3D Shape Analysis

3D Shape Analysis

Fundamentals, Theory, and Applications

Hamid Laga
Murdoch University and University of South Australia, Australia

Yulan Guo
National University of Defense Technology, China

Hedi Tabia
ETIS UMR 8051, Paris Seine University, University of Cergy-Pontoise, ENSEA, CNRS, France

Robert B. Fisher
University of Edinburgh, United Kingdom

Mohammed Bennamoun
The University of Western Australia, Australia

Registered Office
John Wiley & Sons, Inc., 111 River Street, Hoboken, NJ 07030, USA

Editorial Office
111 River Street, Hoboken, NJ 07030, USA

For details of our global editorial offices, customer services, and more information about Wiley products visit us at www.wiley.com.

Wiley also publishes its books in a variety of electronic formats and by print-on-demand. Some content that appears in standard print versions of this book may not be available in other formats.

Library of Congress Cataloging-in-Publication Data

Names: Laga, Hamid, author.
Title: 3D shape analysis : fundamentals, theory, and applications / Hamid
   Laga, Murdoch University and University of South Australia, Australia,
   Yulan Guo, National University of Defense Technology, China, Hedi Tabia,
   ETIS UMR 8051, Paris Seine University, University of Cergy-Pontoise,
   ENSEA, CNRS, France, Robert B. Fisher, University of Edinburgh, United
   Kingdom, Mohammed Bennamoun, The University of Western Australia,
   Australia.
Description: 1st edition. | Hoboken, NJ, USA : Wiley, 2019. | Includes
   bibliographical references and index.
Identifiers: LCCN 2018033203| ISBN 9781119405108 (hardcover) | ISBN
   9781119405191 (epub)
Subjects: LCSH: Three-dimensional imaging. | Pattern recognition systems. |
   Shapes–Computer simulation. | Machine learning.
Classification: LCC TA1560 .L34 2019 | DDC 006.6/93–dc23 LC record available at
   https://lccn.loc.gov/2018033203

Cover design by Wiley
Cover image: © KTSDESIGN/SCIENCE PHOTO LIBRARY/Getty Images

Set in 10/12pt WarnockPro by SPi Global, Chennai, India

Printed in the United States of America

V10006468_112918

Contents

Preface

The primary goal of this book is to provide an in-depth review of 3D shape analysis, which is an important problem and a building block to many applications. This book covers a wide range of basic, intermediate, and advanced topics relating to both the theoretical and practical aspects of 3D shape analysis. It provides a comprehensive overview of the key developments that have occurred for the past two decades in this exciting and continuously expanding field.

This book is organized into 14 chapters, which include an introductory chapter (Chapter 1) and a Conclusions and Perspectives chapter (Chapter 14). The remaining chapters (Chapters 2–13) are structured into three parts. The first part, which is composed of two chapters, introduces the reader to the background concepts of geometry and topology (Chapter 2) that are relevant to most of the 3D shape analysis aspects. It also provides a comprehensive overview of the techniques that are used to capture, create, and preprocess 3D models (Chapter 3). Understanding these techniques will help the reader, not only to understand the various challenges faced in 3D shape analysis but will also motivate the use of 3D shape analysis techniques in improving the algorithms for 3D reconstruction, which is a long-standing problem in computer vision and computer graphics.

The second part, which is composed of two chapters, presents a wide range of mathematical and algorithmic tools that are used for shape description and comparison. In particular, Chapter 4 presents various global descriptors that have been proposed in the literature to characterize the overall shape of a 3D object using its geometry and/or topology. Chapter 5 covers the key algorithms and techniques that are used to detect local features and to characterize the shape of local regions using local descriptors. Both local and global descriptors can be used for shape-based retrieval, recognition, and classification of 3D models. Local descriptors can be also used to compute correspondences between, and to register, 3D objects. This is the focus of the third part of the book, which covers the three commonly studied aspects of the registration and correspondence problem, mainly: rigid registration (Chapter 6), nonrigid registration (Chapter 7), and semantic correspondence (Chapter 8).

The last part and its five chapters are more focused on the application aspects. Specifically, Chapter 9 reviews some of the semantic applications of 3D shape analysis. Chapter 10 focuses on a specific type of 3D object, human faces, and reviews some techniques which are used for 3D face recognition and classification. Chapter 11 focuses on the problem of recognizing objects in 3D scenes. Nowadays, cars, robots, and drones are equipped with 3D sensors, which capture their environments. Tasks such as navigation, target detection and identification, and object tracking require the analysis of the 3D information that is captured by these sensors. Chapter 12 focuses on a classical problem of 3D shape analysis, i.e. how to retrieve 3D objects of interest from a collection of 3D models. It provides a comparative analysis and discusses the pros and cons of various descriptors and similarity measures. Chapter 13, on the other hand, treats the same problem of shape retrieval but this time by using multimodal queries. This is one of the emerging fields of 3D shape analysis and it aims to narrow the gap between the different visual representations of the 3D world (e.g. images, 2D sketches, 3D models, and videos). Finally, Chapter 14 summarizes the book and discusses some of the open problems and future challenges of 3D shape analysis.

The purpose of this book is not to provide a complete and detailed survey of the 3D shape analysis field. Rather, it succinctly covers the key developments of the field in the past two decades and shows their applications in various 3D vision and graphics problems. It is intended to advanced graduate students, postgraduate students, and researchers working in the field. It can also serve as a reference to practitioners and engineers working on the various applications of 3D shape analysis.

May 2018

Hamid Laga
Yulan Guo
Hedi Tabia
Robert B. Fisher
Mohammed Bennamoun

Acknowledgments

The completion of this book would not have been possible without the help, advice, and support of many people. This book is written based on the scientific contributions of a number of colleagues and collaborators. Without their ground breaking contributions to the field, this book would have never matured into this form.

Some of the research presented in this book was developed in collaboration with our PhD students, collaborators, and supervisors. We were very fortunate to work with them and are very grateful for their collaboration. Particularly, we would like to thank (in alphabetical order) Ian H. Jermyn, Sebastian Kurtek, Jonathan Li, Stan Miklavcic, Jacob Montiel, Michela Mortara, Masayuki Nakajima (Hamid Laga's PhD advisor), David Picard, Michela Spagnuolo, Ferdous Sohel, Anuj Srivastava, Antonio Verdone Sanchez, Jianwei Wan, Guan Wang, Hazem Wannous, and Ning Xie. We would also like to thank all our current and previous colleagues at Murdoch University, Tokyo Institute of Technology, National University of Defense Technology, Institute of Computing Technology, Chinese Academy of Sciences, the University of South Australia, the ETIS laboratory, the Graduate School in Electrical Engineering Computer Science and Communications Networks (ENSEA), the University of Western Australia, and The University of Edinburgh.

We are also very grateful to the John Wiley team for helping us create this book.

This work was supported in part by funding from the Australian Research Council (ARC), particularly ARC DP150100294 and ARC DP150104251, National Natural Science Foundation of China (Nos. 61602499 and 61471371), and the National Postdoctoral Program for Innovative Talents of China (No. BX201600172).

Lastly, this book would not have been possible without the incredible support and encouragement of our families.

Hamid Laga dedicates this book to his parents, sisters, and brothers whose love and generosity have always inspired him; to his wife Lan and son Ilyan whose daily encouragement and support in all matters make it all worthwhile.

Yulan Guo dedicates this book to his parents, wife, and son. His son shares the same time for pregnancy, birth, and growth as this book. This would be the first and best gift for the birth of his son.

Hedi Tabia dedicates this book to his family.

Robert B. Fisher dedicates this book to his wife, Miesbeth, who helped make the home a happy place to do the writing. It is also dedicated to his PhD supervisor, Jim Howe, who got him started, and to Bob Beattie, who introduced him to computer vision.

Mohammed Bennamoun dedicates this book to his parents: Mostefa and Rabia, to his children: Miriam, Basheer, and Rayaane and to his seven siblings.

1

Introduction

1.1 Motivation

Shape analysis is an old topic that has been studied, for many centuries, by scientists from different boards, including philosophers, psychologists, mathematicians, biologists, and artists. However, in the past two decades, we have seen a renewed interest in the field motivated by the recent advances in 3D acquisition, modeling, and visualization technologies, and the substantial increase in the computation and storage power. Nowadays, 3D scanning devices are accessible not only to domain-specific experts but also to the general public. Users can scan the real world at high resolution, using devices that are as cheap as video cameras, edit the 3D data using 3D modeling software, share them across the web, and host them in online repositories that are growing in size and in number. Such repositories can include millions of every day objects, cultural heritage artifacts, buildings, as well as medical, scientific, and engineering models.

The increase in the availability of 3D data comes with new challenges in terms of storage, classification, and retrieval of such data. It also brings unprecedented opportunities for solving long-standing problems; First, the rich variability of 3D content in existing shape repositories makes it possible to directly reuse existing 3D models, in whole or in part, to construct new 3D models with rich variations. In many situations, 3D designers and content creators will no more need to scan or model a 3D object or scene from scratch. They can query existing repositories, retrieve the desired models, and fine-tune their geometry and appearance to suit their needs. This concept of context reuse is not specific to 3D models but has been naturally borrowed from other types of media. For instance, one can translate sentences to different languages by performing cross-language search. Similarly, one can create an image composite or a visual art piece by querying images, copying parts of them and pasting them into their own work.

3D Shape Analysis: Fundamentals, Theory, and Applications, First Edition.
Hamid Laga, Yulan Guo, Hedi Tabia, Robert B. Fisher, and Mohammed Bennamoun.
© 2019 John Wiley & Sons, Inc. Published 2019 by John Wiley & Sons, Inc.

Second, these large amounts of 3D data can be used to learn computational models that effectively reason about properties and relationships of shapes without relying on hard-coded rules or explicitly programmed instructions. For instance, they can be used to learn 3D shape variation in medical data in order to model physiological abnormalities in anatomical organs, model their natural growth, and learn how shape is affected by disease progression. They can be also used to model 3D shape variability using statistical models, which, in turn, can be used to facilitate 3D model creation with minimum user interaction.

Finally, data-driven methods facilitate high-level shape understanding by discovering geometric and structural patterns among collections of shapes. These patterns can serve as strong priors not only in various geometry processing applications but also in solving long-standing computer vision problems, ranging from low-level 3D reconstruction to high-level scene understanding.

These technological developments and the opportunities they bring have motivated researchers to take a fresh look at the 3D shape analysis problem. Although most of the recent developments are application-driven, many of them aim to answer fundamental, sometimes philosophical, questions such as: *What is shape? Can we mathematically formulate the concept of shape? How to compare the shape of objects? How to quantify and localize shape similarities and differences?* This book synthesizes the critical mass of 3D shape analysis research that has accumulated over the past 15 years. This rapidly developing field is both profound and broad, with a wide range of applications and many open research questions that are yet to be answered.

1.2   The 3D Shape Analysis Problem

Shape is the external form, outline or surface, of someone or something as opposed to other properties such as color, texture, or material composition.
Source: Wikipedia and Oxford dictionaries.

Humans can easily abstract the form of an object, describe it with a few geometrical attributes or even with words, relate it to the form of another object, and group together, in multiple ways and using various criteria, different objects to form clusters that share some common shape properties. Shape analysis is the general term used to refer to the process of automating these tasks, which are trivial to humans but very challenging to computers. It has been investigated under the umbrella of many applications and has multiple facets. Below, we briefly summarize a few of them.

- **3D shape retrieval, clustering, and classification.** Similar to other types of multimedia information, e.g. text documents, images, and videos, the demand for efficient clustering and classification tools that can organize, automatically or semi-automatically, the continuously expanding collections of 3D models is growing. Likewise, users, whether they are experts, e.g. graphics designers who are increasingly relying on the reuse of existing 3D contents, or novice, will benefit from a search engine that will enable them to search for 3D data of interest in the same way they search for text documents or images.

- **Correspondence and registration.** This problem, which can be summarized as the ability to say which part of an object matches which part on another object, and the ability to align one object onto another, arises in many domains of computer vision, computer graphics, and medical imaging. Probably, one of the most popular examples is the 3D reconstruction problem where usually a 3D object is scanned by multiple sensors positioned at different locations around the object. To build the complete 3D model of the object, one needs to merge the partial scans produced by each sensor. This operation requires a correct alignment, i.e. registration, step that brings all the acquired 3D data into a common coordinate frame. Note also that, in many cases, 3D objects move and deform, in a nonrigid way, during the scanning process. This makes the alignment process even more complex. Another example is in computer graphics where a 3D designer creates a triangulated 3D mesh model, hereinafter referred to as the reference, and assigns to each of its triangular faces some attributes, e.g. color and material properties. The designer then can create additional models with the same attributes but instead of manually setting them, they can be automatically transferred from the reference model if there is a mechanism which finds for each point on the reference model its corresponding points on the other models.

- **Detection and recognition.** This includes the detection of low level features such as corners or regions of high curvatures, as well as the localization and recognition of parts in 3D objects, or objects in 3D scenes. The latter became very popular in the past few years with the availability of cheap 3D scanning devices. In fact, instead of trying to localize and recognize objects in a scene from 2D images, one can develop algorithms that operate on the 3D scans of the scene, eventually acquired using commodity devices. This has the advantage that 3D data are less affected than 2D images by the occlusions and ambiguities, which are inherent to the loss of dimensionality when projecting the 3D world onto 2D images. 3D face and 3D action recognition are, among others, examples of applications that have benefited from the recent advances in 3D technologies.

- **Measurement and characterization** of the geometrical and topological properties of objects on one hand and of the spatial relations between objects on the other hand. This includes the identification of similar regions and finding recurrent patterns within and across 3D objects.
- **Summarization and exploration** of collections of 3D models. Given a set of objects, one would like to compute a representative 3D model, e.g. the average or median shape, as well as other summary statistics such as covariances and modes of variation of their shapes. One would like also to characterize the collection using probability distributions and sample from these distributions new instances of shapes to enrich the collection. In other words, one needs to manipulate 3D models in the same way one manipulates numbers.

Implementing these representative analysis tasks requires solving a set of challenges, and each has been the subject of important research and contributions. The first challenge is **the mathematical representation** of the shape of objects. 3D models, acquired with laser scanners or created using some modeling software, can be represented with point clouds, polygonal soup models, or as volumetric images. Such representations are suitable for storage and visualization but not for high-level analysis tasks. For instance, scanning the same object from two different viewpoints or using different devices will often result in two different point clouds but the shape remains the same. The challenge is in designing mathematical representations that capture the essence of *shape*. A good representation should be independent of (or invariant to) the pose of the 3D object, the way it is scanned or modeled, and the way it is stored. It is also important to ensure that two different shapes cannot have the same representation.

Second, almost every shape analysis task requires a **measure that quantifies shape similarities and differences**. This measure, called dissimilarity, distance, or metric, is essential to many tasks. It can be used to compare the 3D shape of different objects and localize similar parts in and across 3D models. It can also be used to detect and recognize objects in 3D scenes. Shape similarity is, however, one of the most ambiguous concepts in shape analysis since it depends not only on the geometry of the objects being analyzed but also on their semantics, their context, the application, and on the human perception. Figure 1.1 shows a few examples that illustrate the complexity of the shape similarity problem. In Figure 1.1a, we consider human body shapes of the same person but in different poses. One can consider these models as similar since they are of the same person. One may also treat them as different since they differ in pose. On the other hand, the 3D objects of Figure 1.1b are only partially similar. For instance, one part of the centaur model can be treated as similar to the upper body of the human body shape, while the other part is similar to the 3D shape of a horse. Also, one can consider that the

Figure 1.1 Complexity of the shape similarity problem. (a) Nonrigid deformations. (b) Partial similarity. (c) Semantic similarity.

candles of Figure 1.1c are similar despite the significant differences in their geometry and topology. A two-year-old child can easily match together the parts of the candles that have the same functionality despite the fact that they have different geometry, structure, and topology.

Finally, these problems, i.e. representation and dissimilarity, which are interrelated (although many state-of-the-art papers treat them separately), are the core components of and the building blocks for almost every 3D shape analysis system.

1.3 About This Book

The field of 3D shape analysis is being actively studied by researchers originating from at least four different domains: mathematics and statistics, image processing and computer vision, computer graphics, and medical imaging. As a result, a critical mass of research has accumulated over the past 15 years, where almost every major conference in these fields included tracks dedicated to 3D shape analysis. This book provides an in-depth description of the major developments in this continuously expanding field of research. It can serve as a complete reference to graduate students, researchers, and professionals in different fields of mathematics, computer science, and engineering. It could be used for courses of intermediate level in computer vision and computer graphics or for self-study. It is organized into four main parts:

The first part, which is composed of two chapters, provides an in-depth review of the background concepts that are relevant to most of the 3D shape analysis aspects. It begins in Chapter 2 with the basic elements of geometry and topology, which are needed in almost every 3D shape analysis task. We will look in this chapter into elements of differential geometry and into how 3D models are represented. While most of this material is covered in many courses and textbooks, putting them in the broader context of shape analysis will help the reader appreciate the benefits and power of these fundamental mathematical tools.

Chapter 3 reviews the techniques that are used to capture, create, and preprocess 3D models. Understanding these techniques will help the reader, not only to understand the various challenges faced in 3D shape analysis but will also motivate the use of 3D shape analysis techniques in improving the algorithms for 3D reconstruction, which is a long-standing problem in computer vision and computer graphics.

The second part, which is composed of two chapters, presents a range of mathematical and algorithmic tools that are used for shape description and comparison. In particular, Chapter 4 presents the different descriptors that have been proposed in the literature to characterize the global shape of a 3D object using its geometry and/or topology. Early works on 3D shape analysis, in particular classification and retrieval, were based on global descriptors. Although they lack the discrimination power, they are the foundations of modern and powerful 3D shape descriptors.

Chapter 5, on the other hand, covers the algorithms and techniques used for the detection of local features and the characterization of the shape of local regions using local descriptors. Many of the current 3D reconstruction, recognition, and analysis techniques are built on the extraction and matching of feature points. Thus, these are fundamental techniques required in most of the subsequent chapters of the book.

The third part of the book, which is composed of three chapters, focuses on the important problem of computing correspondences and registrations between 3D objects. In fact, almost every task, from 3D reconstruction to animation, and from morphing to attribute transfer, requires accurate correspondence and registration. We will consider the three commonly studied aspects of the problem, which are rigid registration (Chapter 6), nonrigid registration (Chapter 7), and semantic correspondence (Chapter 8). In the first case, we are given two pieces of geometry (which can be partial scans or full 3D models), and we seek to find the rigid transformations (translations, scaling and rotations) that align one piece onto the other. This problem appears mainly in 3D scanning where often a 3D object is scanned by multiple scanners. Each scan produces a set of incomplete point clouds that should be aligned and fused together to form a complete 3D model.

3D models can not only undergo rigid transformations but also nonrigid deformations. Think, for instance, of the problem of scanning a human body. During the scanning process, the body can not only move but also bend. Once it is fully captured, we would like to transfer its properties (e.g. color, texture, and motion) onto another 3D human body of a different shape. This requires finding correspondences and registration between these two 3D objects, which bend and stretch. This is a complex problem since the space of solutions is large and requires efficient techniques to explore it. Solutions to this problem will be discussed in Chapter 7.

Semantic correspondence is even more challenging; think of the problem of finding correspondences between an office chair and a dining chair. While humans can easily match parts across these two models, the problem is very challenging for computers since these two models differ both in geometry and topology. We will review in Chapter 8 the methods that solve this problem using supervised learning, and the methods that used structure and context to infer high-level semantic concepts.

The last part of the book demonstrates the use of the fundamental techniques described in the earlier chapters in a selection of 3D shape analysis applications. In particular, Chapter 9 reviews some of the semantic applications of 3D shape analysis. It also illustrates the range of applications involving 3D data that have been annotated with some sort of meaning (i.e. semantics or labels).

Chapter 10 focuses on a specific type of 3D objects, which are human faces. With the widespread of commodity 3D scanning devices, several recent works use the 3D geometry of the face for various purposes including recognition, gender classification, age recognition, and disease and abnormalities detection. This chapter will review the most relevant works in this area.

Chapter 11 focuses on the problem of recognizing objects in 3D scenes. Nowadays, cars, robots, and drones are all equipped with 3D sensors that capture their environments. Tasks such as navigation, target detection and identification, object tracking, and so on require the analysis of the 3D information that is captured by these sensors.

Chapter 12 focuses on a classical problem of 3D shape analysis, which is how to retrieve 3D objects of interest from a collection of 3D models. Chapter 13, on the other hand, treats the same problem of shape retrieval but this time by using multimodal queries. This is a very recent problem that has received a lot of interest with the emergence of deep-learning techniques that enable embedding different modalities into a common space.

The book concludes in Chapter 14 with a summary of the main ideas and a discussion of the future trends in this very active and continuously expanding field of research.

Readers can proceed sequentially through each chapter. Some readers may want to go straight to topics of their interest. In that case, we recommend to follow the reading chart of Figure 1.2, which illustrates the inter-dependencies between the different chapters.

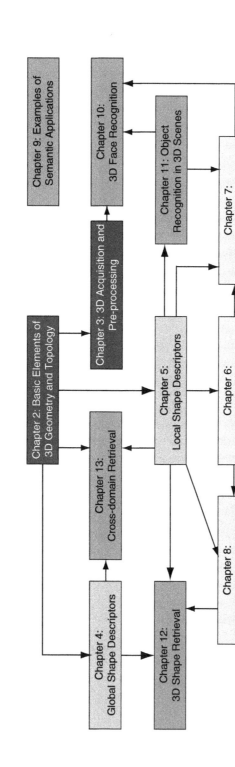

Figure 1.2 Structure of the book and dependencies between the chapters.

1.4   Notation

Table 1.1 summarizes the different notations used throughout the book.

Table 1.1 List of notations used throughout the book.

Symbol	Description
$\mathbb{N}$	Natural numbers
$\mathbb{R}$	Real numbers
$\mathbb{R}_{>0}$	Strictly positive real numbers
$\mathbb{R}_{\geq 0}$	Nonnegative real numbers
$\mathbb{R}^2, \mathbb{R}^3, \ldots, \mathbb{R}^n$	2D, 3D,...,nD Euclidean space, respectively
$\mathbb{S}^2$	The spherical domain or a sphere in $\mathbb{R}^3$
D	A domain of a function. It is also used to refer to a parameterization domain
$\mathbb{C}$	The complex plane, i.e. the space of complex numbers
$p, p_1, p_2, \ldots$	Points in $\mathbb{R}^2$ or $\mathbb{R}^3$
$\mathbf{v}, \mathbf{v}_1, \mathbf{v}_2, \ldots$	Vectors in $\mathbb{R}^2$ or $\mathbb{R}^3$
$\mathbf{n}$	A normal vector
$\tilde{\mathbf{n}}$	A unit normal vector
$\mathbf{v}$	A tangent vector
M	A triangular mesh
V	The set (list) of vertices of a mesh
T	List of (triangular) faces of a mesh
$t, t_1, t_2, \ldots$	Triangular faces of a mesh
$v, v_1, v_2, \ldots$	Vertices of a mesh
m	The number of faces in a mesh
n	The number of vertices in a mesh
G	A graph
V	The set of the nodes of a graph
E	The list of edges in a graph
$v, v_1, v_2, \ldots$	Nodes of a graph
$e, e_1, e_2, \ldots$	Edges of a mesh or a graph
$\mathcal{N}(v)$	The set of vertices (or nodes) that are adjacent or neighbors to the vertex (or node) v in a mesh (or a graph)
$\lambda, \lambda_1, \lambda_2, \ldots$	Eigenvalues
$\Lambda, \Lambda_1, \Lambda_2, \ldots$	Eigenvectors
f	A function
X	A curve in $\mathbb{R}^n$

(Continued)

Table 1.1 (Continued)

Symbol	Description
Δ	A plane
S	A surface in $\mathbb{R}^3$. It is also used to denote the surface of a 3D object
f	The function used to represent the surface of a 3D object
$\mathbf{x}$	A descriptor
κ	Curvature of a curve of a surface at a given point
κ_n	Normal curvature of a surface at a given point
κ_1	Minimum curvature of a surface at a given point
κ_2	Maximum curvature of a surface at a given point
H	Mean curvature of a surface at a given point
K	Gaussian curvature of a surface at a given point
s	Shape index of a surface at a given point
$\mathbf{e}$	Principal direction of a surface at a given point
Δ	The Laplacian operator
∇	The gradient operator
div	The divergence operator
Hess	The Hessian
$SO(3)$	The space of rotations
O	A rotation matrix, which is an element of $SO(3)$
$\mathbf{t}$	A translation vector, which is an element of $\mathbb{R}^3$
Γ	The space of diffeomorphisms
γ	A diffeomorphism, which is an elements of Γ
$\mathbf{t}$	A translation vector in $\mathbb{R}^2$ or $\mathbb{R}^3$

Part I

Foundations

2

Basic Elements of 3D Geometry and Topology

This chapter introduces some of the fundamental concepts of 3D geometry and 3D geometry processing. Since this is a very large topic, we only focus in this chapter on the concepts that are relevant to the 3D shape analysis tasks covered in the subsequent chapters. We structure this chapter into two main parts. **The first** part (Section 2.1) covers the elements of differential geometry that are used to describe the local properties of 3D shapes. **The second** part (Section 2.2) defines the concept of shape, the transformations that preserve the shape of a 3D object, and the deformations that affect the shape of 3D objects. It also describes some preprocessing algorithms, e.g. alignment, which are used to prepare 3D models for shape analysis tasks.

2.1 Elements of Differential Geometry

2.1.1 Parametric Curves

A one-dimensional curve in $\mathbb{R}^3$ can be represented in a parametric form by a vector-valued function of the form:

$$X : [a, b] \subset \mathbb{R} \quad \rightarrow \quad \mathbb{R}^3,$$

$$t \in [a, b] \quad \mapsto \quad X(t) = \begin{pmatrix} x(t) \\ y(t) \\ z(t) \end{pmatrix}. \tag{2.1}$$

The functions $x(\cdot), y(\cdot)$, and $z(\cdot)$ are also called the coordinate functions. If we assume that they are differentiable functions of t, then one can compute the tangent vector, the normal vector, and the curvature at each point on the curve.

3D Shape Analysis: Fundamentals, Theory, and Applications, First Edition.
Hamid Laga, Yulan Guo, Hedi Tabia, Robert B. Fisher, and Mohammed Bennamoun.
© 2019 John Wiley & Sons, Inc. Published 2019 by John Wiley & Sons, Inc.

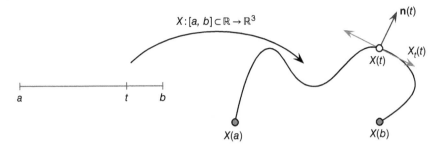

Figure 2.1 An example of a parameterized curve and its differential properties.

For instance, the tangent vector, $X_t(t)$, to the curve at a point $p = X(t)$, see Figure 2.1, is the first derivative of the coordinate functions:

$$X_t(t) = X'(t) = \frac{dX(t)}{dt} = \begin{pmatrix} \frac{\partial x(t)}{\partial t} \\ \frac{\partial y(t)}{\partial t} \\ \frac{\partial z(t)}{\partial t} \end{pmatrix}. \tag{2.2}$$

Note that the tangent vector $X_t(t)$ also corresponds to the *velocity* vector at time t.

Now, let $[c, d] \subset [a, b]$. The length $l(c, d)$ of the curve segment defined between the two points $X(c)$ and $X(d)$ is given by the integral of the tangent vector:

$$l(c, d) = \int_c^d \|X_t(s)\| ds = \int_c^d \langle X_t(s), X_t(s) \rangle^{1/2} ds. \tag{2.3}$$

Here, $\langle \cdot, \cdot \rangle$ denotes the inner dot product. The length L of the curve is then given by $L = l(a, b)$.

Let γ be a smooth and monotonically increasing function, which maps the domain $[a, b]$ onto $[0, L]$ such that $\gamma(t) = l(a, t)$. Reparameterizing X with γ results in another curve $\tilde{X} = X \circ \gamma$ such that:

$$\tilde{X} : [0, L] \to \mathbb{R}^3, \qquad \tilde{X}(s) = (X \circ \gamma)(t) = X(\gamma(t)), \quad \text{where } s = \gamma(t). \tag{2.4}$$

The length of the curve from $\tilde{X}(0)$ to $\tilde{X}(s), s \in [0, L]$, is exactly s. The mapping function γ, as defined above, is called *arc-length parameterization*. In general, γ can be any arbitrary smooth and monotonically increasing function, which maps the domain $[a, b]$ into another domain $[c, d]$. If γ is a bijection and its inverse γ^{-1} is differentiable as well, then γ is called a *diffeomorphism*. Diffeomorphisms are very important in shape analysis. For instance, as shown in Figure 2.2, reparameterizing a curve X with a diffeomorphism γ results in another curve $\tilde{X} = X \circ \gamma$, which has the same shape as X. These two curves are, therefore, equivalent from the shape analysis perspective.

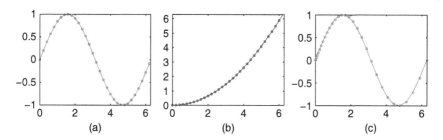

Figure 2.2 Reparameterizing a curve X with a diffeomorphism γ results in another curve $\tilde{X}$ of the same shape as X. (a) A parametric curve $X : [0, 2\pi] \rightarrow \mathbb{R}$. (b) A reparameterization function $\gamma : [0, 2\pi] \rightarrow [0, 2\pi]$. (c) The reparametrized curve $X \circ \gamma$.

Now, for simplicity, we assume that X is a curve parameterized with respect to its arc length. We can define the curvature at a point $p = X(t)$ as the deviation of the curve from the straight line. It is given by the norm of the second derivative of the curve:

$$\kappa(t) = \|X_{tt}\|, \quad \text{where } X_{tt} = X'' = \frac{d^2X(t)}{dt^2} = \begin{pmatrix} \frac{\partial^2 x(t)}{\partial t^2} \\ \frac{\partial^2 y(t)}{\partial t^2} \\ \frac{\partial^2 z(t)}{\partial t^2} \end{pmatrix}. \tag{2.5}$$

Curvatures carry a lot of information about the shape of a curve. For instance, a curve with zero curvature everywhere is a straight line segment, and planar curves of constant curvature are circular arcs.

2.1.2 Continuous Surfaces

We will look in this section at the same concepts as those defined in the previous section for curves but this time for smooth surfaces embedded in $\mathbb{R}^3$.

A parametric surface S (see Figure 2.3) can be defined as a vector-valued function f which maps a two-dimensional domain $D \subset \mathbb{R}^2$ onto $\mathbb{R}^3$. That is:

$$f : D \subset \mathbb{R}^2 \rightarrow \mathbb{R}^3$$
$$(u, v) \mapsto f(u, v) = \begin{pmatrix} x(u, v) \\ y(u, v) \\ z(u, v) \end{pmatrix}, \tag{2.6}$$

where $x(\cdot, \cdot)$, $y(\cdot, \cdot)$, and $z(\cdot, \cdot)$ are differentiable functions with respect to u and v. The domain D is called the *parameter space*, or *parameterization domain*, and the scalars (u, v) are the coordinates in the parameter space.

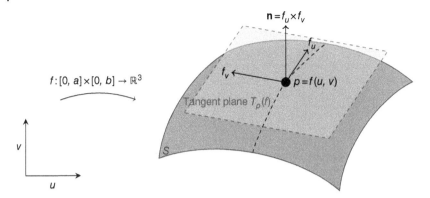

Figure 2.3 Example of an open surface parameterized by a 2D domain $D = [0, a] \times [0, b]$.

As an example, consider S to be the surface of a sphere of radius r and centered at the origin. This surface can be defined using a parametric function of the form

$$f : [0, 2\pi] \times [0, \pi] \to \mathbb{R}^3, \quad (\theta, \phi) \mapsto f(\theta, \phi) = \begin{pmatrix} r \cos \theta \sin \phi \\ r \sin \theta \sin \phi \\ r \cos \phi \end{pmatrix}.$$

In practice, the parameter space D is chosen depending on the types of shapes at hand. In the case of open surfaces such as 3D human faces, D can be a subset of $\mathbb{R}^2$, e.g. the quadrilateral domain $[0, 1] \times [0, 1]$ or the disk domain $[0, 1] \times [0, 2\pi]$. In the case of closed surfaces, such as those shown in Figure 2.4a,b, then D can be chosen to be the unit sphere $\mathbb{S}^2$. In this case, u and v are the spherical coordinates such that $u \in [0, \pi]$ is the polar angle and $v \in [0, 2\pi)$ is the azimuthal angle.

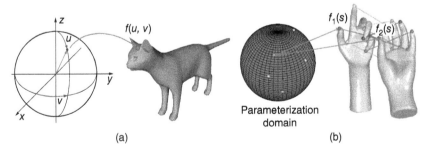

Figure 2.4 Illustration of (a) a spherical parameterization of a closed genus-0 surface, and (b) how parameterization provides correspondence. Here, the two surfaces, which bend and stretch, are in correct correspondence. In general, however, the analysis process should find the optimal reparameterization which brings f_2 into a full correspondence with f_1.

Let $\gamma : D \to D$ be a diffeomorphism, i.e. a smooth invertible function that maps the domain D to itself such that both γ and its inverse are smooth. Let also Γ denote the space of all such diffeomorphisms. γ transforms a surface $f : D \to \mathbb{R}^3$ into $f \circ \gamma : D \to \mathbb{R}^3$. This is a reparameterization process and often γ is referred to as a reparameterization function. Reparameterization is important in 3D shape analysis because it produces registration. For instance, similar to the curve case, both surfaces represented by f and $f \circ \gamma$ have the same shape and thus are equivalent. Also, consider the two surfaces f_1 and f_2 of Figure 2.4b where f_1 represents the surface of one long hand and f_2 represents the surface of the hand of another subject. Let $s \in D$ be such that $f_1(s)$ corresponds to the thumb tip of f_1. If f_1 and f_2 are arbitrarily parameterized, which is often the case since they have been parameterized independently from each other, $f_2(s)$ may refer to any other location on f_2, the fingertip of the index finger, for example. Thus, f_1 and f_2 are not in correct correspondence. Putting f_1 and f_2 in correspondence is equivalent to reparameterizing f_2 with a diffeomorphism $\gamma \in \Gamma$ such that for every $s \in D, f_1(s)$ and $f_2(\gamma(s)) = (f_2 \circ \gamma)(s)$ point to the same feature, e.g. thumb tip to thumb tip. We will use these properties in Chapter 7 to find correspondences and register surfaces which undergo complex nonrigid deformations.

2.1.2.1 Differential Properties of Surfaces

Similar to the curve case, one can define various types of differential properties of a given surface. For instance, the two partial derivatives

$$f_u(u, v) = \frac{\partial f}{\partial u}(u, v) \quad \text{and} \quad f_v(u, v) = \frac{\partial f}{\partial v}(u, v) \tag{2.7}$$

define two tangent vectors to the surface at a point of coordinates (u, v). The plane spanned by these two orthogonal vectors is tangent to the surface. The surface unit normal vector at (u, v), denoted by $\mathbf{n}(u, v)$, can be computed as:

$$\mathbf{n}(u, v) = \frac{f_u(u, v) \times f_v(u, v)}{\|f_u(u, v) \times f_v(u, v)\|}. \tag{2.8}$$

Here, $\times$ denotes the vector cross product.

2.1.2.1.1 First Fundamental Form

A tangent vector $\mathbf{v}$ to the surface at a point $p = f(u, v)$ can be defined in terms of the partial derivatives of f as:

$$\mathbf{v} = \alpha f_u(u, v) + \beta f_v(u, v). \tag{2.9}$$

The two real-valued scalars α and β can be seen as the coordinates of $\mathbf{v}$ in the local coordinate system formed by $f_u(u, v)$ and $f_v(u, v)$. Let $\mathbf{v}_1$ and $\mathbf{v}_2$ be two tangent vectors of coordinates (α_1, β_1) and (α_2, β_2), respectively. Then, the inner product $\langle \mathbf{v}_1, \mathbf{v}_2 \rangle = \|\mathbf{v}_1\| \|\mathbf{v}_2\| \cos \theta$, where θ is the angle between the two vectors, is defined as:

$$\langle \mathbf{v}_1, \mathbf{v}_2 \rangle = (\alpha_1, \beta_1) \begin{pmatrix} f_u^T f_u & f_u^T f_v \\ f_u^T f_v & f_v^T f_v \end{pmatrix} \begin{pmatrix} \alpha_2 \\ \beta_2 \end{pmatrix} = (\alpha_1, \beta_1) g \begin{pmatrix} \alpha_2 \\ \beta_2 \end{pmatrix}, \qquad (2.10)$$

where g is a 2D matrix called *the first fundamental form* of the surface at a point p of coordinates (u, v). Let us write

$$g = \begin{pmatrix} E & F \\ F & G \end{pmatrix}.$$

The first fundamental form, also called *the metric* or *metric tensor*, defines inner products on the tangent space to the surface. It allows to measure:

- **Angles**. The angle between the two tangent vectors $\mathbf{v}_1$ and $\mathbf{v}_2$ can be computed as:

$$\langle \mathbf{v}_1, \mathbf{v}_2 \rangle = \langle (\alpha_1, \beta_1), (\alpha_2, \beta_2) \rangle = (\alpha_1, \beta_1) g (\alpha_2, \beta_2)^T.$$

- **Length**. Consider a curve on the parametric surface. The parametric representation of the curve is $f(u(t), v(t)), t \in [a, b]$. The tangent vector to the curve at any point is given as:

$$\frac{df(u(t), v(t))}{dt} = \frac{\partial f}{\partial u} \frac{du}{dt} + \frac{\partial f}{\partial v} \frac{dv}{dt} = u_t f_u + v_t f_v.$$

Since the curve is on the surface defined by the parametric function f, then f_u and f_v are two orthogonal vectors that are tangent to the surface. Using Eq. (2.3), the length $l(a, b)$ of the curve is given by

$$l(a, b) = \int_a^b \sqrt{\langle (u_t, v_t)^T, (u_t, v_t)^T \rangle} dt = \int_a^b \sqrt{Eu_t^2 + 2Fu_t v_t + Gv_t^2} dt.$$

- **Area**. Similarly, we can measure the surface area a of a certain region $\Omega \subset D$:

$$a = \int \int_\Omega \sqrt{\det g} \, \partial u \, \partial v = \int \int_\Omega \sqrt{EG - F^2} \, \partial u \, \partial v,$$

where $\det g$ refers to the determinant of the square matrix g.

Since the first fundamental form allows measuring angles, distances, and surface areas, it is a strong geometric tool for 3D shape analysis.

2.1.2.1.2 Second Fundamental Form and Shape Operator

The second fundamental form **II** of a surface, defined by its parametric function f, at a point $s = (u, v)$ is a 2×2 matrix defined in terms of the second derivatives of f as follows:

$$\mathbf{II} = \begin{bmatrix} \Pi_{11} & \Pi_{12} \\ \Pi_{21} & \Pi_{22} \end{bmatrix}, \qquad (2.11)$$

where

$$\Pi_{11} = \left\langle \frac{\partial^2 f}{\partial u^2}, \mathbf{n} \right\rangle, \quad \Pi_{12} = \Pi_{21} = \left\langle \frac{\partial^2 f}{\partial u \partial v}, \mathbf{n} \right\rangle, \quad \Pi_{22} = \left\langle \frac{\partial^2 f}{\partial v^2}, \mathbf{n} \right\rangle.$$

$$(2.12)$$

Here, $\langle \cdot, \cdot \rangle$ refers to the standard inner product in $\mathbb{L}^2$, i.e. the Euclidean space. Similarly, the shape operator S of the surface is defined at a point s using the first and second fundamental forms as follows:

$$S(s) = g^{-1}(s)\mathbf{II}(s). \tag{2.13}$$

The shape operator is a linear operator. Along with the second fundamental form, they are used for computing surface curvatures.

2.1.2.2 Curvatures

Let $\mathbf{n}$ be the unit normal vector to the surface at a point p, and $\mathbf{v} = u_t f_u + v_t f_v$ a tangent vector to the surface at p. The intersection of the plane spanned by the two vectors $\mathbf{v}$ and $\mathbf{n}$ with the surface at p forms a curve X called *the normal curve*. This is illustrated in Figure 2.5. Rotating this plane, which is also called *the normal plane*, around the normal vector $\mathbf{n}$ is equivalent to rotating the tangent vector $\mathbf{v}$ around $\mathbf{n}$. This will result in a different normal curve. By analyzing the curvature of such curves, we can define the curvature of the surface.

Figure 2.5 Illustration of the normal vector, normal plane, and normal curve.

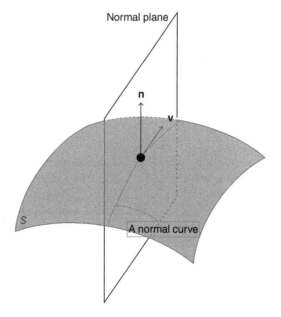

1) **Normal curvature.** The normal curvature κ_n to the surface at a point p in the direction of a tangent vector $\mathbf{v} = u_t f_u + v_t f_v$ is defined as the curvature of the normal curve X created by intersecting the surface at p with the plane spanned by $\mathbf{v}$ and $\mathbf{n}$. Let $\bar{\mathbf{v}}$ denote the representation of $\mathbf{v}$ in the 2D local coordinate system spanned by the two orthogonal vectors f_u and f_v, i.e. $\bar{\mathbf{v}} = (u_t, v_t)^\top$. Then, the normal curvature at p in the direction of $\mathbf{v}$ can be computed using the first and second fundamental forms as follows:

$$\kappa_n(\mathbf{v}) = \frac{\bar{\mathbf{v}}^\top \mathrm{II}\bar{\mathbf{v}}}{\bar{\mathbf{v}}^\top g \bar{\mathbf{v}}}. \tag{2.14}$$

2) **Principal curvatures.** By rotating the tangent vector $\mathbf{v}$ around the normal vector $\mathbf{n}$, we obtain an infinite number of normal curves at p. Thus, $\kappa_n(\mathbf{v})$ varies with $\mathbf{v}$. It has, however, two extremal values called *principal curvatures*; one is the *minimum curvature*, denoted by κ_1, and the other is the *maximum curvature*, denoted by κ_2.

The principal directions, $\mathbf{e}_1$ and $\mathbf{e}_2$, are the tangent vectors to the normal curves, which have the minimum, respectively the maximum, curvature at p. These principal directions are orthogonal and thus they form a local orthonormal basis. Consequently, any vector $\mathbf{v}$ that is tangent to the surface at p can be defined with respect to this basis. For simplicity of notation, let $\mathbf{v} = u_t \mathbf{e}_1 + v_t \mathbf{e}_2$ and let Φ be the angle between $\mathbf{v}$ and $\mathbf{e}_1$. The Euler theorem states that the normal curvature at p in the direction of $\mathbf{v}$ is given by

$$\kappa_n(\theta) = \kappa_1 \cos^2(\theta) + \kappa_2 \sin^2(\theta). \tag{2.15}$$

3) **Mean and Gaussian curvatures.** The mean curvature is defined as the average of the two principal curvatures:

$$H = \frac{\kappa_1 + \kappa_2}{2}. \tag{2.16}$$

The Gaussian curvature, on the other hand, is defined as the product of the two principal curvatures:

$$K = \kappa_1 \kappa_2. \tag{2.17}$$

These two curvature measures are very important and they are extensively used in 3D shape analysis. For instance, the Gaussian curvature can be used to classify surface points into three distinct categories: elliptical points ($K > 0$), hyperbolic points ($K < 0$), and parabolic points ($K = 0$). Figure 2.6 illustrates the relation between different curvatures at a given surface point and the shape of the local patch around that point. For example:

- When the minimum and maximum curvatures are equal and nonzero (i.e. both are either strictly positive or strictly negative) at a given point then the patch around that point is locally spherical.

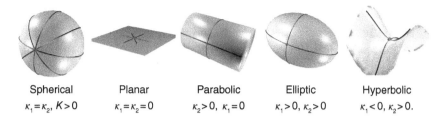

Spherical	Planar	Parabolic	Elliptic	Hyperbolic
$\kappa_1 = \kappa_2,\, K > 0$	$\kappa_1 = \kappa_2 = 0$	$\kappa_2 > 0,\, \kappa_1 = 0$	$\kappa_1 > 0,\, \kappa_2 > 0$	$\kappa_1 < 0,\, \kappa_2 > 0.$

Figure 2.6 Curvatures provide information about the local shape of a 3D object.

- When both minimum and maximum curvatures are zero, then the shape is planar.
- When the minimum curvature is zero and the maximum curvature is strictly positive then the shape of the local patch is parabolic. If every point on the surface of a 3D object satisfies this condition and if the maximum curvature is constant everywhere then the 3D object is a cylinder.
- If both curvatures are strictly positive then the shape is elliptical.
- If the minimum curvature is negative while the maximum curvature is positive then the shape is hyperbolic.

Finally, the curvatures can be also defined using the shape operator. For instance, the normal curvature of the surface at a point p in the direction of a tangent vector $\mathbf{v}$ is defined as

$$\frac{d\mathbf{n}}{d\mathbf{v}} = S(\mathbf{v}). \tag{2.18}$$

The eigenvalues of S are the principal curvatures at p. Its determinant is the Gaussian curvature and its trace is twice the mean curvature at p.

2.1.2.3 Laplace and Laplace–Beltrami Operators

Let f be a function in the Euclidean space. The Laplacian Δf of f is defined as the divergence of the gradient, i.e.;

$$\Delta f = \text{div} \nabla f = \frac{\partial^2 f}{\partial x^2} + \frac{\partial^2 f}{\partial y^2} + \frac{\partial^2 f}{\partial z^2}. \tag{2.19}$$

The Laplace–Beltrami operator is the extension of the Laplace operator to functions defined on manifolds:

$$\Delta_{\mathcal{M}} f = \text{div}_{\mathcal{M}} \nabla_{\mathcal{M}} f, \tag{2.20}$$

where f is a function on the manifold $\mathcal{M}$. If f is a parametric representation of a surface, as defined in Eq. (2.6), then

$$\Delta_{\mathcal{M}} f(u, v) = \text{div}_{\mathcal{M}} \nabla_{\mathcal{M}} f(u, v) = -2K(u, v)\mathbf{n}(u, v). \tag{2.21}$$

Here, $K(u, v)$ and $\mathbf{n}(u, v)$ denote, respectively, the Gaussian curvature and the normal vector at a point $p = f(u, v)$.

2.1.3 Manifolds, Metrics, and Geodesics

The surface of a 3D object can be seen as a domain that is nonlinear, i.e. it is not a vector space. Often, one needs to perform some differential (e.g. computing normals and curvatures) and integral (e.g. tracing curves and paths) calculus on this type of general domains, which are termed *manifolds*. Below, we provide the necessary definitions.

Definition 2.1 (Topological space).   A topological space is a set of points, along with a set of neighborhoods for each point, that satisfy some axioms relating points and neighborhoods.

This definition of topological spaces is the most general notion of a mathematical space, which allows for the definition of concepts such as continuity, connectedness, and convergence. Other spaces are special cases of topological spaces with additional structures or constraints.

Definition 2.2 (Hausdorff space).   A Hausdorff space is a topological space in which distinct points admit disjoint neighborhoods.

Two points p and q in a topological space can be separated by neighborhoods if there exists a neighborhood U of p and a neighborhood V of q such that U and V are disjoint ($U \cap V = \emptyset$). The topological space is a Hausdorff space if all distinct points in it are pairwise neighborhood-separable.

Definition 2.3 (Manifold).   A topological space $\mathcal{M}$ is called a **manifold** of dimension m if it is Hausdorff, it has a countable basis, and for each point $p \in \mathcal{M}$, there is a neighborhood U of p that is homeomorphic to an open subset of $\mathbb{R}^m$.

In our applications of geometry, purely topological manifolds are not sufficient. We would like to be able to evaluate first and higher derivatives of functions on our manifold. A **differentiable manifold** is a manifold with a smoother structure, which allows for the definition of derivatives. Having a notion of differentiability at hand, we can do differential calculus on the manifold and talk about concepts such as tangent vectors, vector fields, and inner products.

Definition 2.4 (Tangent space).   If $\mathcal{M}$ is a smooth manifold of dimension m, then the tangent space to $\mathcal{M}$ at a point $p \in \mathcal{M}$, hereinafter denoted as $T_p(\mathcal{M})$, is the vector space which contains all the possible vectors passing tangentially through p, see Figure 2.3.

If we glue together all the tangent spaces $T_p(\mathcal{M})$, we obtain the tangent bundle $T(\mathcal{M})$.

Definition 2.5 (Riemannian metric). In order to define a notion of distance, we introduce an additional structure called Riemannian metric, which can be thought of as measuring the distance between infinitesimally close points. It does this by assigning an inner product to each tangent space of the manifold, thereby defining a norm on the tangent vectors.

Definition 2.6 (Riemannian manifold). A differentiable manifold with a Riemannian metric on it is called a Riemannian manifold.

Endowing a manifold with a Riemannian metric makes it possible to define on that manifold geometric notions such as angles, lengths of curves, curvatures, and gradients. With a Riemannian metric, it also becomes possible to define the length of a path on a manifold. Let $\alpha : [0, 1] \rightarrow \mathcal{M}$ be a parameterized path on a Riemannian manifold $\mathcal{M}$, such that α is differentiable everywhere on $[0, 1]$. Then, $\dot{\alpha} \equiv \frac{d\alpha}{dt}$, the velocity vector at t, is an element of the tangent space $T_{\alpha(t)}(\mathcal{M})$. Its length is $\|\dot{\alpha}\|$. The length of the path α is then given by:

$$L(\alpha) = \int_0^1 \|\dot{\alpha}(t)\| dt = \int_0^1 \langle \dot{\alpha}(t), \dot{\alpha}(t) \rangle^{1/2} dt. \tag{2.22}$$

This is the integral of the lengths of the velocity vectors along α and, hence, is the length of the whole path α. For any two points $p_1, p_2 \in \mathcal{M}$, one can define the distance between them as the infimum of the lengths of all smooth paths on $\mathcal{M}$ that start at p_1 and end at p_2:

$$d(p_1, p_2) = \inf_{\substack{\alpha : [0,1] \rightarrow \mathcal{M} \\ \alpha(0) = p_1 \\ \alpha(1) = p_2}} L(\alpha) . \tag{2.23}$$

This definition turns $\mathcal{M}$ into **a metric space**, with distance d.

Definition 2.7 If there exists a path α^* that achieves the above minimum, then it is called a **geodesic** between p_1 and p_2 on $\mathcal{M}$.

Very often, the search for geodesics is handled by changing the objective functional from length to an energy functional of the form:

$$E(\alpha) = \int_0^1 \langle \dot{\alpha}(t), \dot{\alpha}(t) \rangle dt. \tag{2.24}$$

The only difference from the path length in Eq. (2.22) is that the square root in the integrand has been removed. It can be shown that a critical point of this functional restricted to paths between p_1 and p_2 is a geodesic between p_1 and p_2. Furthermore, of all the reparameterizations of a geodesic, the one with a constant speed has the minimum energy. As a result, for a minimizing path α^*, we have

$$E(\alpha^*) = L(\alpha^*)^2 . \tag{2.25}$$

2.1.4 Discrete Surfaces

In the previous section, we treated the surface of a 3D object as a continuous function $f : D \to \mathbb{R}^3$. However, most of the applications (including visualization and 3D analysis tasks) require the discretization of these surfaces. Moreover, as will be seen in Chapter 3, most of 3D acquisition devices and 3D modeling software produce discrete 3D data. For example, scanning devices such as the Kinect and stereo cameras produce depth maps and 3D point clouds. CT scanners in medical imaging produce volumetric images. These are then converted into polygonal meshes, similar to what is produced by most of the 3D modeling software, for rendering and visualization. 3D shape analysis algorithms take as input 3D models in one of these representations and convert them into a representation (a set of features and descriptors), which is suitable for comparison and analysis. Thus, it is important to understand these representations (Section 2.1.4.1) and the data structures used for their storage and processing (Section 2.1.4.2).

2.1.4.1 Representations of Discrete Surfaces

Often, complex surfaces cannot be represented with a single parametric function. In that case, one can use a *piecewise* representation. For instance, one can partition a complex surface into smaller subregions, called *patches*, and approximate each patch with one parametric function. The main mathematical challenge here is to ensure continuity between adjacent patches. The most common piecewise representations are spline surfaces, polygonal meshes, and point sets. The spline representation is the standard representation in modern computer aided design (CAD) models. It is used to construct high-quality surfaces and for surface deformation tasks. In this chapter, we will focus on the second and third representations since they are the most commonly used representations for 3D shape analysis. We, however, refer the reader to other textbooks, such as [1], for a detailed description of spline representations.

(1) **Polygonal meshes.** In a polygonal representation, a 3D model is defined as a collection, or union, of planar polygons. Each planar polygon, which approximates a surface patch, is represented with a set of vertices and a set of edges connecting these vertices. In principle, these polygons can be of arbitrary type. In fact, they are not even required to be planar. In practice, however, triangles are the most commonly used since they are guaranteed to be piecewise linear. In other words, triangles are the only polygons that are always guaranteed to be planar. A surface approximated with a set of triangles is referred to as *a triangular mesh*. It consists of a set of vertices

$$V = \{v_1, \dots, v_n\}, \quad v_i \in \mathbb{R}^3$$

and a set of triangular faces

$$T = \{t_k \in V \times V \times V \mid k = 1, \dots, m\},$$

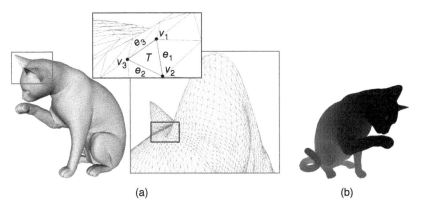

Figure 2.7 Representations of complex 3D objects: (a) triangular mesh-based representation and (b) depth map-based representation.

connecting them, see Figure 2.7a. This representation can be also interpreted as a graph $G = (V, E)$ whose nodes are the vertices of the mesh. Its edges $E = \{e_1, \ldots, e_l\}$ are the edges of the triangular faces. In this graph representation, each triangular face defines, using the barycentric coordinates, a linear parameterization of its interior points. That is, every point p in the interior of a triangle $t = \{v_1, v_2, v_3\}$ can be written as $p = \alpha v_1 + \beta v_2 + \gamma v_3$ where α, β, and $\gamma \in \mathbb{R}_{\geq 0}$ and $\alpha + \beta + \gamma = 1$.

Obviously, the accuracy of the representation depends on the number of vertices (and subsequently the number of faces) that are used to approximate the surface. For instance, a nearly planar surface can be accurately represented using only a few vertices and faces. Surfaces with complex geometry, however, require a large number of faces especially at regions of high curvature.

Definition 2.8 (2-manifold mesh). A triangular mesh is said to be a 2-manifold if the neighborhood around each vertex is homeomorphic to a disk. In other words, the triangular mesh contains neither singular vertices nor singular edges. It is also free from self-intersections.

A singular, or nonmanifold, vertex is a vertex that is attached to two fans of triangles as shown in Figure 2.8a,b. A singular, or nonmanifold, edge is an edge that is attached to more than two faces, see Figure 2.8c. Manifold surfaces are often referred to as *watertight* models.

Manifoldness is an important property of triangular meshes. In particular, a manifold surface S can always be locally parameterized at each point $p \in S$ using a function $f_p : D \to \mathbb{R}^3$, where D is a planar domain, e.g. a disk or a square. As such, differential properties (see Section 2.1.2) can efficiently be computed using some analytical approximations.

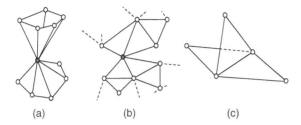

Figure 2.8 Examples of nonmanifold surfaces. (a) A nonmanifold vertex attached two fans of triangles, (b) a nonmanifold vertex attached to patches, and (c) a nonmanifold edge attached to three faces.

When the mesh is nonmanifold, it is often referred to as a *polygon soup model.* Polygon soup models are easier to create since one does not have to worry about the connectivity of the vertices and edges. They are, however, problematic for some shape analysis algorithms since the differential properties of such surfaces cannot be computed analytically. There are, however, efficient numerical methods to compute such differential properties. This will be detailed in Section 2.1.4.3.

(2) **Depth maps**. Depth maps, or depth images, record the distance of the surface of objects in a 3D scene to a viewpoint. They are often the product of some 3D acquisition systems such as stereo cameras or Kinect sensors. They are also used in computer graphics for rendering where the term Z-buffer is used to refer to depth. Mathematically, a depth map is a function $f : [0, 1] \times [0, 1] \rightarrow \mathbb{R}_{\geq 0}$ where $f(u, v)$ is the depth at pixel $s = (u, v)$. In other words, if one projects a ray from the viewpoint through the image pixel $s = (u, v)$, it will intersect the 3D scene at the 3D point of coordinates $p = (u, v, f(u, v))$. Depth maps are compact representations. They, however, represent the 3D world from only one viewpoint, see Figure 2.7b. Thus, they only contain partial information and are often referred to as 2.5D representations. To obtain a complete representation of a 3D scene, one requires multiple depth maps captured from different viewpoints around the 3D scene.

As we will see in the subsequent chapters of this book, depth maps are often used to reduce the 3D shape analysis problem into a 2D problem in order to benefit from the rich image analysis literature.

(3) **Point-based representations**. 3D acquisition systems, such as range scanners, produce unstructured point clouds $\{v_i \in \mathbb{R}^3, i = 1, \dots, n\}$, which can be seen as a dense sampling of the 3D world. In many shape analysis tasks, this point cloud is not used as it is, but converted into a 3D mesh model using some tessellation techniques such as ball pivoting [2] and alpha shapes [3]. Point clouds can be a suitable representation for some types of 3D objects such as hair, fur, or vegetation, which contain dense tiny structures.

(4) **Implicit and volumetric representations**. An implicit surface is defined to be the zero-set of a scalar-valued function $f : \mathbb{R}^2 \rightarrow \mathbb{R}$. In other words, the surface is defined as the set of all points $p \in \mathbb{R}^3$ such that $f(p) = 0$.

Example 2.1 (Implicit sphere). Take the example of a sphere of radius r, centered at the origin. Its implicit representation is of the form:

$$f(x, y, z) = x^2 + y^2 + z^2 - r^2.$$

The implicit or volumetric representation of a solid object is, in general, a function f, which classifies each point in the 3D space to lie either inside, outside, or exactly on the surface S of the object. The most natural choice of the function f is *the signed distance function*, which maps each 3D point to its signed distance from the surface S. By definition, negative values of the function f designate points inside the object, positive values designate points outside the object, and the surface points correspond to points where the function is zero. The surface S is referred to as *the level set* or *the zero-level isosurface*. It separates the inside from the outside of the object.

Implicit representations are well suited for constructive solid geometry where complex objects can be constructed using Boolean operations of simpler objects called primitives. They are also suitable for representing the interior of objects. As such, they are extensively used in the field of medical imaging since the data produced with medical imaging devices, such as CT scans, are volumetric.

There are different representations for implicit surfaces. The most commonly used ones are *discrete voxelization* and *the radial basis functions* (RBFs). The former is just a uniform discretization of the 3D volume around the shape. In other words, it can be seen as a 3D image where each cell is referred to as a *voxel* (voxel element), in analogy to pixels (picture elements) in images.

The latter approximates the implicit function f using a weighted-sum of kernel functions centered around some control points. That is:

$$f(p) = P(p) + \sum_{i=1}^{N} \alpha_i \Phi(\|p - p_i\|), \tag{2.26}$$

where $P(\cdot)$ is a polynomial of low degree and the basis function $\Phi(\cdot)$ is a real-valued function on $[0, \infty)$, usually unbounded and has a noncompact support [4]. The points p_i are referred to as the centers of the radial basis functions.

With this representation, the function f can be seen as a continuous scalar field. In order to efficiently process implicit representations, the continuous scalar field needs to be discretized using a sufficiently dense 3D grid around the solid object. The values within voxels are derived using interpolation. The discretization can be uniform or adaptive. The memory consumption in the former grows cubically if the precision is increased by reducing the size of the voxels. In the adaptive discretization, the sampling density is adapted to the local geometry since the signed distance values are more important in the vicinity of the surface. Thus, in order to achieve better memory efficiency, a higher sampling rate can be used in these regions only, instead of a uniform 3D grid.

2.1.4.2 Mesh Data Structures

Let us assume from now on, unless explicitly stated otherwise, that the boundary S of a 3D object is represented as a piecewise linear surface in the form of a triangulated mesh M, i.e. a set of vertices $V = \{v_1, \ldots, v_n\}$ and triangles $T = \{t_1, \ldots, t_m\}$. The choice of the data structure for storing such 3D objects should be guided by the requirements of the algorithms that will be operating on the data structure. In particular, one should consider whether we only need to render the 3D models, or do we need to have constant-time access to the local neighborhood of the vertices, faces, and edges.

The simplest data structure is the one that stores the mesh as a table of vertices and encodes triangles as a table of triplets of indices into the vertex table. This representation is simple and efficient in terms of storage. While it is very suitable for rendering, it is not suitable for many of the shape analysis algorithms. In particular, most of the 3D shape analysis tasks require access to:

- Individual vertices, edges, and faces.
- The faces attached to a vertex.
- The faces attached to an edge.
- The faces adjacent to a given face.
- The one-ring vertex neighborhood of a given vertex.

Mesh data structures should be designed in such a way that these queries can run in a constant time. The most commonly used data structure is the *halfedge data structure* [5] where each edge is split into two opposing halfedges. All halfedges are oriented consistently in counter-clockwise order around each face and along the boundary, see Figure 2.9. Each halfedge stores a reference to:

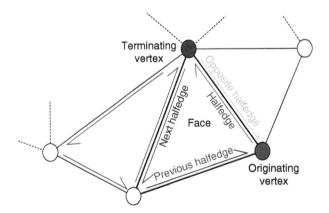

Figure 2.9 A graphical illustration of the half-edge data structure.

- **The head**, i.e. the vertex it points to.
- **The adjacent face**, i.e. a reference to the face which is at the left of the edge (recall that half edges are oriented counter-clockwise). This field stores a zero pointer if this half edge is at the boundary of the surface.
- **The next halfedge** in the face (in the counter-clockwise direction).
- **The pair**, called also the opposite, half-edge.
- **The previous half-edge** in the face. This pointer is optional, but it is used for a better performance.

Additionally, this data structure stores at each face, references to one of its adjacent half-edges. Each vertex also stores a reference to one of its outgoing half-edges. This simple data structure enables to enumerate, for each element (i.e. vertex, edge, half-edge, or face), its adjacent elements in a constant time independently of the size of the mesh.

Other efficient representations include the directed edges [6], which is a memory efficient variant of the half-edge data structure. These data structures are implemented in libraries such as the Computational Geometry Algorithms Library (CGAL)[1] and OpenMesh.[2]

2.1.4.3 Discretization of the Differential Properties of Surfaces

Since, in general, surfaces are represented in a discrete fashion using vertices v_i and faces t_j, their differential properties need to be discretized.

First, the discrete Laplace–Beltrami operator at a vertex v_i can be computed using the following general formula:

$$\Delta_{\mathcal{M}} f(v_i) = \frac{1}{A(v_i)} \sum_{v_j \in \mathcal{N}(v_i)} a_{ij}(f(v_j) - f(v_i)), \tag{2.27}$$

where $\mathcal{N}(v_i)$ denotes the set of all vertices that are adjacent to v_i, a_{ij} are some weight factors, and $A(v_i)$ is a normalization factor. Discretization methods differ in the choice of these weights. The simplest approach is to choose $a_{ij} = 1$ and $A(v_i) = |\mathcal{N}(v_i)|$. This approach, however, is not efficient. In fact, all edges are considered equally, independently of their length. This may lead to poor results (e.g. when smoothing meshes) if the vertices are not uniformly distributed on the surface of the object.

A more accurate way of discretizing the Laplace–Beltrami operator is to account for how the vertices are distributed on the surface. This is done by setting:

$$a_{ij} = \cot \alpha_{ij} + \cot \beta_{ij},$$

and $A(v_i)$ is equal to twice the Voronoi area of the vertex v_i, which is defined as one-third of the total area of all the triangles attached to the vertex v_i. The

---

1 http://www.cgal.org.
2 http://www.openmesh.org.

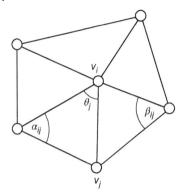

Figure 2.10 Illustration of the cotangent weights used to discretize the Laplace–Beltrami operator.

angles α_{ij} and β_{ij} are the angles opposite to the edge that connects v_i to v_j, see Figure 2.10.

The discrete curvatures can be computed using the discrete Laplace–Beltrami operator as follows:

- The mean curvature:

$$H(v_i) = \frac{1}{2}\|\Delta_{\mathcal{M}}f(v_i)\|. \tag{2.28}$$

- The Gaussian curvature at a vertex v_i is given as

$$K(v_i) = \frac{1}{A(v_i)}\left(2\pi - \sum_j \theta_j\right), \tag{2.29}$$

where $A(v_i)$ is the Voronoi area, and θ_j is the angle between two adjacent edges incident to v_i, see Figure 2.10.
- The principal curvatures:

$$\kappa_1(v_i) = H(v_i) - \sqrt{H(v_i)^2 - K(v_i)}, \tag{2.30}$$

$$\kappa_2(v_i) = H(v_i) + \sqrt{H(v_i)^2 - K(v_i)}. \tag{2.31}$$

2.2 Shape, Shape Transformations, and Deformations

There are several definitions of shape. The most common one is the one that defines shape as *the external form, contour, or outline of someone or something*. Kendall [7] proposed to study shape as that which *is left when the effects asso-ciated with translation, isotropic scaling, and rotation are filtered away*. This is illustrated in the 2D example of Figure 2.11 where although the five plant leaves have different scales and orientations, they all represent the same leaf shape.

Kendall's definition sets the foundation of shape analysis. It implies that when comparing the shape of two objects, whether they are 2D or 3D, the criteria for

Figure 2.11 Similarity transformations: examples of 2D objects with different orientations, scales, and locations, but with identical shapes.

comparison should be invariant to *shape-preserving transformations*, which are translation, rotation, and scale. It should, however, quantify the deformations that affect shape. For instance, 3D objects can bend, e.g. a walking person, stretch, e.g. a growing anatomical organ, or even change their topology, e.g. an eroding bone, or structure, e.g. a growing tree. In this section, we will formulate these transformations and deformations. The subsequent chapters of the book will show how the Kendall's definition is taken into account in almost every shape analysis framework and task.

2.2.1 Shape-Preserving Transformations

Let $\mathcal{F}$ be the space of all surfaces f defined by their parametric function of the form $f : D \to \mathbb{R}^3$. There are three transformations that can be applied to a surface f without changing its shape.

- **Translation.** Translating the surface f with a displacement vector $\mathbf{v} \in \mathbb{R}^3$ will produce another surface $\tilde{f} \in \mathcal{F}$ such that:

$$\forall s, \quad \tilde{f}(s) = f(s) + \mathbf{v}. \tag{2.32}$$

  With a slight abuse of notation, we sometimes write $\tilde{f} = f + \mathbf{v}$.
- **Scale.** Uniformly scaling the surface f with a factor $\alpha \in \mathbb{R}_{>0}$ will produce another surface $\tilde{f} \in \mathcal{F}$ such that:

$$\forall s, \quad \tilde{f}(s) = \alpha f(s). \tag{2.33}$$

- **Rotation.** A rotation is a 3×3 matrix $O \in SO(3)$. When applied to a surface $f \in \mathcal{F}$, we obtain another surface $\tilde{f} \in \mathcal{F}$ such that:

$$\forall s, \quad \tilde{f}(s) = Of(s).$$

  With a slight abuse of notation, we write $\tilde{f} = Of$.

In the field of shape analysis, the transformations associated with translation, rotation, and uniform scaling are termed *shape-preserving*. They are nuisance variables, which should be discarded. This process is called *normalization*. Normalization for translation and scale is probably the easiest one to deal with. Normalization for rotation requires a more in-depth attention.

2.2.1.1 Normalization for Translation

One can discard translation by translating f so that its center of mass is located at the origin:

$$f(s) \rightarrow f(s) - c, \tag{2.34}$$

where c is the center of mass of f. It is computed as follows:

$$c = \frac{\int_D a(s)f(s)}{\int_D a(s)}. \tag{2.35}$$

Here $a(s)$ is the local surface area at s. It is defined as the norm of the normal vector to the surface at s. In the case of objects represented as a set of vertices $V = \{v_i, i = 1, \ldots, n\}$ and faces $T = \{t_i, i = 1, \ldots, m\}$, then the center of mass c is given by:

$$c = \frac{1}{A} \sum_{i=1}^{n} a_i v_i, \quad \text{with } A = \sum_{i=1}^{n} a_i. \tag{2.36}$$

Here, a_i's are weights associated to each vertex v_i. One can assume that all vertices have the same weight and set $a_i = 1, \forall i$. This is a simple solution that works fine when the surfaces are densely sampled. A more efficient choice is to set a_i to be the Voronoi area around the vertex v_i.

2.2.1.2 Normalization for Scale

The scale component can be discarded in several ways. One approach is to scale each 3D object being analyzed in such a way that its overall surface area is equal to one:

$$f(s) \rightarrow \frac{1}{A} f(s), \tag{2.37}$$

where A is a scaling factor chosen to be the entire surface area: $A = \int_D a(s) ds$. When dealing with triangulated surfaces, A is chosen to be the sum of the areas of the triangles which form the mesh.

Another approach is to scale the 3D object in such a way that the radius of its bounding sphere is equal to one. This is usually done in two steps. **First**, the 3D object is normalized for translation. **Then**, it is scaled by a factor d where d is the distance between the origin and the farthest point p on the surface from the origin.

2.2.1.3 Normalization for Rotation

Normalization for rotation is the process of aligning all the 3D models being analyzed in a consistent way. This step is often used after normalizing the shapes for translation and scale. Thus, in what follows, we assume that the 3D models are centered at the origin. There are several ways for rotation normalization. The first approach is to compute the principal directions of the object and then rotate it in such a way that these directions coincide with the three axes of the coordinate system. The second one is more elaborate and uses reflection symmetry planes.

2.2.1.3.1 Rotation Normalization Using Principal Component Analysis (PCA) The first step is to compute the 3×3 covariance matrix C:

$$C = \frac{1}{n}(f(s) - c)(f(s) - c)^{\top}, \tag{2.38}$$

where n is the number of points sampled on the surface f of the 3D object, c is its center of mass, and $X^{\top}$ refers to the transpose of the matrix X. For 3D shapes that are normalized for translation, c is the origin. When dealing with surfaces represented as polygonal meshes, Eq. (2.38) is reduced to

$$C_v = \frac{1}{n} \sum_{v \in V} (v - c)(v - c)^{\top}. \tag{2.39}$$

Here, n is the number of vertices in V. Let $\{\lambda_i, i = 1, 2, 3\}$ be the eigenvalues of C, ordered from the largest to the smallest, and $\{\Lambda_i, i = 1, 2, 3\}$ their associated eigenvectors (of unit length). Rotation normalization is the process of rotating the 3D object such that the first principal axis, Λ_1, coincides with the X axis, the second one, Λ_2, coincides with the Y axis, and the third one, Λ_3, coincides with the Z axis. Note, however, that both Λ_i and $-\Lambda_i$ are correct principal directions. To select the appropriate one, one can set the positive direction to be the one that has maximum volume of the shape. Once this is done, then the 3D object can be normalized for rotation by rotating it with the rotation matrix O that has the unit-length eigenvectors as rows, i.e.:

$$O = [\Lambda_1, \Lambda_2, \Lambda_3]^T \quad \text{and} \quad f(s) \to O(f(s) - c). \tag{2.40}$$

This simple and fast procedure has been extensively used in the literature. Figure 2.12 shows a few examples of 3D shapes normalized using this approach. This figure also shows some of the limitations. For instance, in the case of 3D human body shapes, Principal Component Analysis (PCA) produces plausible

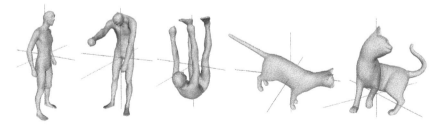

Figure 2.12 PCA-based normalization for translation, scale, and rotation of 3D objects that undergo nonrigid deformations.

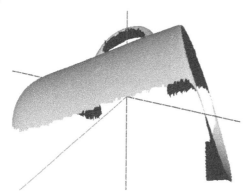

Figure 2.13 PCA-based normalization for translation, scale, and rotation-of partial 3D scans. The 3D model is courtesy of SHREC'15: Range Scans based 3D Shape Retrieval [8].

results when the body is in a neutral pose. The quality of the alignment, however, degradates with the increase in complexity of the nonrigid deformation that the 3D shape undergoes. The cat example of Figure 2.12 also illustrates another limitation of the approach. In fact, PCA-based normalization is very sensitive to outliers in the 3D object and to nonrigid deformations of the object's extruded parts such as the tail and legs of the cat. As shown in Figure 2.13, PCA-based alignment is also very sensitive to geometric and topological noise and often fails to produce plausible alignments when dealing with partial and incomplete data.

In summary, while PCA-based alignment is simple and fast, which justifies the popularity of the method, it can produce implausible results when dealing with complex 3D shapes (e.g. shapes which undergo complex nonrigid deformations, partial scans with geometrical and topological noise). Thus, objects with a similar shape can easily be misaligned using PCA. This misalignment can result in an exaggeration of the difference between the objects, which in turn will affect the subsequent analysis tasks such as similarity estimation and registration.

2.2.1.3.2 *Rotation Normalization Using Planar Reflection Symmetry Analysis*

PCA does not always produce compatible alignments for objects of the same class. It certainly does not produce alignments similar to what a human would select [9]. Podolak et al. [9] used the planar-reflection symmetry transform (PRST) to produce better alignments. Specifically, they introduced two new concepts, the principal symmetry axes (PSA) and the Center of Symmetry (COS), as robust global alignment features of a 3D model. Intuitively, the PSA are the normals of the orthogonal set of planes with maximal symmetry. In 3D, there are three of such planes. The Center of Symmetry is the intersection of those three planes. The first principal symmetry axis is defined as the plane with maximal symmetry. The second axis can be found by searching for the plane with maximal

Figure 2.14 Comparison between (a) PCA-based normalization, and (b) planar-reflection symmetry-based normalization of 3D shapes. Source: Image courtesy of http://gfx.cs .princeton.edu/pubs/ Podolak_2006_APS/ Siggraph2006.ppt and reproduced here with the permission of the authors.

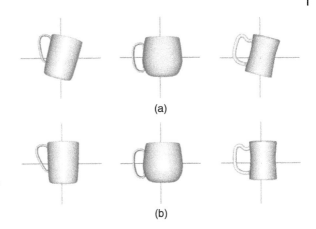

(a)

(b)

symmetry among those that are perpendicular to the first one. Finally, the third axis is found in the same way by searching only the planes that are perpendicular to both the first and the second planes of symmetry.

We refer the reader to [9] for a detailed description of how the planar-reflection symmetry is computed. Readers interested in the general topic of symmetry analysis of 3D models are referred to the survey of Mitra et al. [10] and the related papers. Figure 2.14 compares alignments produced using (a) PCA and (b) planar reflection symmetry analysis. As one can see, the latter produces results that are more intuitive. More importantly, the latter method is less sensitive to noise, missing parts, and extruded parts.

2.2.2 Shape Deformations

In addition to shape-preserving transformations, 3D objects undergo many other deformations that change their shapes. Here, we focus on full elastic deformations, which we further decompose into the bending and the stretching of the surfaces. Let f_1 and f_2 be two parameterized surfaces, which are normalized for translation and scale, and D the parameterization domain. We assume that f_1 deforms onto f_2.

2.2.3 Bending

Bending, also known as *isometric deformation*, is a type of deformations that preserves the intrinsic properties (areas, curve lengths) of a surface. During the bending process, the distances along the surface between the vertices are preserved. To illustrate this, consider the 3D human model example of Figure 2.15. When the model bends its arms or legs, the length of the curves along the

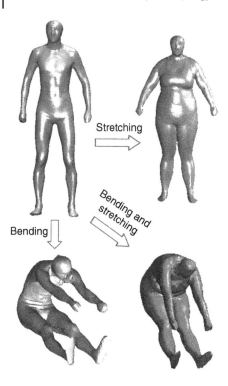

Figure 2.15 Example of nonrigid deformations that affect the shape of a 3D object. A 3D object can bend, and/or stretch. The first one is referred to as *isometric deformations*. Fully elastic shapes can, at the same time, bend and stretch.

Stretching

Bending and stretching

Bending

surface do not change. There are, however, two properties that change. **The first** one is the angle between, or the relative orientation of, the normal vectors of adjacent vertices. **The second** one is the curvature at each vertex. As described in Section 2.1.2, normals and curvatures are encoded in the shape operator S. Hence, to quantify bending, one needs to look at how the shape operator changes when a surface deforms isometrically. This can be done by defining an energy function E_b, which computes distances between shape operators, i.e.:

$$E_b(f_1, f_2) = \int_D \|S_2(s) - S_1(s)\|_F^2 ds, \tag{2.41}$$

where $\| \cdot \|_F$ denotes the Frobenius norm of a matrix, which is defined for a matrix A of elements a_{ij} as $\left(\sum_{ij} a_{ij}^2 \right)^{1/2}$. $S_i(s)$ is the shape operation of f_i at s.

The energy of Eq. (2.41) takes into account the full change in the shape operator. Heeren et al. [11] and Windheuser et al. [12] simplified this metric by only considering changes in the mean curvature, which is the trace of the shape operator:

$$E_b(f_1, f_2) = \int_D \|\text{trace}(S_2(s)) - \text{trace}(S_1(s))\|^2 ds. \tag{2.42}$$

This energy is also known as the *Willmore energy*. In the discrete setup, Heeren et al. [13] approximated the mean curvature by the dihedral angle, i.e. the angle between the normals of two adjacent faces on the triangulated mesh.

Another way of quantifying bending is to look at changes in the orientation of the normal vectors of the surface during its deformation. Let $\tilde{n}(s)$ be the unit normal vector to the surface f at s. Let also $\delta\tilde{n}(s)$ be an infinitesimal perturbation of this normal vector. The magnitude of this perturbation is a natural measure of the strength of bending. It can be defined in terms of inner products as follows:

$$\langle\langle\delta\hat{n}, \delta\hat{n}\rangle\rangle = \int_D \langle\delta\tilde{n}(s), \delta\tilde{n}(s)\rangle ds. \tag{2.43}$$

2.2.4 Stretching

When stretching a surface f, there are two geometric properties of the surface that change, see Figure 2.15. **The first** one is the length of the curves along the surface. Consider two points p_1 and p_2 on f. Consider also a curve on the surface that connects these two points. Unlike bending, when stretching the surface, the length of this curve changes. **The second** property that changes is the local surface area at each surface point.

Both curve lengths and surface areas are encoded in the first fundamental form $g(s)$, or metric, of the surface at each point $p = f(s)$. Formally, let f_1 be a surface and f_2 a stretched version of f_1. Let also $g_i(s)$ be the first fundamental form of f_i at $p_1 = f(s)$. The quantity $G = g_1^{-1}g_2$, called the *Cauchy–Green strain tensor*, accounts for changes in the metric between the source and the target surfaces. In particular, the trace of the Cauchy–Green strain tensor, trace(G), accounts for the local changes in the curve lengths while its determinant, det(G), accounts for the local changes in the surface area [14]. Wirth et al. [15], Heeren et al. [11, 16], Berkels et al. [17], and Zhang et al. [14] used this concept to define a measure of stretch between two surfaces f_1 and f_2 as $E_s(f_1,f_2) = \int_D E_s(G(s))ds$ where:

$$E_s(G(s)) = \frac{\mu}{2} \operatorname{trace}(G(s)) + \frac{\lambda}{4} \det(G(s))$$
$$- \left(\frac{\mu}{2} + \frac{\lambda}{4}\right) \log\det(G(s)) - \mu - \frac{\lambda}{4}. \tag{2.44}$$

Here, λ and μ are constants that balance the importance of each term of the stretch energy. The log det term (the third term) penalizes shape compression.

Jermyn et al. [18], on the other hand, observed that stretch can be divided into two components; a component which changes the area of the local patches, and a component which changes the shape of the local patches while preserving their areas and the orientation of their normal vectors. These quantities can be captured by looking at small perturbations of the first fundamental form of the deforming surface. Let δg be an infinitesimal perturbation of g. Then, stretch can be quantified as the magnitude of this perturbation, which can be defined using inner products as follows:

$$\langle\langle \delta g, \delta g \rangle\rangle = \int_{\mathbb{S}^2} ds \sqrt{|g|}\{\alpha\mathrm{trace}((g^{-1}\delta g)^2) + \beta\mathrm{trace}\ (g^{-1}\delta g)^2\}. \qquad (2.45)$$

The first term of Eq. (2.45) accounts for changes in the shape of a local patch that preserve the patch area. The second term quantifies changes in the area of these patches. α and β are positive values that balance between the importance of each term.

Finally, one can also measure stretch by taking the difference between the first fundamental forms of the original surface f_1 and its stretched version f_2:

$$E_s(f_1, f_2) = \int_D \|g_1(s) - g_2(s)\|_F^2 ds, \qquad (2.46)$$

where $\|\cdot\|_F$ refers to the Frobenius norm and the integration is over the parameterization domain.

2.3 Summary and Further Reading

This chapter covered the important fundamental concepts of geometry and topology, differential geometry, transformations and deformations, and the data structures used for storing and processing 3D shapes. These concepts will be extensively used in the subsequent chapters of the book.

The first part, i.e. the elements of differential geometry, will be extensively used in Chapters 4 and 5 to extract geometrical features and build descriptors for the analysis of 3D shapes. They will be also used to model deformations, which are essential for the analysis of 3D shapes which undergo rigid and elastic deformations (Chapters 6 and 7). The concepts of manifolds, metrics, and geodesics, covered in this part, have been used in two ways. For instance, one can treat the surface of a 3D object as a manifold equipped with a metric. Alternatively, one can see the entire space of shapes as a (Riemannian) manifold equipped with a proper metric and a mechanism for computing geodesics. Geodesics in this space correspond to deformations of shapes. These will be covered in detail in Chapter 7, but we also refer the reader to [19] for an in-depth description of the Riemannian geometry for 3D shape analysis.

The second part of this chapter covered shape transformations and deformations. It has shown how to discard, in a preprocessing step, some of the transformations that are shape-preserving. As we will see in the subsequent chapters, this preprocessing step is required in many shape analysis algorithms. We will see also that, in some situations, one can design descriptors that are invariant to some of these transformations and thus one does not need to discard them at the preprocessing stage.

Finally, this chapter can serve as a reference for the reader. We, however, refer the readers who are interested in other aspects of 3D digital geometry to the rich literature of digital geometry processing such as [1, 19, 20].

The 3D models used in this chapter have been kindly made available by various authors. For instance, the 3D human body shapes used in Figures 2.12 and 2.15 are courtesy of [21], the 3D cat models are part of the TOSCA database [22], and the partial 3D scans of Figure 2.13 are courtesy of [8].

3

3D Acquisition and Preprocessing

3.1 Introduction

Before we can elaborate in detail on 3D shape analysis techniques, we need to understand the process that underpins the creation of 3D models. This is often referred to as 3D reconstruction. It involves the acquisition of 3D data followed by preprocessing steps, which clean, smooth, and optimize the 3D data to facilitate their analysis and usage in various applications. In this chapter, we will review in Section 3.2 the existing 3D acquisition techniques and systems. We will particularly discuss the different characteristics that can influence decision making when it comes to selecting a system or a technique that would best suit a given application.

In general, 3D acquisition systems produce raw 3D data that are not necessarily clean and in a representation that does not suit shape analysis tasks. Section 3.3 introduces some of the basic techniques used for smoothing, cleaning, and repairing 3D data. It also discusses parameterization techniques, which convert discrete representations of 3D shapes into a continuous one.

The material covered in this chapter is a brief summary of a very rich and deep set of topics, traditionally covered in a number of separate books. A more thorough introduction to 3D reconstruction techniques can be found in textbooks on multiview geometry [23]. Preprocessing techniques are usually covered in textbooks on 3D geometry processing such as [1].

3.2 3D Acquisition

The goal of the 3D acquisition task is to collect the geometric samples (i.e. 3D points) and possibly the appearance (i.e. color and texture) of the surface of an object. These points produced by a 3D acquisition system can then be used to represent the shape of the object. The 3D acquisition problem has been extensively investigated, with a number of commercial systems available in the market. 3D acquisition techniques can broadly be divided into (i) contact 3D

3D Shape Analysis: Fundamentals, Theory, and Applications, First Edition.
Hamid Laga, Yulan Guo, Hedi Tabia, Robert B. Fisher, and Mohammed Bennamoun.
© 2019 John Wiley & Sons, Inc. Published 2019 by John Wiley & Sons, Inc.

acquisition and (ii) noncontact (or contactless) 3D acquisition techniques [24]. Most of the research focuses on noncontact 3D acquisition for their practical convenience. These noncontact 3D acquisition methods can further be classified into several different categories, including triangulation, time-of-flight (TOF), structured light, stereo, and shape-from-X, where X can refer to motion, silhouette, shading, photometry, and focus. Each 3D acquisition technique has its specific characteristics, which should be carefully considered when selecting a 3D acquisition system to suit a particular application. Examples of such characteristics include [25]

- **Range**. Contact 3D acquisition systems require physical touch between the probe of the scanner and the object under scanning. On the other hand, the working distance of noncontact 3D acquisition systems can range from few decimeters to several kilometers. Different applications may require different working ranges for the 3D acquisition system.
- **Accuracy**. The accuracy of a 3D acquisition system determines the level of details that can be correctly reconstructed. The requirement for accuracy relies on the particular application. For example, the accuracy requirement for surface defect detection should be higher than that for territory survey.
- **Resolution**. The resolution of a 3D acquisition system determines the smallest element of a surface that the system can differentiate. Sometimes, resolution is also defined as the maximum number of points that the system can obtain in a single frame.
- **Cost**. The cost of different 3D acquisition systems varies significantly. For example, an Optech ILRIS terrestrial laser scanner costs more than one hundred thousand US dollars while a Microsoft Kinect depth sensor just costs about US$150.
- **Efficiency**. The viewing angle and scanning speed of different 3D acquisition systems also vary significantly, resulting in different acquisition efficiencies. Besides, some acquisition systems (e.g. contact 3D acquisition) can produce 3D data in real-time, while other systems may require offline data processing (e.g. large-scale multiview stereo).
- **Portability**. Portability is important for some applications such as the scanning of remote cultural heritage sites and criminal scenes. In these cases, factors such as power supply (e.g. battery), weight, and size of a system should be carefully considered.
- **Flexibility**. The objects under scanning may have various attributes such as materials, topologies, geometrical structures, and dimensions. Consequently, the 3D acquisition system should handle these relevant issues when faced by a specific application.
- **Others**. Several other issues should also be considered for a particular application. Examples include the depth of field and the field of view.

In the following subsections, we will review some of the existing 3D acquisition techniques and discuss their characteristics and their suitability for some applications.

3.2.1 Contact 3D Acquisition

A contact 3D acquisition system produces the 3D coordinates of the surface of a subject by probing the subject through physical touch. There are two typical contact 3D acquisition systems: the coordinate measuring machine (CMM) and the arm-based 3D scanner.

3.2.1.1 Coordinate Measuring Machine (CMM)

A common CMM machine, see Figure 3.1a, is composed of three major components: (i) the main structure, (ii) the probe system, and (iii) the data collection and reduction system. The main structure has three axes of motion and can be made of different materials such as granite, steel, aluminum alloy, and ceramic. The probe, which is attached to the third axis of the main structure, is used to measure the geometric surface of the subject being scanned. A typical probe is a small ball installed at the end of a shaft, but it can also be laser or white light sensors. The probe can glide along each axis through a track. A scale system is installed to each axis of the CMM machine to measure the location of the probe along that axis. During 3D acquisition, the machine, which can be controlled by a user or by a computer, scans a 3D object by moving the touch probe. The 3D

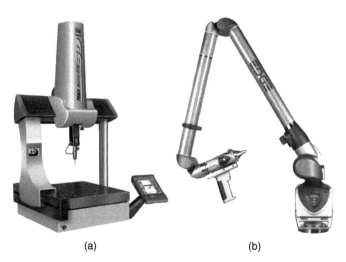

(a)   (b)

Figure 3.1 An illustration of contact 3D acquisition systems. (a) A coordinate measuring machine (CMM). (b) An arm-based 3D scanner.

coordinates of all points are then collected to determine the shape and size of the object being scanned with a micrometer precision.

A CMM machine can produce very precise measurements. It works best for objects with flat profile shapes or simple convex surfaces. The machine has been commonly used in the manufacturing industry. However, it requires physical touch with the object it scans. Thus, it may cause modification or damage to fragile objects such as historical artifacts. Besides, the 3D acquisition efficiency of a CMM machine is very low, as it has to physically move the arm from one position to the next. Note also that, in order to achieve a very high measurement precision (e.g. a few micrometers), the CMM has to be deployed in a controlled environment with a reinforced floor, a strictly controlled temperature and humidity, and a vibration isolation device. The object being scanned should also be fixed to a table to ensure that it remains static during the 3D acquisition.

3.2.1.2 Arm-Based 3D Scanner

Similar to a CMM, an arm-based 3D scanner also requires a touch probe to measure an object. Unlike a CMM, which uses a rigid arm, an arm-based 3D scanner uses an articulated robotic arm in order to provide a high degree of freedom. The articulated arm is usually attached to a table or a stable base, and is held by a user with a hand grip to move around for 3D scanning. The movement of the arm is precisely recorded by the scanner so that the accurate 3D position of each probe point can be calculated.

Arm-based 3D scanners share the same merits and demerits of CMM machines, but the former are more portable. The Faro Arm, which is shown in Figure 3.1b, is one example of arm-based 3D scanners.

3.2.2 Noncontact 3D Acquisition

Numerous noncontact techniques have been developed for 3D acquisition. Examples include TOF cameras, triangulation, stereo, structured light, and shape-from-X. Below, we describe some of these techniques.

3.2.2.1 Time-of-Flight

There are two major types of TOF 3D acquisition systems: pulse-based TOF and phase shift-based TOF.

3.2.2.1.1 *Pulse-Based TOF* For a pulse-based TOF sensor, a laser pulse is first emitted onto the surface of an object. The laser is then reflected back once the surface of the object is reached. The returned laser pulse is finally recorded by a receiver. If c is the speed of light in the air, and Δt is the round-trip flight time of the laser pulse, then the distance d between the scanner and the object surface can be calculated as:

$$d = \frac{1}{2}c\Delta t. \tag{3.1}$$

The accuracy of a pulse-based TOF scanner depends on the accuracy at which time is measured. Specifically, a distance measurement accuracy of 1 mm corresponds to a time counting accuracy of 3.3 ps.

3.2.2.1.2 Phase Shift-Based TOF For a phase shift-based TOF sensor, a laser light is emitted in a continuous beam rather than as a pulse. Besides, the amplitude of the laser is modulated sinusoidally over time. The laser reflected from a surface still has a sinusoidal wave form over time, but the phase is different from the emitted laser. Consequently, the distance between the laser transmitter and the receiver can be calculated by measuring the difference in phases between the received and the emitted laser signals. Usually, phase shift measurement is more accurate than the elapsed time measurement used in pulse-based TOF sensors. Meanwhile, the working range for phase shift-based TOF sensors is usually shorter than pulse-based TOF sensors. In fact, the former works for up to several hundred meters while the latter works for up to a few kilometers.

Note that, for a single-pixel TOF sensor, only a single point can be obtained at each scan. To reconstruct an entire scene, some sort of scanning of the laser beam has to be performed, e.g. by using rotating mirrors. A good alternative is to emit a plane of light (not a light beam or stripe) onto the scene and then collect the reflected light using an array of receivers. In this case, the sensor can acquire the 3D data of the entire scene in one scan.

A major advantage of the TOF sensors is their long working distance. They can acquire about 1 million points per second with a working range of up to few kilometers. Both pulse and phase shift-based TOF sensors are suitable for scanning large objects such as roads, buildings, trees, geographical features, and aircrafts. Phase shift-based TOF sensors are also suitable for medium or short range applications such as the scanning of vehicles, rooms, and industrial parts.

The measurement accuracy of TOF sensors is relatively low (usually at the scale of a few millimeters). This is because measuring the elapsed time is very challenging as the speed of light is too high. This accuracy can dramatically be low when the laser shines on the edge of a surface. This is because when a laser ray hits a surface edge, the reflected laser, which is collected by the receiver, comes from two different locations on the surface for one laser, and the average distance is considered as the measurement of the point on the edge, which is incorrect. A possible solution to this problem is to reduce the width of the laser beam.

3.2.2.2 Triangulation

A triangulation scanner, see Figure 3.2a, consists of a light projector and a digital camera. The projector projects a light (usually a laser) spot onto the surface of an object. The reflected light is then recorded by the camera. Consequently, the center of the projector, the light dot on the object, and the center of the camera form a triangle. The distance between the camera and the projector

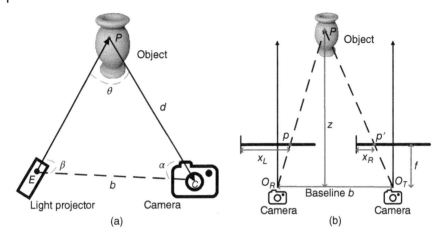

Figure 3.2 An illustration of (a) triangulation and (b) stereo-based 3D acquisition systems. A triangulation-based system is composed of a light projector and a camera, while a stereo-based systems uses two cameras. In both cases, the depth of a point on the 3D object is inferred by triangulation.

is known as the baseline, the angle at the projector is known in priori, and the angle at the camera can be obtained by looking at the light dot on the object surface in the field of view of the camera. Given these two angles and the length of the baseline b, the shape and the size of this triangle are fully determined. Consequently, the distance from the projector to the light spot can be calculated. That is, the coordinates of the sample point on the surface where the light spot lies can be obtained.

For simplicity, let us consider the 2D example of Figure 3.2a, which illustrates the triangulation principle. In this example, the projector center E, the camera center C, and the point p on the surface of the 3D object form a triangle. With the known baseline b and angle β, we have:

$$\frac{b}{\sin(\theta)} = \frac{d}{\sin(\beta)}, \tag{3.2}$$

where $\theta = 180 - \alpha - \beta$ is the angle formed by the lines PE and PC, β is the angle formed by the lines EP and EC, and d is the distance from the camera center C to the point P on the surface. From this equation, we can derive that:

$$d = \frac{b\sin(\beta)}{\sin(\theta)} = \frac{b\sin(\beta)}{\sin(\alpha + \beta)}, \tag{3.3}$$

where α is the angle formed by the lines CE and CP. It can be obtained by the projection geometry as it is determined by the position of the projected point P in the image. Finally, the 2D coordinates of a point P can be obtained as $(d\cos(\alpha), d\sin(\alpha))$.

By illuminating a light spot onto a surface, only one single point can be reconstructed each time. To reconstruct a large surface, the light spot can be projected over the whole surface using mirrors. However, this approach is relatively slow. An alternative approach is to use a plane of light rather than a single light spot. That is, a light stripe is projected onto the surface of the object. The reflected light stripe is then recorded by a camera. Consequently, a range profile can be obtained from each projection. The entire surface of the object can efficiently be reconstructed by sweeping the light stripe over the surface [26]. The triangulation-based scanner with a light stripe is also called *a slit scanner* [27].

Triangulation-based 3D acquisition systems are able to scan tough surfaces (such as shiny or dark surfaces). They are less sensitive to illumination variations and ambient light, easy to use, and usually portable. These systems usually have a working range of several meters with a submillimeter accuracy. However, they are often less accurate on shiny and transparent surfaces [25]. Triangulation-based 3D acquisition systems are not suitable for scanning moving objects since several spots or stripes need to be swept over the scene. During this time, the objects in the scene are required to remain static. Finally, due to the separation between the projector and the camera, several regions may be occluded, i.e. cannot be seen at the same time by the camera and the projector, and thus, they cannot be reconstructed [27].

3.2.2.3 Stereo

Stereo vision systems, which are inspired by the human binocular vision system, use two images (hereinafter referred to as the left and right images), captured by two cameras located slightly apart from each other, to estimate the scene depth at each image pixel. The process is fairly simple. It start by finding for each pixel in the left image, its corresponding pixel in the right image. Two pixels, one in the left image and the other in the right image, are in correspondence if they are the image of the same point in the 3D scene. This is referred to as *stereo matching*. Once the correspondences are computed, depth can be estimated by triangulation.

In practice, correspondences are unknown a priori, and computing them automatically without any constraints is not an easy task. Rather than searching for each pixel in the left image its corresponding pixel in the right image over the entire 2D image space, epipolar constraints can be enforced to restrict the search to just along the epipolar lines. Moreover, instead of working directly on the original images, one can work on rectified images, which are rectified in such a way that the epipolar lines are horizontal. In this case, for each pixel s in the left image, its corresponding point in the right image is located on the same row as s. Epipolar constraints significantly reduce the search space and thus simplify the stereo matching process.

Consider the simplest case shown in Figure 3.2b where the two cameras have the same focal length f, and their imaging planes are located in the same plane. Let P be a point in the 3D scene. P is unknown and the goal of 3D reconstruction is to retrieve its coordinates from its two projections p and p' on the left and right image respectively. If O_L and O_R are the projection centers of the left and right cameras, respectively, then the two triangles $\triangle Ppp'$ and $\triangle PO_LO_R$ are similar. Thus, one can write:

$$\frac{b}{z} = \frac{(b + x_R) - x_L}{z - f}. \text{Thus, } z = \frac{bf}{d}. \tag{3.4}$$

Here, b is the baseline of the stereo rig, which is defined as the distance between the centers of the two cameras, z is the depth from the camera to point P on the scene, and $d = x_L - x_R$ is the difference between the x coordinates of the two corresponding points. d, which is referred to as *the disparity*, is inversely proportional to the depth z.

This procedure assumes that the parameters of the two cameras (e.g. focal lengths and center of projections) are known. This is, however, not the case in practice. Thus, 3D reconstruction by stereo vision usually consists of four main steps:

- **Calibration.** The offline calibration step estimates the intrinsic parameters (e.g. the focal lengths, the image centers, and the lens distortion) and extrinsic parameters (rotation and translation) of the two cameras. This is usually done by processing a set of stereo image pairs of a checkerboard.
- **Rectification.** The rectification step removes the perspective and lens distortions, and transforms the stereo rig into a basic form where the two cameras have parallel optical axes. Thus, the epipolar lines of the two cameras become horizontal.
- **Stereo matching or correspondence.** Stereo correspondence finds for each pixel in the left image, its corresponding pixel in the right image.
- **Triangulation.** Once point correspondence is known, the disparity map can be estimated, using Eq. (3.4), to generate the depth map.

In practice, finding correspondences between the stereo pair of images, even when rectified, is very challenging, especially for scenes with noise, photometric distortions, perspective distortions, specular surfaces, and bland regions. Most of the existing stereo-based 3D reconstruction algorithms perform the stereo correspondence, and thus the disparity estimation, using all or part of the following four steps [28]; (i) matching cost computation, (ii) cost aggregation, (iii) disparity computation/optimization, and (iv) disparity refinement. Local methods usually use a simple method to perform matching cost computation and then reduce the ambiguity by aggregating the matching costs in the support region. In contrast, global methods usually do not have the cost aggregation step. Instead, they obtain the optimal disparity map by minimizing an energy function, which is defined using pixel-based matching costs over the entire stereo image pair. Below, we detail these steps.

- **Matching cost computation.** Commonly used matching cost calculation methods include pixel-based matching costs and area-based matching costs. Examples of pixel-based matching costs include the absolute intensity differences (ADs) and the Squared intensity Difference (SD). Examples of area-based matching costs include normalized cross-correlation (NCC), sum of absolute differences (SAD), and sum of squared differences (SSDs) [28, 29].

- **Cost aggregation.** For local methods, the matching costs are aggregated by summing or averaging these costs in a support region in the disparity space image. Different types of 2D support regions have been investigated for cost aggregation. Examples include square windows, Gaussian convolution windows, windows with adaptive sizes, shiftable windows, and windows based on connected components of constant disparity. Several 3D support regions have also been studied. Examples include limited disparity difference and limited disparity gradient. A comprehensive evaluation of the cost aggregation strategies for stereo matching can be found in [30].

- **Disparity computation/optimization.** For local-based methods, a local winner-take-all (WTA) strategy is usually performed at each pixel. That is, the disparity with the minimum cost is selected for each pixel. The problem is that a point in one image may correspond to multiple points in the other image, resulting in incorrect disparity estimation. Global methods usually minimize a global energy function, which contains both a data term and a smoothness term. The data term measures the degree of agreement between the disparity and the stereo image pair. The smoothness term is used to enforce piecewise disparity continuity between neighboring pixels. Once the global energy function has been defined, several methods, such as simulated annealing, max-flow, graph-cut, and dynamic programming, can be used to optimize them [28].

- **Disparity refinement.** The raw disparity maps produced by most stereo correspondence algorithms contain outliers and are represented in discrete pixel levels (such as integer disparities or splines). In order to obtain more accurate disparity maps, a disparity refinement step is required. Examples of refinement methods include subpixel interpolation and image filtering [28].

Recently, deep neural networks have also been extensively investigated to learn a disparity map from two stereo images in an end-to-end manner. For instance, iResNet aggregates the four steps of stereo matching into a single network. It achieves the state-of-the-art performance on the KITTI 2012 [31] and KITTI 2015 [32] benchmarks with fast running times [33].

In summary, stereo sensors have many advantages. They are cheap, highly portable, can reconstruct 3D objects and scenes with their color and textures, and can be used for real-time applications. In contrast, the spatial resolution and accuracy of the 3D point clouds reconstructed by stereo sensors are usually low [25]. Besides, stereo-based reconstruction of scenes with no texture or periodic textures is still challenging [27].

3.2.2.4 Structured Light

One of the main challenges in stereo-based reconstruction is the computation of correspondences between the left and right images. To solve this problem, structured light sensors use light patterns (or codes) to facilitate the correspondence problem. A structured light sensor consists of a camera and a projector, see Figure 3.3. The projector shines a single pattern or a set of patterns onto the surface of an object. The camera then records the patterns on the surface. If the surface of the object under scanning is planar, then the pattern acquired by the camera would be similar to the pattern illuminated by the projector. However, if the object has some variations on the surface, the pattern acquired by the camera would be distorted compared to the projected pattern. Therefore, the 3D shape of the object can be reconstructed by comparing the projected patterns and the distorted patterns acquired by the camera [34].

Specifically, the patterns are designed such that each coded pixel has an associated codeword. That is, a mapping from the designed codewords to the coordinates of the pixels in the pattern is provided using the gray or color values of the pattern. Compared to stereo matching, the projected pattern helps to distinguish pixels using a local coding strategy [35]. Consequently, correspondences between the projected pattern and the pattern perceived by the camera can easily be established. Once correspondences are established, the distance from the sensor to each point on the surface of the 3D object under scanning can be calculated by triangulation, as described in Section 3.2.2.2.

Structured light sensors have several advantages. They are fast and can be used for large areas. They are also able to reconstruct the geometry and texture of the 3D objects, at high resolution, and with a high accuracy. However, they are sensitive to ambient illumination. They are also not suitable for scanning reflective and transparent surfaces.

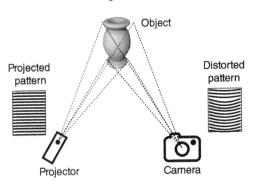

Figure 3.3 An illustration of a structured light 3D acquisition sensor. The projector shines a single pattern or a set of patterns onto the surface of an object. The camera then records the patterns on the surface. The 3D shape of the surface is estimated by comparing the projected patterns and the distorted patterns acquired by the camera.

Note that, the design of light patterns is critical for structured light-based 3D acquisition. Several methods have been proposed in the literature to design various light patterns, which can roughly be divided into temporal coded patterns (also called time-multiplexing), spatial coded patterns (also called spatial codification), and direct coded patterns (also called direct codification) [35]. Below, we discuss each of these patterns.

3.2.2.4.1 Temporal Coded Patterns In this type of coding, which is also called time multiplexing, a series of patterns are sequentially projected onto the surface of the object under reconstruction. The codeword of each pixel is then formulated as the sequence of values in these patterns on that pixel. That is, the bits in the codeword of each pixel are multiplexed in time.

There are several types of temporal coded patterns that can be used. Examples include binary patterns, gray patterns, phase shifting patterns, and hybrid patterns [34, 35].

- **Binary patterns**. These patterns are composed of two illumination levels, e.g. black and white, to represent the codes 0 and 1. Thus, when the projector projects a sequence of binary patterns onto the object, each pixel in the images captured by the camera will be assigned a sequence of zeros and ones. This sequence of binary values will form a unique codeword for each image pixel. Figure 3.4a shows an example of binary patterns. Since only two illumination levels are used, the segmentation of the binary patterns is relatively easy and straightforward. However, a large number of patterns are required to uniquely encode these pixels. Besides, binary pattern-based 3D acquisition sensors usually assume that the object under scanning has a uniform albedo. Otherwise, surface acquisition becomes very hard [35]. That is because if patterns are projected with high illumination intensities, regions with high reflectance are hard to scan due to pixel saturation. In contrast, if the patterns are projected with low illumination intensities, regions with low reflectance are challenging to be reconstructed since the signal-to-noise ratio in these regions is low [35].
- **Gray patterns**. To reduce the number of patterns that are required to be projected on the surface, one can use $n_i > 2$ intensity levels (instead of 0 and 1). Consequently, $n_i^{n_p}$ codes can be represented with n_p patterns. Gary patterns are more efficient than binary patterns, which can only represent 2^{n_p} codes. Note that, in gray pattern-based structured light systems, the gray patterns should be designed in such a way that the difference among all unique codewords (i.e. stripes) is maximized. The system should also be carefully calibrated so that different gray levels can easily be differentiated. Figure 3.4b shows an example of gray patterns.
- **Phase shifting patterns**. These methods project the same periodic pattern with different phase shifts. Take, for example, sinusoidal patterns with three

phase shifts. The intensities of each pixel (x, y) in the three patterns P_1, P_2, and P_3 are defined as:

$$P_1(x, y) = P_0(x, y) + P_{mod}(x, y) \cos(\phi(x, y) - \theta). \tag{3.5}$$

$$P_2(x, y) = P_0(x, y) + P_{mod}(x, y) \cos(\phi(x, y)). \tag{3.6}$$

$$P_3(x, y) = P_0(x, y) + P_{mod}(x, y) \cos(\phi(x, y) + \theta). \tag{3.7}$$

Here, $P_0(x, y)$ is the background component, $P_{mod}(x, y)$ defines the modulation of the signal amplitude, $\phi(x, y)$ gives the phase in pixel (x, y), and θ is the angle for phase shifting. An example of phase shifting patterns is shown in Figure 3.4c. Using the intensities of these phase-shifted patterns received by the camera, the phase $\phi(x, y)$ can be calculated for the estimation of depth value. A critical point for phase shifting methods is how to correctly estimate the absolute phase $\phi(x, y)$. Note that, only the relative phase (not the absolute phase) can be obtained using unwarping methods [34]. That is, if the phase is larger than 2π, the phase cannot correctly be estimated, resulting in a problem called *phase ambiguity.*

- **Hybrid patterns**. The binary pattern and gray pattern-based sensors can achieve a high 3D acquisition accuracy. Their spatial resolution, however, is limited by the width of the stripes and the optical quality. In contrast, the phase shifting pattern-based sensor can achieve a high spatial resolution (e.g. a fraction of the stripe width), but they are challenged by the phase ambiguity problem. In order to address these problems, a combination of binary/gray patterns and phase shifting patterns has been proposed. As shown in Figure 3.4d, the hybrid pattern consists of four binary patterns and four phase shifting patterns. The binary patterns are used to obtain the range of the absolute phase without any ambiguity, while the phase shifting patterns are employed to achieve subpixel resolution (which is better than that achieved by binary patterns). Note that, the hybrid patterns improve the 3D acquisition accuracy at the cost of additional patterns.

Temporal coded patterns are able to achieve a very high 3D acquisition accuracy. This is because the codeword basis is small since multiple patterns are projected, and a coarse-to-fine approach is adopted during the successive projection of more patterns [35]. Besides, methods based on temporal coded patterns are easy to implement and they achieve high spatial resolution. However, they require a long computational time. Thus, they cannot be used for scanning moving objects since several patterns have to be projected for 3D acquisition.

3.2.2.4.2 Spatial Coded Patterns

Unlike temporal coded patterns, 3D sensors with spatial coded patterns can acquire the 3D shape of an object by projecting only a single pattern. The codeword for each pixel in the pattern is represented by the neighboring points of that pixel. Thus, methods based on these types of patterns are suitable for scanning moving objects. The main challenge,

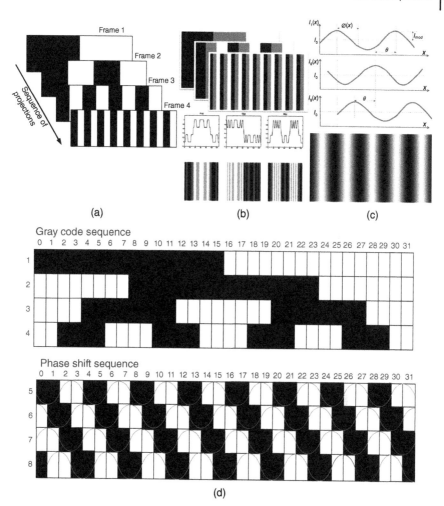

Figure 3.4 An illustration of different temporal coded patterns for structured light sensors. (a) Sequential binary pattern. (b) Sequential gray pattern. (c) Phase shift with three projection patterns. Source: Reprinted with permission from OSA Publishing. (d) Hybrid pattern consisting of gray pattern and phase shift. Source: Panel (b,c) – Reprinted with permission of Geng [34]. ©2011, OSA Publishing. Panel (d) is based on Figure 3 from Ref. [36]. ©1998, SPIE.)

however, is in the decoding process since the spatial neighborhood of a point cannot always be correctly extracted. In many cases, local smoothness of the underlying surface is assumed to achieve correct decoding of the spatial neighborhood. The main disadvantage of the spatial coded pattern-based sensors is that their spatial resolution is relatively low since they only use one spatial pattern.

A number of spatial coded patterns have been developed in the past few decades. In this section, we only briefly describe two representative methods, namely, the 1D stripe patterns and the 2D grid patterns.

- **1D stripe patterns**. A typical 1D stripe pattern is proposed based on the De Bruijn sequence [37]. Given an alphabet with a size of k, a De Bruijn sequence of rank l gives a cyclic word where each subword of length n is different from any of the remaining $k^l - 1$ subwords in the cyclic word. Figure 3.5a shows an example of De Bruijn patterns generated with an alphabet of $\{0, 1\}$, and with $k = 2$ and $l = 3$. It can be observed that each of the 8, i.e. 2^3, sub-words with length 3 (which are $111, 110, 100, 000, 001, 010, 101, 011$) is all different. The fact that each sub-word, i.e. a pattern, is unique and different from any other sub-words in the sequence, reduces the difficulty for pattern decoding. The De Bruijn sequence can therefore be used to combine the R, G, and B channels to obtain a color stripe pattern. Figure 3.5b shows an example of a

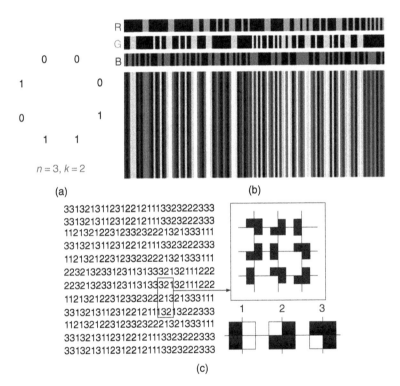

(a)   (b)

(c)

Figure 3.5 An illustration of different spatial coded patterns for structured light sensors. (a) A De Bruijn sequence. (b) A color stripe pattern generated by the De Bruijn sequence. Source: Geng 2011 [34]. ©2011. Reprinted with permission of OSA Publishing. (c) An M-array with three symbols.

color stripe pattern generated by the De Bruijn sequence with $k = 5$ and $l = 3$ [38]. Note that all neighboring stripes in Figure 3.5b have different colors, which is helpful for the decoding of the spatial patterns.

- **2D grid patterns.** For these methods, each position in the 2D pattern is uniquely defined by a 2D subwindow. A number of 2D grid patterns have been proposed based on the theory of *perfect maps* [35]. Let A be an $a \times b$ matrix with each of its elements coming from an alphabet of k symbols. The matrix A is considered as a *perfect map* if each sub-matrix of size $l \times m$ is different from all of the remaining sub-matrices of A. That is, A is a *perfect map* if each of its sub-matrices appears just once. Further, if all types of matrices of size $l \times m$, except the one with zeros everywhere, are contained in A, then A is called an *M-array*, or a *pseudorandom array* [35]. Based on the theory of *perfect maps*, an M-array with three symbols (i.e. 1–3) is proposed by Grinffin et al. [39]. Here, these symbols (i.e. 1–3) are represented by different shape primitives for pattern projection, as shown on the right part in Figure 3.5c. Specifically, a unique codeword is assigned to each element of the array using the values of that element and its left, right, up and down neighbors. An example of the M-array proposed in [39] is shown in Figure 3.5c. In this example, each symbol is represented by a shape primitive.

3.2.2.4.3 Direct Coded Patterns
These methods represent each pixel in the pattern using only the information at that pixel. That is, the codeword for a pixel is only contained in that pixel. There are mainly two groups of approaches for the generation of direct coded patterns. The first group of methods uses a set of gray levels to encode the pixels in the pattern, while the second group of methods uses a number of colors to encode the pixels in the pattern. The direct codification-based sensors require a reduced number of patterns and can achieve a high-resolution reconstruction. However, these methods achieve a low 3D reconstruction accuracy, and are usually sensitive to noise. Specifically, to achieve high resolution, a large number of color or gray values have to be used to represent the pattern. In that case, the distance between different codewords is very small, and the pattern decoding process becomes very sensitive to noise. Besides, the color/gray values perceived by the camera not only depend on the projected light but also on other factors such as the property of the surface under scanning. Consequently, the direct coded patterns are only suitable for neutral color or pale objects [35].

3.2.2.5 Shape from X
Shape from X techniques use information, such as motion, shading, photometry, silhouette, texture, focus, and defocus, to reconstruct 3D objects. Below, we give a brief overview of these methods.

- **Shape from motion.** The 3D information of an object/scene is reconstructed from a sequence of images of the scene captured by a moving camera. This method is usually low in cost, highly portable, and rich in texture. However, its spatial resolution and accuracy are low [25].

- **Shape from shading**. The 3D information of an object is reconstructed from a set of images captured by a fixed camera under different illuminations. This method usually assumes orthographic projection, a Lambertian surface, a known light source at infinity, no shadows, and uniform surface material. This method is also inexpensive, highly portable, and rich in texture. Its resolution and accuracy, however, are relatively low.
- **Shape from photometry**. Similar to shape from shading, this method recovers the 3D information from a set of images acquired under different illuminations. This method can achieve a precision of submillimeters, but operates in controlled environments, and requires calibrated reference objects or calibrated reference light sources.
- **Shape from silhouette**. The 3D shape of an object is constructed using the silhouette images of the object [40]. This method can work on reflective surfaces or even transparent/translucent surfaces.
- **Shape from texture**. The shape of a textured surface is estimated from the deformations of texture elements. This method usually assumes a smooth closed surface and a homogeneous texture.
- **Shape from focus**. The shape of an object is reconstructed from a sequence of images acquired from the same point of view but with different focus levels. This method usually uses a microscope, and is used to reconstruct small objects.

3.3   Preprocessing 3D Models

In general, the output of most of the 3D scanning devices and the 3D modeling software is a point cloud or a polygonal soup model. The performance, in terms of accuracy and computational efficiency, of the subsequent 3D shape analysis tasks depends on the quality of the input data. For instance:

- Raw 3D data produced by 3D scanners can be very noisy containing undesirable holes and spikes. Such noise, which can be geometric (e.g. displaced vertices and surface points) and/or topological (e.g. holes and missing parts), affects shape normalization. It also affects the quality of the features, which can be extracted from such data. In fact, some features, such as normal vectors, curvatures, geodesic paths, and geodesic distances, cannot be reliably computed on such noisy data.
- The size of a 3D model, in terms of number of points or vertices, can be very large. This will significantly impact on the computation time needed to extract features and to compute descriptors, see Chapters 4 and 5.

In this section, we will first look at a few preprocessing algorithms which are used to reduce noise in 3D models prior to the analysis stage (Section 3.3.1). In Section 3.3.2, we will look at parameterization techniques, which convert 3D models represented as triangulated meshes into continuous surfaces with a well-defined parameterization domain.

3.3.1 Surface Smoothing and Fairing

3.3.1.1 Laplacian Smoothing

One simple way to reduce noise from a 3D mesh M is by using Laplacian smoothing, which is an iterative process. At each iteration t, the Laplacian smoothing algorithm takes each vertex of the mesh and moves it with a small amount in the direction of the gradient at that vertex. The new location of each vertex v_i is given by:

$$v_i^{t+1} \leftarrow v_i^t + \lambda \Delta_M(v_i^t), \text{ with } 0 < \lambda < 1. \tag{3.8}$$

In this equation, $\Delta_M(v_i)$ is the discrete Laplacian given by Eq. (2.27). Equation (3.8) is first applied to all the vertices of the mesh and then repeated a few times until a satisfactory result is attained.

Laplacian smoothing is fast and straightforward to implement. Its main limitation, however, is that, after a few iterations, the Laplacian smoothing rapidly results in a significant shrinkage of the shape. In other words, it does not preserve the volume of the model being smoothed as shown in the first row of Figure 3.6. In this figure, we take the input model of Figure 3.6a, simulate some noise by randomly displacing the mesh vertices (Figure 3.6b), and then try to recover the original surface by applying several iterations of the Laplacian smoothing. As shown in Figure 3.6c, the more iterations we

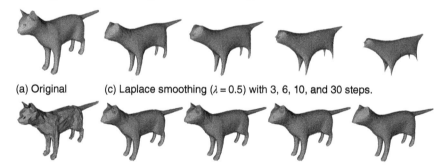

(a) Original   (c) Laplace smoothing ($\lambda = 0.5$) with 3, 6, 10, and 30 steps.

(b) Added noise   (d) Taubin smoothing ($\lambda = 0.5$, $\mu = -0.53$) with 3, 6, 10, and 30 steps.

Figure 3.6 Comparison between the Laplacian smoothing and and Taubin smoothing. (a) Original shape. (b) Input shape after adding random noise. (c) Laplacian smoothing of the 3D shape in (b). (d) Taubin smoothing of the 3D shape in (b). Observe how the Laplacian smoothing results in volume shrinkage.

apply, the smoother the surface becomes. However, the volume of the object becomes smaller and we can clearly see that elongated parts, such as the tail and the four legs, become thinner like needles. Of course, this behavior is not desirable and ideally, we would like a method that preserves the shape and volume of the 3D object being smoothed.

3.3.1.2 Taubin Smoothing

In order to reduce the shrinkage effect of the Laplacian smoothing, Taubin [41] proposed a two-step algorithm. **First**, a Laplacian smoothing step with a positive scale factor λ (see Eq. (3.8)) is applied to all the vertices of the shape. **Second**, another Laplacian smoothing step is applied to all the vertices, but this time with a negative scale factor μ, which is greater in magnitude than the first scale factor, i.e. $0 < \lambda < -\mu$ (e.g. $\lambda = 0.33$, $\mu = -0.34$). By using a negative scale factor, the goal is to compensate for the volume loss, which happened in the first smoothing step. Taubin's algorithm alternates between these two steps a number of times until a satisfactory result is obtained.

Taubin's method [41] produces a low pass filter effect where the surface curvature plays the role of frequency. In fact, a noisy surface can be seen as a smooth surface plus a perturbation of each of its vertices (or points on the surface) along the normal direction at that vertex or point. This perturbation, which needs to be filtered out, is treated as a zero mean high curvature noise. The two scale factors λ and μ determine the pass-band and stop-band curvatures. They must be tuned by the user to satisfy the nonshrinking property. In general, however, μ should be slightly larger in magnitude than λ. The amount of noise attenuation is then determined by the number of iterations.

Figure 3.6d shows some results obtained using Taubin smoothing. Compared to Laplacian smoothing (Figure 3.6c), Taubin smoothing preserves the volume of the shapes.

3.3.1.3 Curvature Flow Smoothing

Curvature flow [42] smooths the surface by moving each vertex v_i along the surface normal $\mathbf{n}_i$ with a speed equal to the mean curvature H_i at v_i:

$$v_i^{t+1} \leftarrow v_i^t - H_i^t \mathbf{n}_i^t. \tag{3.9}$$

An effective approximation of $-H_i\mathbf{n}_i$, proposed in [42], is by using the cotangent weights of Figure 2.10:

$$-H_i\mathbf{n}_i = \frac{1}{4A_i} \sum_{j \in \mathcal{N}_i} (\cot \alpha_j + \cot \beta_j)(v_j - v_i).$$

Here, A_i is the sum of the areas of the triangles having v_i as a common vertex.

3.3.2 Spherical Parameterization of 3D Surfaces

In general, the surface S of a 3D object is represented as a piecewise linear surface in the form of a triangulated mesh M, i.e. a set of vertices $V = \{v_1, \ldots, v_n\}$

Figure 3.7 The concept of spherical mapping.

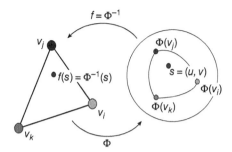

and triangles $T = \{t_1, \ldots, t_m\}$ such that the triangles intersect only at common vertices or edges [43]. The goal of parameterization is to find a suitable continuous parameterization domain D and then estimate the underlying continuous function $f : D \to \mathbb{R}^3$ such that $\forall s \in D, f(s) \in S$. If S is a disk-like surface, then D can be chosen to be a simply-connected region of the plane, e.g. the unit disk. When dealing with closed genus-0 surfaces then the unit sphere $\mathbb{S}^2$ is the natural parameterization domain. In what follows, we will discuss a few methods, which have been used in the literature for parameterizing closed genus-0 surfaces. We refer to these methods as *spherical parameterization* methods. Spherical parameterization is a two-step process;

- **First**, find a continuous mapping Φ that maps the mesh M onto the surface of a unit sphere, i.e. $\Phi : M \to \mathbb{S}^2$. This mapping should be uniquely determined by the images $\Phi(v) \in \mathbb{S}^2$, where $v \in V$ are the vertices of the triangulated mesh M.
- **Second**, for each point $s = (u, v) \in \mathbb{S}^2$, estimate its corresponding surface point $p = f(s) = \Phi^{-1}(s)$, which is uniquely determined by the vertices $v_i, v_j, v_k \in V$ of a triangle $t \in T$ such that the polygon formed by $\Phi(v_i)$, $\Phi(v_j)$, and $\Phi(v_k)$ encloses p, see Figure 3.7.

Spherical parameterization proves to be very challenging in practice. In fact, parameterization algorithms should be robust by preventing parametric fold overs and thus guaranteeing one-to-one spherical maps. Also, a good parameterization should adequately sample all surface regions with minimum distortion. This is particularly difficult since genus-0 surfaces can contain complex extruding parts that are highly deformed. A good parameterization is a one-to-one map that minimizes these distortions in some sense [43]. Here, we are particularly interested in conformal spherical parameterization, which has nice mathematical properties. Before laying down the parameterization algorithms, we will first introduce some definitions.

Definition 3.1 (Gauss map). A Gauss map $\Phi_N : M \to \mathbb{S}^2$ is defined as

$$\Phi_N(p) = \mathbf{n}(p), p \in M, \tag{3.10}$$

where $\mathbf{n}(p)$ is the unit normal vector to M at p.

In other words, a Gauss map maps each point of M onto its unit normal vector at that point. Naturally, this is a spherical mapping but it is one-to-one only in the case where M represents a star-shaped surfaces. By a star-shaped surface, we mean a surface S which, up to a translation, can be represented with a parametric function f of the form $f(u, v) = \rho(u, v)\mathbf{n}(u, v)$, where $\rho(u, v) \in \mathbb{R}_{\geq 0}$, (u, v) are spherical coordinates, and $\mathbf{n}(u, v)$ is a point on the unit sphere $\mathbb{S}^2$. It is also the unit normal vector to $\mathbb{S}^2$ at (u, v).

Definition 3.2 (Conformal mapping). A mapping $\Phi : M \rightarrow \mathbb{S}^2$ is conformal if the angle of intersection of every pair of intersecting arcs on $\mathbb{S}^2$ is the same as that of the corresponding preimages on M at the corresponding point.

In other words, conformal mapping preserves angles and it is often referred to as angle-preserving.

Definition 3.3 (String energy). The string energy of a mapping $\Phi : M \rightarrow \mathbb{S}^2$ is defined as:

$$E(\Phi) = \sum_{e_{ij}=(v_i,v_j)\in E} w_{ij}\|\Phi(v_i) - \Phi(v_j)\|^2, \tag{3.11}$$

where v_i and v_j are vertices of M, E is the set of edges $e_{ij} = (v_i, v_j)$ of the mesh, and w_{ij} is the weight associated to the edge e_{ij}. When $w_{ij} = 1$ for every edge e_{ij} then the String energy is known as the *Tuette energy*.

Definition 3.4 (Tuette mapping). Tuette mapping is the mapping that minimizes the Tuette energy obtained by setting the weights of Eq. (3.11) to one.

Definition 3.5 (Harmonic energy). The harmonic energy is a special case of the String energy where the weights w_{ij} are constructed using the cotangent of the angles α_{ij} and β_{ij} opposite to the edge e_{ij} (see Figure 2.10):

$$w_{ij} = \cot \alpha_{ij} + \cot \beta_{ij}. \tag{3.12}$$

A mapping Φ with zero harmonic energy is called *harmonic mapping*. Geometrically, Φ is harmonic if at each point $p \in M$, the Laplacian $\Delta_M \Phi(p)$ is orthogonal to the tangent plane to $\mathbb{S}^2$ at $\Phi(p)$ [43]. In other words, the tangential component of the Laplacian vanishes.

An important key observation made in [44] is that in the case of closed spherical (i.e. genus-0) surfaces, harmonic mapping and conformal mapping are the same. Gu et al. [44] used this property and proposed an iterative method, which builds an approximate conformal map of closed genus-0 surfaces without splitting or cutting the surfaces. The algorithm, summarized in Algorithm 3.1, starts with an initial mapping Φ_0 and iteratively updates it

Algorithm 3.1 Spherical conformal mapping.

Input: A triangulated mesh M, a step size δt, and an energy difference threshold δE.

Output: A mapping $\Phi : M \to \mathbb{S}^2$ that minimizes the harmonic energy and satisfies the zero-mass center constraint.

1: Compute the Tuette map $\Phi_t : M \to \mathbb{S}^2$.
2: Let $\Phi = \Phi_t$. Compute the harmonic energy E_0.
3: For each vertex v of M, compute the absolute derivative $D\Phi(v)$.
4: Update $\Phi(v)$ with $\delta\Phi(v) = -D\Phi(v)\delta_t$.
5: Ensure that Φ satisfies the zero-mass condition:

- Compute the mass center $c = \int_{\mathbb{S}^2} \Phi(s)ds$, where $ds = d\phi d\theta \sin(\theta)$ is the standard Lebesgue norm on $\mathbb{S}^2$.
- For all $v \in M, \Phi(v) \leftarrow \Phi(v) - c$.
- For all $v \in M, \Phi(v) \leftarrow \Phi(v)/\|\Phi(v)\|$.

6: Compute the harmonic energy E.
7: If $\|E - E_0\| < \delta E$, return Φ. Otherwise, assign E to E_0 and repeat from Steps 3–7.

---

using a gradient descent-like approach. At each iteration, the estimated map Φ is updated by subtracting a proportion δ_t, referred to as the step size, of the tangential component $D\Phi$ of its Laplacian. The tangential component of the Laplacian, also called the absolute derivative, is computed as follows:

$$D\Phi(p) = \Delta\Phi(p) - \langle\Delta\Phi(p), \mathbf{n}(\Phi(p))\rangle\mathbf{n}(\Phi(p)), \tag{3.13}$$

where $\mathbf{n}(\Phi(p))$ is the unit normal vector to $\mathbb{S}^2$ at $\Phi(p)$. Finally, Gu et al. [44] enforces an additional zero-mass condition to improve the convergence and to ensure the uniqueness of the solution. This procedure is summarized in Algorithm 3.1.

The speed of convergence of Algorithm 3.1 depends on the initial map Φ_0 and on the step size δ_t. Gu et al. [44] showed that by carefully choosing the step length, the energy can be decreased monotonically at each iteration. For the initial map Φ_0, one can use the Gauss map Φ_N or the Tuette map. The former will require more iterations for the conformal mapping algorithm to converge. Since conformal mapping iterations are computationally expensive, initializing the algorithm with a Tuette map would significantly reduce the computation time. Algorithm 3.2 details the procedure for computing the Tuette mapping of a surface S represented with a mesh M. Note that the Tuette algorithm converges rapidly, and more importantly, it has a unique minimum.

Figure 3.8 shows an example of three genus-0 surfaces and their corresponding spherical conformal parameterizations.

Algorithm 3.2 Tuette mapping.

Input: A triangulated mesh M, a step size δt, and an energy difference threshold δE.

Output: A mapping $\Phi : M \rightarrow \mathbb{S}^2$ that minimizes the Tuette energy.

1: Compute the Gauss map $\Phi_N : M \rightarrow \mathbb{S}^2$.
2: Let $\Phi = \Phi_N$. Compute the Tuette energy E_0.
3: For each vertex v of M, compute the absolute derivative $D\Phi(v)$.
4: Update $\Phi(v)$ with $\delta\Phi(v) = -D\Phi(v)\delta_t$.
5: Compute the Tuette energy E.
6: If $\|E - E_0\| < \delta E$, return Φ. Otherwise, assign E to E_0 and repeat steps 3–6.

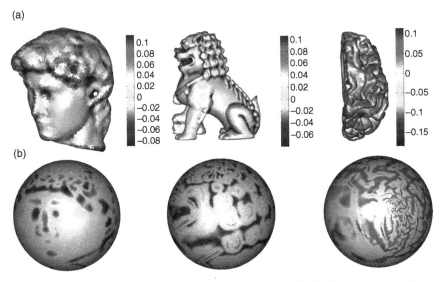

Figure 3.8 Example of genus-0 surfaces (a) and their conformal spherical parameterizations (b). The color at each point on the surfaces indicates the mean curvature of the surface at that point.

3.4 Summary and Further Reading

As mentioned at the beginning of this chapter, the material covered here is a brief summary of a very rich and deep set of topics, which are traditionally covered in a number of separate books. A more thorough introduction to 3D reconstruction techniques can be found in textbooks on multiview geometry

[23] and on computer vision in general, e.g. [45]. Preprocessing techniques are usually covered in textbooks on 3D geometry processing, e.g. [1]. The spherical parameterization procedures described in this chapter have been originally proposed by Gu et al. [44]. In this chapter, we only discussed conformal spherical parameterization. There are, however, many other techniques, which dealt with nonconformal mapping [46], and with surfaces, which are not necessarily genus-0. We refer the reader to [47] for a detailed survey on the topic.

Part II

3D Shape Descriptors

4

Global Shape Descriptors

4.1 Introduction

The main building block and core component of any 3D shape analysis task is a mechanism for measuring the similarity, or the dissimilarity, between 3D models. This is often achieved using:

- **Descriptors**, which characterize the geometry, topology, and semantics of 3D objects,
- **Dissimilarity functions**, which measure distances between descriptors, and
- **Algorithms**, which use the descriptors and the dissimilarity measures for the matching, classification, or retrieval of 3D models.

In short, a descriptor is a compact numerical representation of the shape of a 3D object. Distances in the descriptor space should reflect the dissimilarities between the objects that these descriptors represent. Such dissimilarities can be geometric, topological, or semantic. A good descriptor is the one that is

- **Concise, compact, quick to compute, and efficient to compare**. Ideally, a descriptor should be composed of a few numbers, which capture the *essence* of the shape so that it will be easy to store and to compare.
- **Discriminative**. Objects of different shapes should have different descriptors while objects of similar shapes should have similar descriptors.
- **Insensitive to geometric and topological noise** such as acquisition noise (see Chapter 2) and missing parts.
- **Invariant to similarity transformations**. A good shape descriptor should capture properties that characterize the shape of objects, by representing key shape features, while remaining invariant to transformations, such as translation, scale, and rotation, which do not affect shape.
- **Independent of the representation of the 3D objects**. Whether a 3D model is represented as a polygon soup, a parameterized surface, or a voxelized 3D volume, its shape descriptor should remain the same since shape is invariant to such representations.

3D Shape Analysis: Fundamentals, Theory, and Applications, First Edition.
Hamid Laga, Yulan Guo, Hedi Tabia, Robert B. Fisher, and Mohammed Bennamoun.
© 2019 John Wiley & Sons, Inc. Published 2019 by John Wiley & Sons, Inc.

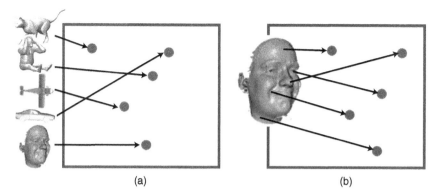

(a)   (b)

Figure 4.1 (a) Global descriptors describe whole objects with a single real-valued vector. (b) Local descriptors describe the shape of regions (or neighborhoods) around feature points. In both cases, the descriptors form a space, which is not necessarily Euclidean, with a meaningful distance metric.

In practice, it is unlikely that any single descriptor will satisfy all these properties. Also, some of these properties such as the invariance and robustness can be achieved either (i) in a preprocessing stage, by normalizing, cleaning, and/or repairing the 3D models before computing their descriptors, (ii) at the descriptor computation level, by designing descriptors that satisfy these properties, or (iii) at the dissimilarity computation level.

This chapter, as well as Chapter 5, will review the rich literature of descriptors, which have been proposed for the analysis of the shape of 3D models, and the dissimilarity measures, which have been used to compare them. Descriptors can be global or local (Figure 4.1). Historically, global descriptors have been proposed before the local descriptors. They describe a 3D object as a whole, while local descriptors characterize local regions of the 3D objects, e.g. semantic parts, small regions extracted around key points, or randomly sampled local regions. For instance, one can take a small region of a 3D model, e.g. the 3D face extracted from a full 3D human body, and compute its global geometric descriptor. In other words, global descriptors become local when applied to specific regions of the 3D object. As such, many local descriptors are adapted from their global counterparts. Inversely, local descriptors can be made global using some aggregation techniques. We will elaborate more on this in Chapter 5.

In this chapter, we focus on global descriptors, which we group into four different categories, namely, distribution-based (Section 4.2), view-based (Section 4.3), spherical representation-based (Section 4.4), and deep learning-based (Section 4.5). In what follows, we assume that a 3D object is represented as a polygonal mesh M, which is composed of a set of vertices $V = \{v_i, i = 1, \ldots, n\}$ and a set of triangulated faces $T = \{t_i, i = 1, \ldots, m\}$.

Recall that, in general, a 3D object can have any arbitrary representation. However, one can always convert any representation into a polygonal mesh representation.

4.2 Distribution-Based Descriptors

The first class of methods that we are going to explore represents the shape of a 3D object using probability distributions of some geometric properties computed at different locations on the 3D model. This is a three-step process, which involves:

- Sampling a set of points on the surface or in the interior of the 3D model.
- Measuring some geometric properties at each sample point, and
- Summarizing these properties into a histogram, which approximates their distribution on the 3D model. The histogram is then used as a global shape signature.

Existing techniques differ in the way points are sampled, the type of geometric properties that are computed at each point, and the way these histogram-based signatures are compared. Below, we discuss these steps.

4.2.1 Point Sampling

The first step for building distribution-based descriptors is to sample N points $\{p_1, \ldots, p_N\}$ on the surface M. There are two issues to consider when choosing these points; the first one is the sampling density and the second one is the sample generation process.

The simplest way to generate these samples is to consider each vertex v_i as one sample point p_i. This choice, however, has many limitations. For instance, in 3D shape analysis, we are generally dealing with arbitrary 3D models, which come from various sources. These 3D models can have an arbitrary number of vertices, which often are not uniformly distributed on the shape's surface. As such, the resulting distribution-based signature would be biased and sensitive to changes in the tessellation of the 3D model.

A more efficient approach is to sample random points on the surface of a 3D model following a uniform distribution. The sampling algorithm can be summarized as follows:

- Let A be an array of size m where $m + 1$ is the number of triangular faces in the 3D model. Set $A[0] = 0$.
- **First,** for each triangle t_i, compute its area area(t_i) and set $A[i] = A[i - 1] +$ area(t_i).
- **Next,** select a triangle with a probability proportional to its area. This is done by first generating a random number a between 0 and $A[m]$,

the total cumulative area. Then, select the triangular face t_s such that $A[s] \leq a < A[s+1]$. This can be efficiently done using a binary search on the array A.

- **Finally,** if (v_i, v_j, v_k) are the vertices of the selected triangle then one can construct a point p on its surface by generating two random numbers, r_1 and r_2, between 0 and 1, and evaluating the following equation:

$$p = (1 - \sqrt{r_1})v_i + \sqrt{r_1}(1 - r_2)v_j + \sqrt{r_1}r_2 v_k. \tag{4.1}$$

Here, $\sqrt{r_1}$ sets the percentage from vertex v_i to the opposing edge, while r_2 represents the percentage along that edge. Taking the square-root of r_1 gives a uniform random point with respect to the surface area.

This method ensures that the points are uniformly sampled with respect to the surface area. The procedure is also very fast and works for watertight meshes as well as for polygonal soup models.

4.2.2 Geometric Features

Once a set of N points have been sampled on the surface of the input 3D shape M, the next step is to measure some shape properties at these points and build a signature per property type by taking the histogram of the measured property over the sampled points. We refer to these properties as *shape functions* following Osada et al. [48]. In general, any shape property can be used to form a shape distribution. Below, we describe some of those that have been commonly used in the literature. Some of them are based on geometric attributes such as angles, distances, areas, and volumes. Others are based on differential properties such as normal vectors and curvatures.

4.2.2.1 Geometric Attributes
We consider seven geometric attributes, some of them have been introduced in [48], see Figure 4.2 for a 2D illustration of each of these measures.

- **A3**: measures the angle between every three random points on the surface of a 3D model.
- **D1**: measures the distance between a fixed point, e.g. the centroid or the center of mass, of the boundary of the model, and every random point on the surface.
- **D2**: measures the distance between every two random points on the surface of the 3D model.
- **D3**: measures the square root of the area of the triangle between three random points on the surface of the 3D model.
- **D4**: measures the cubic root of the volume of the tetrahedron between four random points on the surface of the 3D model.

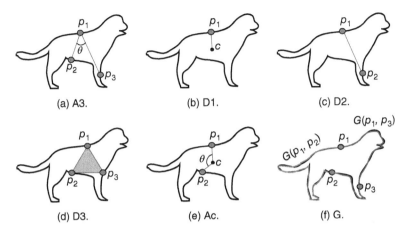

Figure 4.2 Some shape functions based on angles (A3), lengths (D1, D2, Ac, G), and areas (D3), see the text for their detailed description. Here, we use a 2D contour for illustration purposes only. The functions are computed on 3D models.

- **Ac:** first, consider ray segments, called cords, that connect each point on the mesh to the barycenter of the mesh. The Ac measure is defined as the angle between two arbitrary segments [49].
- **G:** measures the geodesic distance between two random points on the surface of the 3D model. This measure can only be computed on watertight surfaces, which have a manifold structure.

These shape functions are invariant to translations and rotations of the 3D models. In addition, A3 and Ac shape functions are invariant to scale, while the others have to be normalized, using one of the procedures described in Chapter 2. Finally, the G shape function is invariant to bending of the shapes and thus it is suitable for the computation of bending-invariant dissimilarity between 3D shapes.

4.2.2.2 Differential Attributes
Instead of, or in addition to, the geometric attributes described above, one can define attributes using the differential properties, such the normals and curvatures, of the surfaces being analyzed. Examples of such attributes include

- **The angles** between the surface normal at each point and the axes of the coordinate frame [49].
- **The mean and Gaussian curvatures** at each point on the surface of the 3D model.
- **The shape index** s at each surface point [50], which is defined as

$$s = \frac{2}{\pi} \arctan\left(\frac{\kappa_2 + \kappa_1}{\kappa_2 - \kappa_1}\right). \tag{4.2}$$

Here, κ_1 and κ_2 are, respectively, the minimum and maximum curvatures at the current surface point. The shape index represents salient elementary shape features (convex, concave, rut, ridge, saddle, and so on).

The curvatures and the shape index are invariant to the translation, scale, and rotation of the 3D objects. The angles between the surface normal and the axes of the coordinate frame are only invariant to translation and scale. Thus, it requires normalization for rotation.

4.2.3 Signature Construction and Comparison

Having chosen a shape function, the next step is to compute its distribution. This is done by evaluating the shape function at N samples and constructing a histogram by counting how many samples fall into each of the B fixed-size bins. The histogram can be further smoothed or approximated using, for example, a piecewise linear function with $b \leq B$ equally spaced vertices [48]. At the end, the shape distribution of a 3D model M is a 1D real-valued vector h of size B. It is constructed once for each model, for each shape function, and then stored and used for shape analysis tasks. Note that computing the shape distribution using the D1 function results in a shape descriptor called SHELLS [51].

Now, let M_1 and M_2 be two input 3D models, and h_1 and h_2 their shape distributions, respectively. The remaining task is to estimate the dissimilarity between the two models by comparing their respective shape functions using some dissimilarity measure or metric $d(h_1, h_2)$. Several metrics can be used, e.g. the $\mathbb{L}^2$ metric. However, since these descriptors represent distributions, it is more appropriate to use metrics that are more suitable for comparing distributions. Examples of such metrics include

- The χ^2 statistics:

$$d(h_1, h_2) = \sum_{i=1}^{B} \frac{(h_1[i] - h_2[i])^2}{h_1[i] + h_2[i]}. \tag{4.3}$$

Here, $h_1[i]$, respectively $h_2[i]$, denotes the ith component of the 1D vector h_1, respectively, h_2.

- Bhattacharyya distance:

$$d(h_1, h_2) = 1 - \sum_{i=1}^{B} \sqrt{h_1[i]h_2[i]}. \tag{4.4}$$

- PDF L_a norm: The Minkowski L_a norm of the probability distribution functions:

$$d(h_1, h_2) = \left(\sum_{i=1}^{B} \|h_1[i] - h_2[i]\|^a \right)^{\frac{1}{a}}, \tag{4.5}$$

where $a \in \mathbb{N}$ and $a \geq 1$.

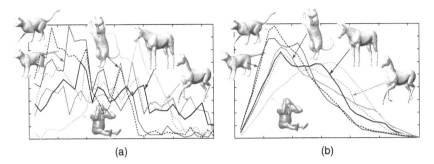

Figure 4.3 Illustration of the (a) D1 and (b) D2 shape distributions computed from six different 3D models.

- CDF L_a: The Minkowski L_a norm of the conditional density functions:

$$d(h_1, h_2) = \left(\sum_{i=1}^{B} \|\hat{h}_1[i] - \hat{h}_2[i]\|^a \right)^{\frac{1}{a}}, \quad \text{where } \hat{h}[i] = \sum_{k=1}^{i} h[k]. \quad (4.6)$$

- The Kullback–Leibler divergence:

$$D_{KL}(h_1 \| h_2) = \sum_{i=1}^{B} h_1[i] \log \frac{h_1[i]}{h_2[i]}. \quad (4.7)$$

Note that the performance of the shape distributions depends on the choice of the shape function, the number of sample points N, and the number of bins B. Osada et al. [48] found that $N = 1024^2$ and $B = 1024$ provide a good trade-off between accuracy and computation time. This choice, however, is experimental and may vary with the geometric complexity of the 3D models being analyzed.

While shape distribution-based descriptors are compact and easy to compute and compare, they are not sufficiently discriminative. As such, they have been often used in combination with other methods to improve results. For instance, shape distributions can serve as an active filter after which more detailed comparisons can be performed using more powerful and computationally demanding descriptors.

Figure 4.3 shows examples of the D1 and D2 shape distributions computed on six different 3D models.

4.3 View-Based 3D Shape Descriptors

Probably, the most straightforward approach for comparing 3D models is to represent them as a collection of images rendered from different viewpoints, called viewers, carefully positioned around the objects. The motivation behind such viewer-centered representations is threefold; the first one stems from the fact that in the human visual system, 3D objects are represented by a set of 2D views rather than by the full 3D representation. Second, if two 3D models are similar, they also look similar from all viewing angles. Finally, 2D image analysis problems have been investigated for decades prior to the widespread of 3D technologies. As such, many efficient techniques have been developed for indexing, classification, and retrieval of images. The rationale is that 3D shape analysis would benefit from this rich literature if one can represent a 3D model using 2D projections.

To build view-based descriptors, we take a 3D model M and place around it a system of N virtual cameras. To ensure that the 3D model is entirely enclosed in the view field of the cameras, it is first translated so that its center of mass is at the origin of the coordinate system, and then scaled to fit inside a bounding sphere of radius one. Each virtual camera $c_i, i = 1, \ldots, N$, can then capture:

- A 2D projection I_i of the 3D model from which one can extract various 2D features such as the object's silhouette and its internal regions.
- A depth map D_i of the 3D model as viewed from the camera's viewpoint.

Let us consider two shapes M_A and M_B. To compare them, we first place a camera system C of N cameras: $C = \{c_i, i = 1, \ldots, N\}$, which captures 2D projections $A = \{A_i, i = 1, \ldots, N\}$, respectively $B = \{B_i, i = 1, \ldots, N\}$, of M_A, respectively M_B. (These projections can be 2D images, depth maps, or a combination of both.) We then rotate the camera system around M_B with a rotation $O \in SO(3)$ and capture another set of 2D projections of M_B, hereinafter denoted by $B_O = \{B_{O(i)}, i = 1, \ldots, N\}$. This process is then repeated for every possible rotation $O \in SO(3)$. The dissimilarity between the two shapes can be defined as the minimum distance between A and B_O over all possible rotations $O \in SO(3)$:

$$D(M_A, M_B) = \min_{O \in SO(3)} \sum_{i=1}^{N} d_{img}(A_i, B_{O(i)}), \qquad (4.8)$$

where $d_{img}(\cdot, \cdot)$ is a certain measure of distance between 2D projections, or more precisely, between the descriptors of the 2D projections. There are various types of descriptors that one can compute. These will be discussed in Section 4.3.2.

4.3.1 The Light Field Descriptors (LFD)

The light field descriptor (LFD), introduced by Chen et al. [52], is an efficient view-based descriptor computed from the silhouettes of a 3D object captured from 20 cameras placed at the vertices of a regular dodecahedron. Since two silhouettes of a 3D model taken from two opposite points are similar, only $N = 10$ views from one hemisphere are considered. With this configuration, there are

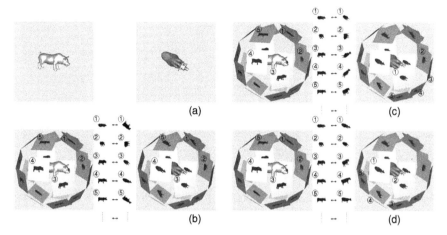

Figure 4.4 Comparing the light field descriptors (LFDs) of the two 3D models in (a). The LFD of the first shape (top row, first shape) is compared to every LFD of the second shape (top row, second shape), obtained by all possible rotations of the camera system around the second shape (b–d). The dissimilarity between the two shapes is the minimum over all possible rotations of the camera system. Source: The image is courtesy of [52].

60 possible rotations of such a camera system, see Figure 4.4. Thus, Eq. (4.8) becomes:

$$D(M_A, M_B) \equiv D(A, B) = \min_{r=1}^{60} \sum_{i=1}^{N} d(A_i, B_{\text{rot}(r,i)}), \qquad (4.9)$$

where rot(r, i) is the image B_i aligned to A_i using the rth rotation. Equation (4.9) finds the minimum distance over 60 possible rotations of the camera system and thus the dissimilarity measure it defines accounts for rotation variability. However, since the cameras are placed at discrete locations, one can increase the chances of finding the right alignment by considering multiple camera systems. For instance, Chen et al. [52] used 10 different camera systems, see Figure 4.5. The LFD dissimilarity measure is then defined as:

$$D_{\text{LFD}}(M_A, M_B) = \min_{j,k=1}^{10} D(A^j, B^k), \qquad (4.10)$$

where $D(\cdot, \cdot)$ is given by Eq. (4.9) and $A^j = \{A_i^j, i = 1, \dots, N\}$, respectively, $B^k = \{B_i^k, i = 1, \dots, N\}$, is the set of 2D views of M_A, respectively, M_B, captured by the jth, respectively, the kth, camera system.

4.3.2 Feature Extraction

From the 2D projections of the 3D model, one can extract various types of features, which can be used to index the 3D model. Below, we discuss a few of them.

Figure 4.5 A set of 10 light field descriptors for a 3D model. Each descriptor is obtained with a system of 10 cameras placed at the vertices of the half hemisphere of a regular dodecahedron. Source: Image courtesy of [52].

- **Contour-based descriptor.** Given a 2D image I_i of a 3D model, one can extract the boundary points of the shape it represents through a contour tracing algorithm, and then convert the contour points to a centroid distance function $f_c : [0, 2\phi) \to \mathbb{R}_{\geq 0}$ such that $f(\theta)$ is the extent of the shape in the direction of the ray which forms an angle θ with the first axis of the coordinate system. One can then apply Fourier transform to the function f_c and take the first (e.g. 10) Fourier coefficients to form a descriptor that characterizes the shape of the silhouette [52].
- **Region-based descriptor.** In addition to the descriptors that capture the outer contour of the 2D projections, one can also use region-based descriptors to capture the interior features. Chen et al. [52] used 35 Zernike moment coefficients on each 2D projection. Similar descriptors can also be extracted from the depth images. Note that the Zernike moment descriptor has been also used in MPEG-7 under the name of *RegionShape* descriptor.

The final descriptor can be obtained by concatenating the contour and the region-based descriptors. The distance between two images is then given by the Euclidean distance between their respective descriptors.

4.3.3 Properties

The LFD descriptor has very nice properties, which makes it suitable for global shape analysis. In particular:

- It is reasonably quick to compute since the 2D projections can be efficiently rendered with graphics programming units.
- It has a good discrimination power. In fact, it ranked first among 12 other descriptors benchmarked on the Princeton Shape Benchmark dataset [53].

- It is invariant to rigid transformations. The invariance to translation and scale is achieved by normalizing the 3D shapes in a preprocessing step. The invariance to rotations is achieved by using an exhaustive search for the best alignment during the process of computing the dissimilarity between LFD descriptors (Eq. (4.8)).
- It is robust to noise, mesh topology, and degeneracies since the mesh is discarded once the 3D models are projected onto the 2D images.

One of the major limitations of the LFD descriptor is the fact that it is not compact. In the original implementation [52], the LFD descriptor of each 3D model is a vector of size 100×45 coefficients.

4.4 Spherical Function-Based Descriptors

One of the main challenges for comparing 3D models is the lack of a proper parameterization domain. For instance, an image can be represented as a function from a quadrilateral domain onto $\mathbb{R}^d$, where $d > 0$ is a natural number. For example, in the case of RGB images, we have $d = 3$, while in the case of grayscale images, we have $d = 1$. Thus, one can use plenty of tools, e.g. Fourier and Wavelet transforms, to extract features that can be used for their comparison. View-based techniques for 3D shape analysis (Section 4.3) overcome the parameterization issue by projecting the 3D models onto 2D images and thus reducing the shape analysis task into a 2D (image) analysis problem.

Instead of using 2D projections, one can parameterize a 3D model by functions defined on a spherical domain $\mathbb{S}^2$ [54, 55]. The rational behind the choice of the spherical domain is that rotations in the Euclidean space are equivalent to translations if one equips the spherical domain with the spherical coordinate system. Thus, a number of methods have been proposed to compute one-to-one and onto spherical parameterizations [43]. Although these approaches have been mainly intended for graphics applications, e.g. texture mapping and shape morphing, they can be used to build effective descriptors for shape analysis [54, 55]. They are, however, limited to genus-0 manifold surfaces, i.e. surfaces that are topologically equivalent to a sphere. As a result, these methods can only be applied to a very restricted subset of the 3D models that are currently available. In this section, we will first review the various spherical representations that have been proposed in the literature for the analysis of generic 3D models of arbitrary topology and genus (Section 4.4.1). We will then discuss a few methods and strategies for comparing them (Section 4.4.2).

In what follows, we consider the sphere $\mathbb{S}^2$ parameterized by its spherical coordinates $\theta \in [0, \pi]$ and $\phi \in [0, 2\pi)$.

4.4.1 Spherical Shape Functions

Figure 4.6 illustrates different types of spherical functions that have been used to represent the shape of 3D objects. Below we detail each of them.

1) **Extended Gaussian image**s (EGIs) [56]. The Gaussian image of a 3D model is the mapping of all the normal vectors (called also surface orientations) of the 3D object onto a sphere of radius one. In this mapping, all the tails of the normal vectors are set to be on the center of the sphere while the heads lie on the surface of the unit sphere. The EGI is a histogram that records the variation of surface area with surface orientation. It is constructed as follows:
   - **First**, decompose the sphere into equidistant cells, called bins, along the θ and ϕ directions.
   - **Next**, for each bin b, find all the surface patches for which the orientations of their surface normals fall within b, sum their surface area, and store the result in the bin b. In practice, since objects are represented as polygon meshes, we only need to look at the normals of the surface faces.
   - **Finally**, normalize the result by dividing each bin by the total surface area of the 3D model.

   This procedure produces a spherical function $h : \mathbb{S}^2 \to [0, 1]$, which is a two-dimensional distribution since the bins are obtained by partitioning the sphere along the θ and ϕ directions.
2) **Complex extended Gaussian image**s (CEGIs) [57]. In the EGI representation, the weights associated with the normals of the object are scalars that

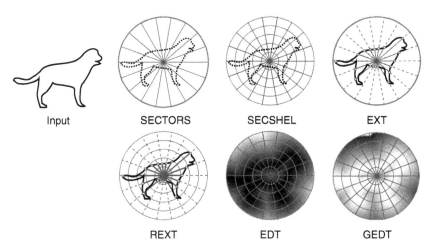

Input	SECTORS	SECSHEL	EXT

REXT	EDT	GEDT

Figure 4.6 Examples of methods that have been used to convert a 3D shape into a set of spherical functions, which in turn are used to build compact descriptors, see the text for more details.

represent the associated visible face area. The CEGIs, introduced in [57], extend this representation by adding the normal distance (i.e. the distance in the normal direction) of that face to the center of mass of the shape. Let $s = (\theta, \phi)$ be a point on the unit sphere $\mathbb{S}^2$. The value of the spherical shape function h at s is computed as follows:

- Compute a vector $\mathbf{n} = (\sin\theta\cos\phi, \sin\theta\sin\phi, \cos\theta)^\mathsf{T}$.
- Find the surface faces whose normals are $\mathbf{n}$. Let $A_1, \ldots, A_k$ be the areas of these faces, and let $d_1, \ldots, d_k$ be the normal distances of these faces.
- The spherical function h at $s = (\theta, \phi)$ is then given as:

$$h(\theta, \phi) = \sum_{i=1}^{k} A_i \exp(j d_k), \quad \text{where } j^2 = -1. \tag{4.11}$$

In other words, the value of the CEGI at each spherical point is a complex number whose magnitude is the visible surface area along that direction, and whose phase is the normal distance.

3) **Shape Histogram (SECTORS)** [51] is a spherical function that represents the distribution of the model area as a function of the spherical angle. To build this descriptor, Shilane et al. [53] first sample uniformly a large number of points on the surface of the input 3D model. Then, they divide the spherical volume into sectors by sampling the spherical grids into 64×64 along the θ and ϕ directions, see Figure 4.6. The method then constructs the shape histogram by counting the number of surface points that fall inside each sector.

4) **Shape Histogram (SECSHEL)** [51] is a collection of spherical functions that represent the distribution of the model area as a function of the radius and the spherical angle. Similar to SECTORS, the spherical grid is discretized into 64×64 along the spherical angles and into 32 equidistant shells along the radial direction, see Figure 4.6. The histogram is then constructed by counting the number of surface points that fall inside each spherical cell.

5) **Spherical Extent Function (EXT)** [58] is a spherical function that represents the maximal distance from the shape's center of mass as a function of the spherical angle. Here also, the resolution of the spherical grid is set to 64×64, see Figure 4.6.

6) **Radialized Spherical Extent Function (REXT)** [59] is a collection of spherical functions which represent the maximal distance from the center of mass as a function of spherical angle and radius. To compute this descriptor, the spherical volume is discretized into a grid of 64×64 along the spherical angles and 32 along the radial direction, see Figure 4.6.

7) **Gaussian Euclidean Distance Transform** (GEDT) [60] is a 3D function whose value at each point is given by composition of a Gaussian with the Euclidean distance transform (EDT) of the surface, see Figure 4.6.

To compute the EDT, the 3D volume is first discretized into a grid of size $64 \times 64 \times 64$, translated so that the origin is at the point of coordinates $(32, 32, 32)$, scaled by a factor of 32, and then represented by 32 spherical descriptors representing the intersection of the voxel grid with concentric spherical shells. The values within each shell are scaled by the square-root of the corresponding area [53].

4.4.2 Comparing Spherical Functions

Once 3D models are represented using spherical functions, the next step is to define a mechanism for comparing these functions.

4.4.2.1 Spherical Harmonic Descriptors

The spherical harmonic transform is the equivalent of the Fourier transform but for functions which are defined on the sphere. Let $h : \mathbb{S}^2 \to \mathbb{R}$ be a spherical function. Its spherical harmonic decomposition is given by

$$h(\theta, \phi) = \sum_{l=0}^{\infty} h_l(\theta, \phi), \quad \text{such that } h_l(\theta, \phi) = \sum_{m=-l}^{l} a_{lm} Y_l^m(\theta, \phi), \qquad (4.12)$$

where $\theta \in [0, \pi]$, $\phi \in [0, 2\pi)$, and

$$Y_l^m(\theta, \phi) = \sqrt{\frac{(2l+1)}{4\pi} \frac{(l-|m|)!}{(l+|m|)!}} P_{lm}(\cos \theta) e^{im\phi}. \qquad (4.13)$$

Y_l^m are orthogonal functions, which form a complete orthonormal basis of the sphere $\mathbb{S}^2$. Here, P_{lm} are Legendre polynomials.

Similar to the Fourier transform, one can use the coefficients $a_{lm}, l \geq 0, |m| \leq l$, to construct various types of shape descriptors. For example, one can gather all the coefficients up to a frequency $l = b$ (where b is referred to as the bandwidth) to form one shape descriptor and then use the Euclidean distance to compare them. These types of descriptors require the normalization of the input 3D models for translation, scale, and rotation prior to harmonic decomposition. While normalization for translation and scale is straightforward, normalization for rotation is usually very challenging since existing techniques such Principal Component Analysis (see Chapter 2) are very sensitive to small surface noise.

Instead of using directly the coefficients of the spherical harmonic decomposition, Kazhdan et al. [60] proposed to represent 3D shapes using the energies, $SH(h) = \{\|h_0(\theta, \phi)\|, \|h_1(\theta, \phi)\|, \ldots\}$, of their spherical functions, where

$$\|h_l(\theta, \phi)\|^2 = \sum_{m=-l}^{l} \|a_{lm}\|^2. \qquad (4.14)$$

The main motivation is that this representation is independent of the orientation of the spherical function. In other words, applying a rotation $O \in SO(3)$ to a 3D model does not change its energy representation. Kazhdan et al. [60] studied the properties of the energy-based representation of spherical functions for 3D shape analysis and showed that:

- The $\mathbb{L}^2$-difference between the energy representations of two spherical functions h and g is a lower bound for the minimum of the $\mathbb{L}^2$-difference between the two functions, taken over all possible rotations:

$$\|SH(h) - SH(g)\| \leq \|h - Og\|.$$

- The space of spherical functions with bandwidth b is of dimension $O(b^2)$. Its energy, however, is of dimension $O(b)$. Thus, a full dimension of information is lost.

The first property makes spherical harmonic energy-based descriptors suitable for rotation-invariant shape comparison without prealignment using PCA. The second property, however, implies that the performance of such descriptors is lower than when using the harmonics coefficients since a full dimension of information is lost.

Using this framework, several types of spherical descriptors can be constructed. For example:

- In [53], all the spherical descriptors of Section 4.4.1 are represented by their spherical harmonic coefficients up to order 16 and compared using the $\mathbb{L}^2$ metric. Since the harmonic coefficients are not invariant to translation, scale, and rotation, the 3D shapes should be first aligned to their principal axes and scaled to fit within a bounding sphere of radius one prior to the computation of the descriptors. Hereinafter, we also denote the spherical harmonic transformations of these descriptors as EGI, CEGI, SECTORS, SECSHELL, EXT, RECT, and GEDT.
- Shilane et al. [53] defined the SHD descriptor as a rotation invariant representation of the GEDT obtained by computing the restriction of the function to concentric spheres and storing the norm (or energy) of each (harmonic) frequency [60]. This representation does not require the normalization of the 3D models for rotation.

Note that energy-based descriptors are compact and thus are very fast to compare. We will discuss the performance of these descriptors in Chapter 12.

4.4.2.2 Spherical Wavelet Transforms

Instead of using spherical harmonic transforms, Laga et al. [61–63] proposed to use spherical wavelet transforms to build global descriptors from the spherical representations of 3D shapes. There are two ways of constructing the wavelet transform of a spherical function $h : \mathbb{S}^2 \rightarrow \mathbb{R}^3$. The first one is the second-generation wavelets, introduced in [64], and which operates on the vertices of the geodesic sphere. The second one uses the standard image-based wavelet transforms. It takes a spherical function h, maps it onto a geometry image [46, 65], and then applies standard image wavelet transform. The entire procedure can be summarized as follows:

1) **Initialization:**
   - Discretize the spherical function h into an image I of size $2^{n+1} \times 2^n$. The integer n can be set to any value, but in general, it should be chosen in such a way that all the shape details are captured while keeping the storage requirement and computation time reasonable.
   - Set $A^{(n)} \leftarrow I$ and $l \leftarrow n$.
2) **Forward transform:** repeat the following steps until $l = 0$:
   - Apply the forward wavelet transform on $A^{(l)}$, which will result in an approximation image, hereinafter denoted by $LL^{(l-1)}$ and three detail coefficient images $C^{(l-1)} = \{LH^{(l-1)}, HL^{(l-1)}, HH^{(l-1)}\}$. These images are of size $2^l \times 2^{l-1}$ each.
   - Set $A^{(l-1)} \leftarrow LL^{(l-1)}$ and $l \leftarrow l - 1$.
3) **Coefficient collection:**
   - Collect the approximation $A^{(0)}$ and the coefficients $C^{(0)}, \dots, C^{(n-1)}$ into a vector F of wavelet coefficients.

In principle, any type of wavelet transform can be used to build the vector F of coefficients. Laga et al. [61–63] used the Haar wavelet whose scaling function is designed to take the rolling average of the data, and its wavelet function is designed to take the difference between every two samples in the signal.

4.4.2.2.1 Wavelet Coefficients as a Shape Descriptor Once the spherical wavelet transform is performed, one may use the wavelet coefficients for shape description. Using the entire coefficients is computationally expensive. Instead, Laga et al. [61, 63] have chosen to keep the coefficients up to level d. The resulting shape descriptor is denoted by SWC_d, where $d = 0, \dots, n - 1$. When $d = 3$, we obtain two-dimensional feature vectors of size $N = 2^{d+2} \times 2^{d+1} = 32 \times 16 = 512$.

Similar to spherical harmonic coefficients, comparing directly wavelet coefficients requires an efficient alignment of the 3D models, using, for example, PCA as discussed in Chapter 2.

4.4.2.2.2 Spherical Wavelet Energy Signatures Instead of using the entire wavelet coefficients as a shape descriptor, Laga et al. [61, 63] used their energies to build compact rotation-invariant shape signatures. There are two types of energies:

- The L_2 energy, which is defined as:

$$F_l^{(2)} = \left(\frac{1}{k_l} \sum_{j=1}^{k_l} x_{l,j}^2 \right)^{\frac{1}{2}}.$$

(4.15)

- The L_1 energy, which is defined as:

$$F_l^{(1)} = \frac{1}{k_l} \sum_{j=1}^{k_l} \|x_{l,j}\|.$$

(4.16)

Here, $x_{l,j}, j = 1 \dots k_l$ are the wavelet coefficients of the lth wavelet subband. Laga et al. [61] used $n = 7$. Thus, the size of the final energy descriptor is 19. We refer to the L_1 and L_2 energy-based descriptors by $SWEL_1$ and $SWEL_2$, respectively. The main benefits of these two descriptors are (i) their compactness, which reduces the storage and computation time required for their comparison and (ii) their rotation invariance. However, similar to the power spectrum or energy of the spherical harmonic transform, information about feature localization is lost in the energy spectrum.

Finally, since 3D shapes are now represented using multidimensional vectors with real-valued components (SWC_d is two-dimensional, while $SWEL_1$ and $SWEL_2$ are one-dimensional descriptors), a natural way for comparing them is by using the $\mathbb{L}^2$, i.e. Euclidean, distance.

4.5 Deep Neural Network-Based 3D Descriptors

The global descriptors described in the previous sections have been used in various 3D shape analysis tasks such as retrieval, recognition, and classification. They often produce acceptable performances on relatively small datasets. Their performance, however, remains far from what can be achieved in image analysis when used for the analysis of large scale 3D datasets. We argue that the main reason for this limited performance is the fact that each of these *hand-crafted* descriptors has been designed to capture specific shape properties. Large datasets often exhibit large interclass variability and large geometric and topological similarities across some classes. There has been some attempts in the past to boost the performance of these hand-crafted descriptors by learning either the best combination of descriptors [66], the best similarity measures which maximize performance [67], or by incorporating user feedback (also called relevance feedback). Nevertheless, their performance still remains far from what humans can achieve on similar tasks.

In the recent literature, Deep Neural Networks have revolutionized image analysis. They have been used to learn general-purpose image descriptors which outperform hand-crafted features on a number of vision tasks such as object detection, texture and scene recognition, and fine-grained classification [68]. The most straightforward way to extend these techniques to the 3D case is by using a multiview representation of the 3D objects (Section 4.3). Along this line, Su et al. [68] proposed the concept of multiview Convolutional Neural Networks (MVCNN), which aggregate into a single global descriptor the output of multiple image-based Convolutional Neural Networks (CNNs), each one applied to a 2D projection of the 3D model. Sections 4.5.1 and 4.5.2 describe the details of this approach. Another approach is to treat 3D shapes as volumetric images and thus one can design deep networks which operate directly on the 3D grid rather than on the 2D projections of the 3D models. This type of techniques will be covered in Section 4.5.3.

4.5.1 CNN-Based Image Descriptors

A CNN is a neural network which takes as input an image I and produces a set of scores where each score is the likelihood that the image belongs to one of the image classes. A CNN is generally composed of five types of neuron layers (see Figure 4.7):

- **The input layer**. The neurons in this layer hold the raw pixel values of the input image. In the case of an RGB image of size $w \times h$, this layer is a 3D volume of size $w \times h \times 3$.
- **The convolution layer**. Each neuron in this layer computes the dot product between their weights and a small region in the input volume to which they are connected. This is equivalent to performing a convolution of the input with the filter whose elements are the weights of the neuron.
- **The ReLU layer**. This applies an element-wise activation function, e.g. the $\max(0, x)$, to its input.
- **The POOL layer**. This layer down-samples the input along the spatial dimensions.
- **The Fully-Connected (FC) layer**. This is a classification layer. It computes the class scores, i.e. the likelihood that the input to the network belongs to one of the classes.

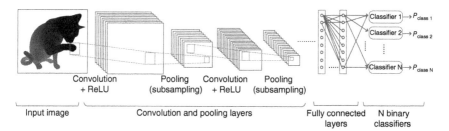

Figure 4.7 Illustration of the typical architecture of a Convolutional Neural Network (CNN) for the analysis of 2D images.

The main property of CNNs is that the weights of the neurons in the convolutional layers are learnable. They can be seen as a set of learnable filters. Each filter is small spatially, and extends through the full depth of the input volume. During the forward pass, the input volume is convolved with each filter producing 2D activation maps, one per filter. The activation maps are then stacked together to form one output volume whose depth is equal to the number of convolution filters. During the convolution, the stride s specifies the number of pixels with which we slide the convolution filter. A stride of one, i.e. $s = 1$, means that the filters slide one pixel at a time. $s = 2$ means that the filters move two pixels at a time as we slide them around. If $s > 1$ then this procedure will produce spatially smaller output volumes. Thus, the convolution layer:

- Accepts a volume of size $w \times h \times d$, where w is the width, h is the height, and d is the depth ($d = 3$ in the first layer since it takes as input the input RGB image).
- Requires as parameters the number of filters n_F, their spatial extent e, the stride s, and the amount of zero padding p.
- Produces a volume of size $w_o \times h_o \times d_o$ where:

$$w_o = \frac{1}{s}(w - e + 2p) + 1, \qquad (4.17)$$

$$h_o = \frac{1}{s}(h - e + 2p) + 1, \qquad (4.18)$$

$$d_o = n_F. \qquad (4.19)$$

At runtime, an input image fed to the CNN transits through the different layers to produce at the end some scores (one score per neuron in the last layer). In the case of image classification, these scores can be interpreted as the probability of the image to belong to each of the classes, e.g. chairs, horses, cars, and so on. The goal of the training process is to learn the weights of the filters at the various layers of the CNN. Note that, often, the output of one of the layers before the last layer, which is fully connected, can be used as a global descriptor for the input image. The descriptor can then be used for various image analysis tasks including classification, recognition, and retrieval.

4.5.2 Multiview CNN for 3D Shapes

CNNs can be used to analyze the shape of 3D objects as follows:

- **First**, render 2D views of a 3D shape from arbitrary or carefully chosen view points;
- **Second**, pass each 2D view through a pretrained CNN and take the neuron activations of one of the layers, not necessarily the last one, as a descriptor for the input 2D view;
- **Finally**, aggregate the descriptors of the different views into a single global descriptor.

Such CNNs can be pretrained on large databases of 2D images. However, for better accuracy, one can fine-tune the network using a training set of 2D views rendered from 3D shapes. Below, we describe some of the network architectures used for this purpose and the methods used for the aggregation of the view descriptors.

4.5.2.1 Network Architecture

Su et al. [68] used the VGG-M network from [69] whose architecture is similar to the one shown in Figure 4.7. It consists of:

- Five convolutional layers, which we denote by $conv_{1,...,5}$,
- Three fully connected layers $fc_{6,...,8}$, and
- A softmax classification layer.

The output of the second last fully connected layer, i.e. fc_7, which is a 4096-dimensional vector, is used as a view descriptor. The network is first pretrained on ImageNet images [70], which has 1000 image categories, and then fine-tuned on 2D views of 3D shapes.

4.5.2.2 View Aggregation using CNN

Once a descriptor is computed for each view, these descriptors need to be aggregated so that 3D shapes can be compared. This can be done is several ways. For instance, when comparing two shapes, each one represented with N views, one can compute $N \times N$ pairwise distances and take either the minimum or the average pairwise distance as the measure of dissimilarity between the two shapes. Alternatively, one can concatenate the descriptors of each view to form a single long descriptor for the 3D model. This would require an accurate alignment of the input 3D models using one of the methods discussed in Chapter 2. Efficient alignment of 3D models is not straightforward to solve, especially when dealing with arbitrary shapes such as those found in the public domain. Su et al. [68] overcome these difficulties by using a MVCNN on top of the image-based CNNs as follows (see also Figure 4.8):

- Take the output signal of the last convolutional layer of the base network (CNN1) from each view, and combine them, element-by-element, using a max-pooling operation. The view-pooling layer can be placed anywhere in the network. Su et al. [68] suggested that it should be placed close to the last convolutional layer ($conv_5$) for an optimal performance.
- Pass this view-pooled signal through a second CNN (MVCNN), which can be trained or fine-tuned in exactly the same manner as CNN1 using back-propagation and gradient descent.
- Finally, use the output of fc_7 of the MVCNN as an aggregated shape descriptor.

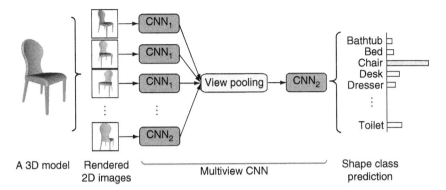

Figure 4.8 Illustration of the MVCNN architecture for view-based 3D shape analysis. A set of cameras are placed around the 3D model. Each camera captures a single image that is fed to a view-based CNN. The output of the view-based CNNs is aggregated using a view pooling and then fed into another CNN that produces the class scores.

This aggregated descriptor is readily available for a variety of tasks, e.g. shape classification and retrieval, and offers significant speed-ups against multiple image descriptors.

The second camera setup does not make any assumption about the upright orientation of the 3D shapes. In this case, images are rendered from more viewpoints by placing 20 virtual cameras at the 20 vertices of an icosahedron enclosing the shape. All the cameras point toward the centroid of the mesh. Four images are then rendered from each camera, using $0, 90, 180, 270$ degrees rotation along the axis passing through the camera and the object centroid, yielding a total of $N = 80$ views. One can also generate more views by using different shading coefficients or illumination models. Su et al. [68], however, found that this did not affect the performance of the output descriptors.

4.5.3 Volumetric CNN

To leverage the full power of 3D data, several papers, e.g. 3DShapeNets [71] and VoxNet [72], proposed to extend 2D CNN architectures, which operate on 2D projections of 3D models, to 3D (or volumetric) CNNs, which operate on volumetric representations of the 3D models. In 3DShapeNet of Wu et al. [71], for example, a 3D model is represented using a volumetric occupancy grid of size $32 \times 32 \times 32$. Each voxel is a binary variable that encodes whether the voxel is inside or outside the 3D shape. This 3D binary voxel grid is then used as input to a CNN with volumetric filter banks. The network architecture is composed of a CNN part with three convolutional layers ($conv_{1,...,3}$), and a multilayer perceptron (MLP) part with two fully connected layers ($fc_{1,2}$):

- $conv_1$ uses 48 volumetric filters of size $6 \times 6 \times 6$ and stride 2;
- $conv_2$ uses 160 filters of size $5 \times 5 \times 5$ and stride 2;
- $conv_3$ uses 512 filters of size 4;
- fc_1 and fc_2 have 1200 and 400 units, respectively.

Wu et al. [71] used the output of the fifth layer, i.e. fc_2, as a global descriptor for the indexation of 3D shapes. Recall that, in general, one can also use the output of the intermediate layers as descriptors.

To train this model, Wu et al. [71] selected 40 common object categories from ModelNet [73] with 100 unique CAD models per category. These data are then augmented by rotating each model every certain angle θ (Wu et al. used $\theta = 30$ degrees along the gravity direction, which generates 12 poses per model) resulting in models in arbitrary poses.

Interestingly, while this approach achieves good results, the MVCNN achieves a significantly better performance, see [68]. This indicates that the simple 3D CNN architecture described above is unable to fully exploit the power of 3D representations. Qi et al. [74] proposed two distinct volumetric CNNs architectures, which significantly improve state-of-the-art CNNs on 3D volumetric data. The first network is designed to mitigate overfitting by

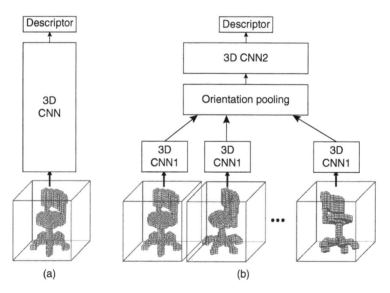

Figure 4.9 Illustration of orientation pooling strategies in volumetric CNN architectures. (a) Volumetric CNN (VCNN) with single orientation input. (b) MultiOrientation Volumetric Convolutional Neural Network (MO-VCNN), which takes in various orientations of the 3D input, extracts features from shared CNN1 and then pass pooled feature through another network CNN2 to make a prediction. Source: Qi et al. 2016 [74]. ©2016. Reprinted with permission of IEEE.

introducing auxiliary training tasks, which encourage the network to predict object class labels from partial subvolumes. The second network mimics MVCNNs, but instead of using rendering routines from computer graphics, the network projects a 3D shape to 2D by convolving its 3D volume with an anisotropic probing kernel. This kernel is capable of encoding long-range interactions between voxels.

Qi et al. [74] also used various data augmentation techniques. For instance, new training samples can be generated by rotating the training data with different azimuth and elevation rotations. Finally, in the same work, Qi et al. [74] proposed to add an orientation pooling stage that aggregates information from different orientations, see Figure 4.9. The descriptor obtained with this method, which is called *MultiOrientation Volumetric Convolutional Neural Network* (MO-VCNN), is less sensitive to changes in the shape orientation. This idea is conceptually similar to the MVCNN architecture of Figure 4.8, which has a view pooling layer (which is the same as orientation pooling) followed by another CNN.

4.6   Summary and Further Reading

This chapter presented the main global descriptors that have been used for the analysis of the shape of 3D objects. Table 4.1 summarizes some of their properties. In summary, early works used hand-crafted global descriptors to map a 3D model into a compact representation, which characterizes the geometry and topology of the 3D model. These descriptors perform relatively well on small datasets, e.g. the Princeton Shape Benchmark [53]. Their performance, however, drops significantly when applied to large databases such as ShapeNet [75] and ModelNet [73]. With the advent of deep learning techniques and CNNs, trained on large collections of images and 3D models, most of the recent global descriptors are deep-learning based. They are achieving a remarkable performance as will be discussed in Chapter 12. Table 4.1 summarizes the main properties of the global descriptors discussed in this chapter. The reader can also find in Chapter 12 a detailed evaluation of the performance of these descriptors on various 3D shape benchmarks.

The techniques presented in this chapter appeared in various research papers. For instance, shape distribution-based descriptors have been published in [48]. Spherical harmonic and power spectrum-based descriptors appeared in [58, 60, 76]. View-based descriptors have been very popular. For instance, the LFD of Chen et al. [52] is one of the most popular descriptors in this category as it achieved the best performance on the Princeton Shape Benchmark [53]. As discussed earlier in Section 4.5, deep learning techniques are becoming increasingly popular in 3D shape analysis. While Section 4.5 provides the fundamental concepts, this topic will be revisited a few more times in the subsequent chapters.

Table 4.1 A summary of the characteristics of the global descriptors presented in this chapter.

Category	Acronym	Input	Size (bytes)	Invariance			
				Trans.	Scale	Rotation	Isometry
Distribution	A3 [48]	Polygon soup	136	✓	✓	✓	—
	D1 [48]	Polygon soup	136	✓	—	✓	—
	D2 [48]	Polygon soup	136	✓	—	✓	—
	D3 [48]	Polygon soup	136	✓	—	✓	—
	D4 [48]	Polygon soup	136	✓	✓	✓	—
	Ac [48]	Polygon soup	136	✓	—	✓	—
	G	Watertight mesh	136	✓	—	✓	✓
View-based	LFD [52]	Polygon soup	4 700	—	✓	—	—
Spherical	EGI [53, 56, 60]	Polygon soup	1 032	✓	✓	—	—
	CEGI [53, 57, 60]	Polygon soup	2 056	✓	—	—	—
	SECTORS [51, 53, 60]	Polygon soup	552	✓	—	—	—
	SECSHELL [51, 53, 60]	Polygon soup	32 776	✓	—	—	—
	EXT [53, 58, 60]	Polygon soup	552	✓	—	—	—
	REXT [53, 59, 60]	Polygon soup	17 416	✓	—	—	—
	GEDT [53, 60]	Polygon soup	32 776	✓	—	—	—
	SHD [53, 60]	Polygon soup	2 184	✓	—	—	—
	SWC [55, 63]	Polygon soup	2 048	✓	—	✓	—
	SWEL1 [55, 63]	Polygon soup	76	✓	—	✓	—
	SWEL2 [55, 63]	Polygon soup	76	✓	—	✓	—
Deep learning	Multi-view CNN (MVCNN) [68]	Polygon soup	16 384	—	—	✓	—
	Multi-orientation vol. CNN (MO-VCNN) [74]	Polygon soup	—	—	—	✓	—

Note that when a descriptor is not invariant to a specific type of transformations, then either one needs to normalize the 3D models prior to the computation of the descriptor or use a distance measure that is invariant to such transformations.

Finally, the descriptors presented in this chapter are just a sample from the wide range of descriptors that have been proposed in the literature. Other examples include symmetry descriptors [9, 77] and graph-based descriptors [78]. For an additional discussion of the state-of-the-art of hand-crafted global descriptors, we refer the reader to some of the recent survey papers, e.g. [53, 79, 80]. For shape analysis using deep learning techniques, many methods are based on either the multiview representation presented in Section 4.5, e.g. [68, 81, 82], or the volumetric representations [71, 74]. We refer the reader to [83] for a comprehensive survey of deep learning techniques for 3D shape analysis.

5

Local Shape Descriptors

5.1   Introduction

Local shape description is an essential component for many 3D computer vision and graphics applications. For example, in 3D shape registration, two partial scans of the same object, but captured from two different viewpoints, can be merged together by matching keypoints on the two scans. In 3D object recognition, we may wish to detect and recognize an object occluded in a cluttered 3D scene by matching 3D exemplar models to patches or parts of the 3D scene. In partial 3D shape retrieval, we may wish to retrieve a list of 3D objects that are similar to a given partial scan or a given part of another 3D model. The global shape descriptors described in Chapter 4 are usually not suitable for such tasks. In fact, a global descriptor abstracts the shape of an entire 3D object into one real-valued vector. As such, the information they convey is not localized to specific parts of the object. This makes them not suitable for tasks such as registration, recognition under occlusions or in cluttered 3D scenes, and partial 3D shape matching and retrieval.

In this chapter, we focus on local descriptors, which, in contrast to global descriptors, encode the shape of small patches around a set of specific keypoints. Thus, they are more robust to occlusions and clutter and are more suitable for partial shape matching and retrieval. Local descriptors can be also aggregated to form global descriptors with a significantly higher descriptive power than the global descriptors described in Chapter 4.

The process of computing local shape descriptors usually consists of two main modules: the 3D keypoint detection module and the local feature description module, see Figure 5.1. During the 3D keypoint detection, keypoints (also called feature points, salient points [SPs], or interest points) with rich information contents are located, and a characteristic scale is defined for each keypoint. The location and scale of each keypoint provide a local surface patch for the subsequent feature description module [84]. During the local feature description, the geometric information of the patch around the

3D Shape Analysis: Fundamentals, Theory, and Applications, First Edition.
Hamid Laga, Yulan Guo, Hedi Tabia, Robert B. Fisher, and Mohammed Bennamoun.

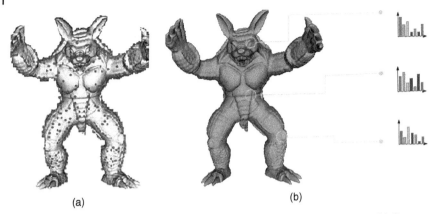

Figure 5.1 An illustration of keypoint detection (a) and local feature description (b). Source: Panel (a) Reprinted with permission from Ref. [84]. ©2013, Springer.

keypoint is encoded into a compact vector called *local descriptor* [85, 86]. These local descriptors, extracted from the 3D shape, can then be used for various high-level tasks, e.g. 3D shape registration [87], 3D object recognition [85], 3D face recognition [88], robot grasping [89], and 3D shape retrieval [90].

In this chapter, we first present the challenges and requirements for the design of a 3D keypoint detector and a feature descriptor (Section 5.2). We then introduce the basic concepts, the taxonomy, and several representative methods for 3D keypoint detection (Section 5.3) and feature description (Section 5.4). Finally, we discuss a few fundamental techniques, which have been used to aggregate local descriptors into global descriptors, which in turn can be used for 3D shape retrieval and classification tasks.

5.2 Challenges and Criteria

In this section, we will introduce the challenges for local feature extraction and the criteria for the design of 3D keypoint detection and local feature description.

5.2.1 Challenges

Considering the application scenarios for local shape descriptors, several major challenges should be considered when designing or selecting a local shape descriptor for 3D shape-related applications. For example:

- **Occlusion**. Due to the shape complexity of an object under observation, self-occlusion exists naturally in 3D data.

- **Clutter**. 3D scenes usually contain multiple objects. As such, the clutter caused by other objects does not only introduce occlusions but also brings in additional undesired points to the shape of the object of interest.
- **Noise**. A 3D shape may be affected by different types of geometrical noise, which can be Gaussian, or shot noise. It can be even affected by color noise and topological noise.
- **Data resolution**. There are various types of 3D shape representations (see Chapter 2). While mesh and point cloud representations are the most commonly used, a 3D shape can be represented with different levels of mesh/point-cloud resolutions. On the one hand, a shape with a low resolution, which is computationally efficient, may have several missing geometric details. On the other hand, the analysis of a shape with a high resolution, which captures fine geometric details, is usually computationally expensive.
- **Pose variation**. 3D acquisition systems scan 3D shapes from arbitrary view points. The scanned 3D shapes can also be in different poses and scales. Such rigid transformations introduce additional nuisance that affect the performance of descriptors.
- **Deformation**. For a nonrigid object (e.g. a human body which bends and stretches), the shape may be deformed to a certain degree, making the extrinsic shapes of an object significantly different.

Besides, several other challenges, e.g. scale variation, missing data, holes, and topology change, should also be considered [91].

5.2.2 Criteria for 3D Keypoint Detection

Given the challenges listed in Section 5.2.1, the following criteria should be taken into consideration when designing 3D keypoint detection algorithms [92, 93]:

- **Repeatability**. Keypoints detected on the overlapped surface of two different scans of an object (e.g. observed from different viewpoints or after a specific transformation) should have a high repeatability. That is, the majority of the keypoints detected on one scan should have their corresponding keypoints on another scan with a small error tolerance.
- **Distinctiveness**. The local patch around each keypoint should contain sufficiently distinctive information to uniquely characterize that point. By doing so, the subsequently extracted feature descriptors can be correctly distinguished and matched.
- **Accuracy**. The location and scale of each keypoint should be accurately detected to ensure that exactly the same local patch can be extracted on different scans of an object.

- **Quantity**. The number of detected keypoints should be large enough to encode more information of the shape. Meanwhile, it cannot be too large to avoid high computational costs. Thus, there should be a trade-off between descriptiveness and computation cost for a specific application.

5.2.3 Criteria for Local Feature Description

Similarly, the following criteria should also be considered for local feature description methods:

- **Descriptiveness**. A good local feature descriptor should be able to encapsulate the predominant information about the shape of the underlying surface and to provide sufficient descriptive richness for the discrimination of different local surface shapes.
- **Robustness**. A good local feature descriptor should not vary too much when different nuisances, such as noise, holes, occlusion, clutter, and deformation, are added to the 3D shape.

Besides, other issues such as computational and memory cost are also important for some applications.

5.3   3D Keypoint Detection

Keypoint detection is a prerequisite module prior to the local feature description module. In the most trivial cases, feature descriptors are computed at all points of a 3D shape, thereby bypassing the module of keypoint detection. However, extracting local features at all points is computationally too expensive because the redundancy in areas with small shape variations is not removed [94]. Therefore, detecting only a small subset of points over the entire shape is essential for efficient feature description. Alternatively, random sampling, uniform sampling, and mesh decimation are used to reduce the number of points for feature description [95–97]. However, both sparse sampling and mesh decimation cannot produce optimal keypoints in terms of repeatability and discriminativenss, as no consideration is given to the richness of the discriminative information of these selected points [93]. Therefore, efficient methods for detecting representative keypoints from a 3D surface are critical.

The task of 3D keypoint detection is to determine the location and scale of each keypoint. Note that, the notion of "scale" used in 3D data is different from the one used in 2D images [84]. In 2D images, scale is used to determine the image structures that are invariant to size changes during the projection from the 3D space to the 2D imaging plane. For 3D data, since metric information (i.e. physical size) is already included in the data, scale is used to determine the most discriminative neighborhood (i.e. support region) around a 3D keypoint.

The scale can be predetermined or adaptively detected. Consequently, existing 3D keypoint detection methods can be divided into two categories: fixed-scale keypoint detection methods and adaptive-scale keypoint detection methods [84, 85].

- Fixed-scale keypoint detection methods consider a point as a keypoint if it is distinctive within a predetermined neighborhood. The neighborhood size is determined by the scale, which is a parameter of the method.
- Adaptive-scale keypoint detection methods first construct a scale-space for a 3D shape and then select as keypoints the points with extreme distinctiveness measures in both the spatial and scale neighborhoods.

An important issue for these two types of methods is the definition of a measure of distinctiveness, which determines the specific type of 3D structures that are selected by the detector, and subsequently the robustness and repeatability of the keypoints [84].

A comprehensive survey on 3D keypoint detection methods can be found in [85], which presents the taxonomy of these methods and describes several representative methods within each category. Tombari et al. [84] have tested eight 3D keypoint detectors in terms of repeatability, disctinctiveness, and computational efficiency on five datasets under different nuisances including noise, occlusion, clutter, and viewpoint changes. Besides, a few experimental evaluation papers are also available in [98, 99].

5.3.1 Fixed-Scale Keypoint Detection

The fixed-scale keypoint detection methods define a keypoint as the point that is distinctive within a predetermined neighborhood (scale). Here, fixed-scale implies that these methods detect all the keypoints at a specific scale without any consideration to the intrinsic spatial extent of the underlying local geometric structure [84]. Specifically, these methods first compute a distinctiveness score for each point, and then select keypoints by maximizing the distinctiveness score in a spatial neighborhood defined by the scale. Naturally, different types of curvatures and surface variations are used as distinctiveness measures.

5.3.1.1 Curvature-Based Methods

As shown in Figure 5.2, these methods assume that the majority of points on a 3D shape lie on regions with small curvatures. Therefore, most points with low curvatures are redundant and not qualified to serve as keypoints. In contrast, points with high curvature can be considered as unique and rich in information about the shape.

Several different types of curvatures, e.g. the mean curvature, the Gaussian curvature, the principal curvatures, and the shape index (see Section 2.1), have been used to measure the distinctiveness. For instance, in [100], a point p is

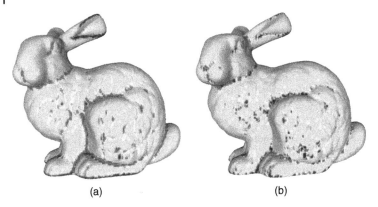

(a)   (b)

Figure 5.2 Curvature on the Bunny model. (a) Mean curvature. (b) Gaussian curvature. Light gray denotes low curvature values and dark gray denotes high curvature values.

selected as a keypoint if its Gaussian or mean curvature value is higher than the curvature values of its neighboring points.

5.3.1.1.1 Local Surface Patch (LSP) Instead of the Gaussian and mean curvatures, Chen and Bhanu [101] consider a point as a keypoint if its shape index value is a maximum or minimum in a local neighborhood, and exceeds a threshold defined by the average shape index value of the neighboring points. Specifically, a point p is marked as a keypoint only if its shape index $s_I(p)$ satisfies one of the following two conditions:

$$s_I(p) = \max_{j=1}^{n}(s_I(p_j)) \quad \text{and} \quad s_I \geq (1 + \alpha) \cdot \mu, \tag{5.1}$$

$$s_I(p) = \min_{j=1}^{n}(s_I(p_j)) \quad \text{and} \quad s_I \leq (1 - \beta) \cdot \mu, \tag{5.2}$$

where n is the number of points in the neighborhood of point p, $p_j, j = 1, \ldots, n$ are points in the neighborhood of p, μ is the average shape index in this neighborhood, and α and β are parameters to control the keypoint detection. An illustration of the keypoints detected by this method on the Armadillo model is shown in Figure 5.3a. It can be observed that the detected keypoints are uniformly distributed on the 3D shape, i.e. keypoints can be detected not only on highly protruded areas, but also on areas with few features, just avoiding the planar surfaces (e.g. the chest of Armadillo).

5.3.1.2 Other Surface Variation-Based Methods

Instead of different types of curvatures, other surface variation measures are also used to detect keypoints from a 3D shape. Since the eigenvalues λ_1, λ_2, and λ_3 of the covariance matrix of points on a local surface explicitly represent the variation of that surface, they can be used to detect 3D keypoints.

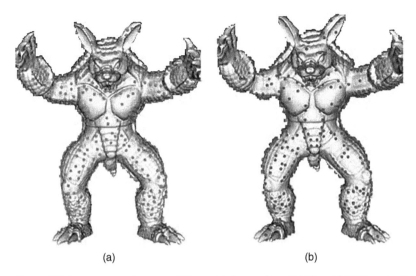

(a)   (b)

Figure 5.3 Keypoints on the Armadillo model detected by (a) LSP and (b) ISS methods. Source: Reprinted with permission from Ref. [84]. ©2013, Springer.

5.3.1.2.1 *Matei's Method* For example, Matei et al. [102] use the smallest eigenvalue λ_1 of the local surface around a point p to measure the distinctiveness of that point. All points are sorted in decreasing order of their distinctiveness measures, and keypoints are then selected greedily from the sorted list. In order to reduce redundancy, unchecked points within the neighborhood of a selected keypoint are removed from the candidate list.

5.3.1.2.2 *Intrinsic Shape Signatures (ISS)* Similarly, Zhong [103] proposes an ISS method to detect keypoints using the smallest eigenvalue λ_1. In addition, Zhong [103] uses the ratio of two successive eigenvalues to further prune the candidate keypoints. Only the points satisfying the constraints $\frac{\lambda_2}{\lambda_1} < \tau_{12}$ and $\frac{\lambda_3}{\lambda_2} < \tau_{23}$ are retained, where τ_{12} and τ_{23} are two thresholds for the selection of keypoints. Figure 5.3b shows the keypoints detected on the Armadillo model using this method.

5.3.1.2.3 *Harris 3D* Inspired by the success of Harris detector for 2D images, Sipiran and Bustos [104] propose a Harris 3D detector for 3D meshes. Given a point p, the neighboring points are first translated into a canonical coordinate system by subtracting the value of the centroid. These points are then rotated to make the normal at p aligned with the Z axis of the coordinate system. Next, these transformed points are fitted into a quadratic surface of the form $f(u, v) = a_1 u^2 + a_2 uv + a_3 v^2 a_4 u + a_5 v + a_6$, where $a_i, i = 1, \ldots, 6$, are real values. A symmetric matrix E is then defined as:

$$E = \frac{1}{\sqrt{2\pi}\sigma} \int_{\mathbb{R}^2} e^{-\frac{u^2+v^2}{2\sigma^2}} \begin{pmatrix} f_u^2(u,v) & f_u(u,v)f_v(u,v) \\ f_u(u,v)f_v(u,v) & f_v^2(u,v) \end{pmatrix} du\, dv, \qquad (5.3)$$

where σ is the Gaussian variance. The Harris 3D operator value at a point p is calculated as:

$$h(p) = \det(E) - \alpha(\text{trace}(E))^2, \qquad (5.4)$$

where $\det(E)$ and $\text{tr}(E)$ are, respectively, the determinant and the trace of the matrix E. α is a parameter that needs to be tuned experimentally. Finally, a fixed percentage of points with the largest values of $h(p)$ are selected as keypoints.

The Harris 3D detector is robust to several nuisances such noise, change in the tessellation of the 3D shape, local scaling, shot noise, and holes [104].

Since only a fixed-scale is used to find keypoints, the implementation of fixed-scale 3D keypoint detection algorithms is relatively straightforward. However, these methods have also a few limitations. For example:

- These methods may detect too few keypoints, particularly on the less-curved parts of the 3D object. This could be problematic for object recognition. For example, no keypoint can be detected on the belly of the chicken of Figure 5.4a with a neighborhood size of 20 mm. However, if we reduce the

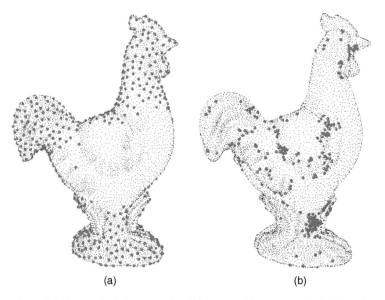

(a)   (b)

Figure 5.4 Keypoints detected on the Chicken model at two scales. (a) Keypoints with a neighborhood size of 20 mm. (b) Keypoints with a neighborhood size of 5 mm. The sampling rate of the model is 2 mm. Source: Reprinted with permission from Ref. [93]. ©2010, Springer.

neighborhood size to 5 mm, a sufficient number of keypoints can be detected on the belly, as shown in Figure 5.4b.

- These methods determine the scale empirically. They do not fully exploit the scale information encoded in the local geometric structures for the estimation of an inherent scale of a keypoint. Therefore, the neighborhood size cannot be adaptively determined [105].

5.3.2 Adaptive-Scale Keypoint Detection

The geometric structure on a surface has its inherent scale. The inherent scale represents the intrinsic spatial extent (neighborhood size) of the underlying local structure. For example, the keypoints detected at small scales represent the subtle geometric characteristics of the underlying surface. In contrast, keypoints detected at large scales represent the dominant features. Therefore, besides its location, the inherent scale of a keypoint should be adaptively determined.

Adaptive-scale keypoint detection methods first construct a scale-space for a 3D shape, and then consider a point as keypoint if it has the local extreme distinctiveness measure in both its spatial and scale neighborhood. Both the location and scale of a keypoint can be determined in the scale-space. Different from our previous review [85], we classify the existing adaptive-scale keypoint detection methods into two categories: extrinsic scale-space and intrinsic scale-space based methods.

5.3.2.1 Extrinsic Scale-Space Based Methods

These methods detect keypoints by constructing an extrinsic scale-space which directly works on the coordinates of a 3D shape. This can be achieved either by performing filtering on the 3D shape or by calculating surface variations at multiple scales.

5.3.2.1.1 3D Shape Filtering

Salient Points A straightforward approach to construct a scale-space is by successively smoothing the surface of a 3D shape, and then detecting the locations and scales of the keypoints from the constructed scale-space. Following this paradigm, Castellani et al. [106] apply a series of Gaussian filters to the mesh M and then detect, as keypoints, the points which had the maximum displacement along the direction of the normals. Specifically, the mesh M is first resampled at n_o levels to yield a set of octave meshes $M^j, j = 1, \ldots, n_o$. Here, M^j is called an *octave-j* mesh since the variation of resolution is a jump of an octave in the scale space [106]. The keypoint detection method then consists of two major phases: intraoctave and interoctave phases.

Intraoctave Phase A set of n_s Gaussian filters are first applied to each octave mesh M^j to obtain a set of multidimensional filtering maps $\mathcal{F}^j_i (i = 1, \dots, n_s)$. Given the Gaussian kernel $g(p, \sigma)$ with a standard deviation σ, the Difference-of-Gaussians (DoG) operator is given as:

$$\mathcal{F}^j_i(p) = g(p, \sigma_i) - g(p, 2\sigma_i), \tag{5.5}$$

where $g(p, \sigma_i)$ is the Gaussian operator applied to the point p, and σ_i is the standard deviation of the Gaussian kernel for the ith scale. Six values of σ_i, mainly $\{1\varepsilon, 2\varepsilon, 3\varepsilon, 4\varepsilon, 5\varepsilon, 6\varepsilon\}$, are used to obtain six scales of filtering. Here, ε is set to 0.1% of the length of the main diagonal of the bounding box of the 3D shape. Note that, the DoG calculation in Eq. (5.5) approximates the scale-normalized Laplacian of Gaussian (LOG), which is important to achieve scale invariance [107].

The filtering map $\mathcal{F}^j_i(p)$ for a point p is a 3D vector representing the displacement of p from its original location after filtering, and the most significant displacement is along the normal direction of that point. Therefore, the filtering map $\mathcal{F}^j_i(p)$ is further projected to the normal $\mathbf{n}(p)$ of p to obtain a scalar scale map $M^j_i(p)$. That is:

$$M^j_i(p) = |\mathbf{n}(p) \cdot \mathcal{F}^j_i(p)|. \tag{5.6}$$

The scale map $M^j_i(p)$ is further normalized using the Itti's approach [108] to obtain the final normalized scale map $\hat{M}^j_i$. That is, the values in $M^j_i(p)$ are first mapped to a fixed range $[0, 1, \dots, R]$. Then, the global maximum T and the average $\hat{t}$ of the remaining local maxima are obtained. Finally, the map is multiplied by $(T - \hat{t})^2$ to produce the final normalized scale map $\hat{M}^j_i$.

Next, an adaptive inhibition process is enforced on each normalized scale map. Specifically, for a point p, if its normalized scale map value $\hat{M}^j_i(p)$ is larger than 85% of the values of its neighboring points, the value of $\hat{M}^j_i(p)$ remains unchanged. Otherwise, it is set to 0, resulting in an inhibited scale map. Finally, the inhibited saliency map is generated by adding the values of all the inhibited scale maps. A candidate keypoint is then selected if its inhibited scale map value is a local maximum and it is larger than 30% of the global maximum. The process of the intraoctave phase is shown in Figure 5.5.

Interoctave Phase Interoctave phase is a validation phase, which is used to further improve the robustness of the keypoint detectors to varying mesh resolutions. That is, five octave meshes are generated using five mesh decimation levels $j \in \{0, h, 2h, 3h, 4h\}$, where h is set to 0.2. The intraoctave phase is then applied to each octave mesh to detect candidate keypoints. The candidate keypoints detected in at least three octaves are considered as final keypoints.

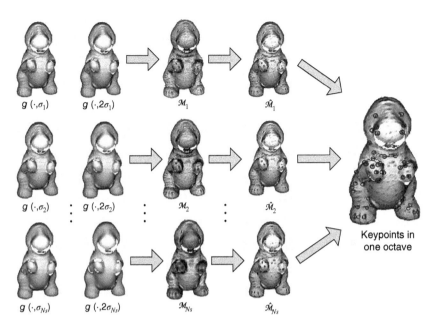

Figure 5.5 The process of the intraoctave phase. A set of Gaussian filters are applied to an octave mesh M^j to obtain several multidimensional filtering maps, which are further used to produce several scalar scale maps M_i^j. These scale maps M_i^j are normalized to obtain the normalized scale maps $\hat{M}_{i}^j$, which are finally used to produce 3D keypoints in an octave.

Figure 5.6a shows an example of keypoints detected on the Armadillo model. We can see that this method is able to detect the most significant parts such as the eyes and the fingers of the paws of the Armadillo model [106].

Note that, this method directly applies the 2D scale-space theory to 3D geometric data by replacing pixel intensities with 3D point coordinates. It is argued

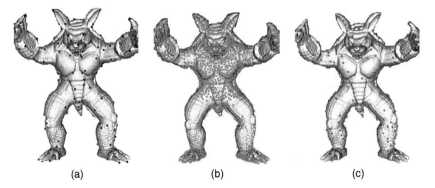

Figure 5.6 Keypoints detected on the Armadillo model by Castellani et al. [106], Mian et al. [93] and Unnikrishnan and Hebert [109]. Source: Reprinted with permission from Ref. [84]. ©2013, Springer.

that the extrinsic geometry and topology of the 3D shape are modified (e.g. shape shrinkage), and the causality property of a scale-space is violated [105, 109–111]. The causality property is an important axiom for a scale-space representation, which means that any structure (feature) in a coarse scale should be able to find its cause in finer scales [112].

5.3.2.1.2 Multiscale Surface Variation Another approach to generate a scale-space is to calculate surface variations at a set of varying neighborhood sizes. Methods in this category assume that the neighborhood size can be interpreted as a discrete scale parameter. Therefore, the process of increasing the local neighborhood size is similar to filter smoothing [113]. Compared to shape filtering, this approach avoids the direct modification of the 3D surfaces.

Pauly's Method Given a point p on a 3D mesh and a neighborhood size ρ, Pauly et al. [113] measure the surface variation δ using the eigenvalues λ_1, λ_2, and λ_3 of the covariance matrix of its neighboring points. That is:

$$\delta(p, \rho) = \frac{\lambda_1}{\lambda_1 + \lambda_2 + \lambda_3}, \tag{5.7}$$

where $\lambda_1 < \lambda_2 < \lambda_3$. By increasing the neighborhood size ρ, a set of surface variations $\delta(p, \rho)$ can be calculated for a point p. Note that, the neighborhood size of a point can be interpreted as a discrete scale parameter. It is also demonstrated that the surface variation corresponded well to the smoothed surface using the standard Gaussian filtering. Finally, keypoints are detected as those points having local maximum values in the surface variation across the scale axis.

Keypoint Quality-Adaptive Scale (KPQ-AS) Following the approach in [113, 114], Mian et al. [93] first crop the neighboring points from a 3D shape using a sphere of radius ρ centered at point p. The neighboring points are rotated to align the surface normal at p with the positive Z axis, and then the principal component analysis (PCA) is performed on the covariance matrix of the neighboring points. The ratio between the length along the first two principal axes (i.e. X and Y) is used to measure the surface variation $\delta(p, \rho)$:

$$\delta(p, \rho) = \frac{\max(x) - \min(x)}{\max(y) - \min(y)}. \tag{5.8}$$

Candidate keypoints are detected by comparing their surface variations $\delta(p, \rho)$ with a predefined threshold. Then, in order to rank these candidate keypoints, a keypoint quality measure is defined using a combination of the Gaussian and principal curvatures.

Since the principal curvatures correspond to the second-order derivatives of a surface, they are sensitive to noise. Instead of calculating the principal curvatures from the raw data of the 3D shape, a surface S is fitted to the raw local patch. An $n_s \times n_s$ lattice is then used to sample the fitted surface S. The principal curvatures are finally estimated at each point on the surface S. The distinctiveness (keypoint quality) Q_k of a point is calculated as:

$$Q_k = \frac{1000}{n_s^2} \sum_{p \in S} |K(p)| + \max_{p \in S}(100K(p)) + |\min_{p \in S}(100K(p))|$$
$$+ \max_{p \in S}(100\kappa_1(p)) + |\min_{p \in S}(1\kappa_2(p))|, \tag{5.9}$$

where K is the Gaussian curvature, κ_2 and κ_1 are the maximum and minimum curvatures, respectively. Note that, the summation, maximization, and minimization operations are defined over the $n_s \times n_s$ lattice of the surface S. Since the closely located keypoints usually represent similar local surfaces, the information contained in the feature descriptors extracted at these keypoints is highly redundant. Therefore, the candidate keypoints are sorted according to their distinctiveness values. Final keypoints are selected from the sorted candidate keypoints starting from the one with the highest distinctiveness. Once a keypoint is selected, its neighboring candidate keypoints within a particular radius are discarded. The scale for a keypoint is finally determined as the neighborhood size ρ for which the surface variation δ reaches the local maximum.

This method can detect sufficient keypoints. It is demonstrated that the distinctiveness measure Q_k correctly relates to the repeatability of keypoints [93]. The detected keypoints are repeatable and are also robust to noise. However, one major limitation of this method is its computational inefficiency [84]. An illustration of keypoints detected on the Armadillo model by this method is shown in Figure 5.6b.

Laplace–Beltrami Scale-Space (LBSS) Rather than using eigenvalues, Unnikrishnan and Hebert [109] construct a scale-space using the Laplace–Beltrami operator Δ_M. Given a neighborhood size ρ, an integral operator $B(p, \rho)$ is used to capture the surface variation around the neighborhood of a point p. In fact, $B(p, \rho)$ displaces a point p along its normal direction $\tilde{n}$, and the magnitude of displacement is proportional to the mean curvature H. That is

$$\|p - B(p, \rho)\| \approx \frac{\rho^2}{2} \Delta_M p. \tag{5.10}$$

In order to handle nonuniform sampling of a point cloud, a normalized integral operator $\tilde{B}(p, \rho)$ is proposed to replace $B(p, \rho)$, where

$$\|p - \tilde{B}(p, \rho)\| = H\frac{\rho^2}{2}. \tag{5.11}$$

The surface variation $\delta(p, \rho)$ is finally defined as:

$$\delta(p, \rho) = \frac{2}{\rho}\|p - \tilde{B}(p, \rho)\| \exp\left(-\frac{2}{\rho}\|p - \tilde{B}(p, \rho)\|\right). \tag{5.12}$$

A point is considered as a keypoint if it has an extreme value of $\delta(p, \rho)$ in both its geodesic neighborhood of radius ρ_k and over a range of neighboring scales (ρ_{k-1}, ρ_{k+1}). An illustration of the keypoints detected on the Armadillo model is shown in Figure 5.6c. From Eq. (5.11), we can see that the distinctiveness measure used in this method is proportional to the mean curvature. In the case of a perfect spherical surface, its characteristic scale is defined by the radius of the sphere. This method is very effective in determining the characteristic scales of 3D structures even in the presence of noise. However, its repeatability is relatively low and the number of detected keypoints is small, with most keypoints detected around bumps and arched surfaces [84].

5.3.2.2 Intrinsic Scale-Space Based Methods

These methods detect adaptive-scale keypoints by constructing a scale-space using the geometric attributes of a 3D shape. They use three different approaches to generate scale-spaces, namely, (i) scale-space on 2D parameterized images, (ii) scale-space on 3D shapes, and (iii) scale-space on a transformed domain.

5.3.2.2.1 *Scale-Space on 2D Parameterized Images* These methods first map a 3D shape to a 2D planar vector/scalar image, then construct a scale-space in the mapped 2D image and detect keypoints in this 2D scale-space. The motivation behind these methods is that the mapping makes the analysis of a 3D surface much easier since the 3D surface lacks a regular and uniform parameterization. In contrast, the resulting 2D image has a regular structure [111].

Geometric Scale-Space Given a triangular mesh of a 3D shape (Figure 5.7a), Novatnack et al. [110] first parameterize the 3D mesh M on a 2D plane D using a bijective mapping $\phi : M \rightarrow D$. They then encode the surface normals in the 2D image to represent the original shape, resulting in a normal map

| (a) | (b) | (c) | (d) |

Figure 5.7 The scheme of the keypoint detection method in [110]. (a) Original lion vase model. (b) Dense 2D normal map. (c) Keypoints detected in 2D normal map. (d) Color-coded keypoints on the 3D models. Larger spheres represent keypoints at coarser scales. Source: Reprinted with permission from Ref. [110]. ©2006, IEEE.

(Figure 5.7b). Since the parameterization is not isometric and the normal map does not preserve the angle and surface area of the original 3D shape, they compute a dense distortion map ε to account for this distortion. Analogous to the construction of the scale-space of 2D images [107], a discrete scale-space of the original 3D shape is then constructed by successively convolving the 2D normal map with distortion-adapted Gaussian kernels of increasing standard deviations. Specifically, given a Gaussian filter centered at a point p with a standard deviation σ, the value of the distortion-adapted Gaussian kernel at a point q is defined as:

$$g_\varepsilon(q,p) = \frac{1}{2\pi\sigma^2} \exp^{-\left[\frac{1}{8\sigma^2}(\varepsilon(q)^{-1}+\varepsilon(p)^{-1})^2((q_x-p_x)^2+(q_y-p_y)^2)\right]}. \tag{5.13}$$

The distortion-adapted Gaussian kernel is then independently applied to the x, y, and z components of the normal map. The vector at each point of the normal map is further normalized to ensure that each vector has a unit length. Gaussian kernels with different standard deviations are applied to the normal map to obtain a scale-space of the 3D shape. Since the derivative of the normals is represented by vectors rather than scalar gradient values, the gradients of the normal map in the horizontal (f_h) and vertical (f_v) directions are defined using the angle α between the surface normals. That is:

$$f_h(p) = \begin{cases} \dfrac{\varepsilon(p)}{2}\alpha(p_{h-1},p_{h+1}) & \text{if } \alpha(p_{h+1},p_h) > \alpha(p_{h-1},p_h) \\ -\dfrac{\varepsilon(p)}{2}\alpha(p_{h-1},p_{h+1}) & \text{if } \alpha(p_{h+1},p_h) < \alpha(p_{h-1},p_h), \end{cases} \tag{5.14}$$

and

$$f_v(p) = \begin{cases} \dfrac{\varepsilon(p)}{2}\alpha(p_{v-1},p_{v+1}) & \text{if } \alpha(p_{v+1},p_v) > \alpha(p_{v-1},p_v) \\ -\dfrac{\varepsilon(p)}{2}\alpha(p_{v-1},p_{v+1}) & \text{if } \alpha(p_{v+1},p_v) < \alpha(p_{v-1},p_v), \end{cases} \tag{5.15}$$

where $\varepsilon(p)$ is the distortion at point p, and $\alpha(p_1,p_2)$ represents the angle between the normals of two points p_1 and p_2. That is:

$$\alpha(p_1,p_2) = \arccos(\tilde{\mathbf{n}}(p_1) \cdot \tilde{\mathbf{n}}(p_2)). \tag{5.16}$$

Similar to Harris detector [115], given a window $\mathcal{W}$ centered at the point of interest p, the Gram matrix for p is calculated as:

$$C(p) = \begin{bmatrix} \sum_{q\in\mathcal{W}} g_\varepsilon(q,p)|f_h(q)|^2 & \sum_{q\in\mathcal{W}} g_\varepsilon(q,p)|f_h(q)||f_v(q)| \\ \sum_{q\in\mathcal{W}} g_\varepsilon(q,p)|f_h(q)||f_v(q)| & \sum_{q\in\mathcal{W}} g_\varepsilon(q,p)|f_v(q)|^2 \end{bmatrix}. \tag{5.17}$$

The corner response is then defined as:

$$\mathcal{R}(p) = \det(C(p)) - \alpha\text{Trace}(C(p))^2, \tag{5.18}$$

where α is a tuning parameter. Finally, the points with maximum corner responses along both the spatial and scale axes in the scale-space are considered as keypoints (Figure 5.7c,d). The procedure of this method is illustrated in Figure 5.7. This method is able to detect a sufficient number of keypoints [116]. One limitation of this method is that it requires an accurate estimation of the surface normals to construct the 2D scale-space [109, 117]. The normals, however, can only be reliably constructed from a 3D mesh with connectivity information. In their later work, Bariya et al. [105] extend this method to range images and define all operators in terms of geodesic distances instead of Euclidean distances to avoid the computation of a distortion map.

Diffusion Space Hua et al. [111] generate shape vector images by mapping a 3D shape to a canonical 2D domain, and then detect 3D keypoints from the shape vector images via linear diffusion. Specifically, conformal mapping is used to map a 3D shape to a 2D representation, namely, a shape vector image. Given two surfaces M_a and M_b, a diffeomorphism $f : M_a \rightarrow M_b$ is determined as conformal if it preserves angles. That is, if two intersecting curves X_1^a and X_2^a on M_a are mapped onto X_1^b and X_2^b on M_b by a conformal mapping function f, the angle between X_1^a and X_2^a should be equal to the angle between X_1^b and X_2^b. For a 3D shape M, both the conformal factor $\lambda(u, v)$ and the mean curvature $H(u, v)$ are encoded into the rectangular 2D domain D to obtain the shape vector images $I = (I_1, I_2, \ldots, I_{n_c})^\top$, where n_c is the number of 2D channels. The vector diffusion is then performed by:

$$\frac{\partial I_j}{\partial t} = \mathrm{div}\left(g\left(\sum_{n=1}^{n_c} \|\nabla f_{I_n}^2\| \right) \nabla I_j \right), \qquad j = 1, \ldots, n_c, \qquad (5.19)$$

where div is the divergence operator, ∇ denotes the gradient operator, I_j is the actual diffused image, f_{I_n} is the original image, and g is the diffusivity function:

$$g(x) = \frac{1}{\sqrt{1 + \left(\frac{x}{l}\right)^2}}. \qquad (5.20)$$

The diffusion space can then be obtained by solving Eq. (5.19). That is:

$$\begin{pmatrix} I_1 \\ I_2 \\ \vdots \\ I_{n_c} \end{pmatrix} = \left(I^{t_0} \; I^{t_1} \; \ldots \; I^{t_{n_s}} \right) = \begin{pmatrix} I_1^{t_0} & I_1^{t_1} & \cdots & I_1^{t_{n_s}} \\ I_2^{t_0} & I_2^{t_1} & \cdots & I_2^{t_{n_s}} \\ \vdots & \vdots & \ddots & \vdots \\ I_{n_c}^{t_0} & I_{n_c}^{t_1} & \cdots & I_{n_c}^{t_{n_s}} \end{pmatrix}, \qquad (5.21)$$

where n_c is the number of channels, and n_s is the number of scales. Each row of the matrix represents the sequence images of a specific channel at different scales t. Each column of the matrix represents the vector image at a specific scale. The difference of diffusion (DoD) is then calculated as

$$\text{DoD}^{t_i} = I^{t_{i+1}} - I^{t_i}, \quad i = 0, 1, \ldots, n_s - 1. \tag{5.22}$$

Once the DoD images are obtained, the points with maxima/minima DoD values across the scales are detected as keypoints. Specifically, for each DoD_j of a particular channel, if a point in $\text{DoD}_j^{t_i}$ with scale t_i has the maximum or minimum value compared to its 8 neighbors at the same scale t_i and 18 neighbors at the neighboring scales t_{i-1} and t_{i+1}, it is then considered as a keypoint and the scale t_i is determined as its associated scale. The keypoints detected in all of the channels are considered as the final keypoints.

Although it is possible to leverage the well-established keypoint detection methods for 2D images to detect keypoints on 3D shapes by surface mapping from a 3D domain to a 2D domain, several issues are still unresolved. **First,** these methods may introduce inevitable and large distortions when mapping a large and topologically complex surface to a 2D canonical domain. These unwanted distortions may consequently change the characteristics of the actual 3D shape and adversely affect the performance of the keypoint detection method [118, 119]. **Second,** these methods cannot be used to high-genus surfaces (i.e. surfaces with a large number of holes) [111, 119] since most of the existing parameterization methods operate only on surfaces of low genus.

5.3.2.2.2 Scale-Space on 3D Shapes These methods first calculate the geometric attributes (e.g. normals and curvatures) of a 3D shape, and then construct a scale-space by successively smoothing the geometric attributes. Keypoints are finally detected in this geometric attribute scale-space. Since filtering is applied to the geometric attributes rather than the coordinates of the 3D shape, no modification is made to the extrinsic geometry of the 3D shape. Therefore, the causality property of the scale-space is preserved.

MeshDoG Zaharescu et al. [91] propose the MeshDoG keypoint detector by constructing a scale-space of a specific scalar function f. The scalar function can either be a photometric attribute (e.g. texture) or a geometric attribute (e.g. curvature) defined on the vertices of a 3D mesh. The scale-space is constructed by progressive convolutions between a Gaussian kernel $g_{\sigma(t)}$ and the scalar function f. That is:

$$F_0(p_i) = f(p_i), \tag{5.23}$$

$$F_t(p_i) = F_{t-1} * g(p_i, \sigma(t)), \tag{5.24}$$

$$= \frac{1}{K_i} \sum_{q_j \in \mathcal{N}(p_i)} F_{t-1}(q_j) \exp\left(\frac{-d_g^2(p_i, q_j)}{2\sigma^2(t)}\right), \tag{5.25}$$

where $\mathcal{N}(p_i)$ represents the neighborhood of the point p_i, $d_g(p_i, q_j)$ is the geodesic distance between the two points p_i and q_j, and $t = \{1, \ldots, s \cdot c\}$. Here, s is the number of octaves and is set to $s = 3$, and c is the number of steps in each octave and is set to $c = 6$. K_i is a normalization term, which is defined as:

$$K_i = \sum_{q_j \in \mathcal{N}(p_i)} \exp\left(\frac{-d_g^2(p_i, q_j)}{2\sigma^2(t)}\right). \tag{5.26}$$

The standard deviation $\sigma(t)$ of the Gaussian kernel at iteration t is defined as:

$$\sigma(t) = 2^{\frac{1}{c-1}\lceil\frac{t}{c}\rceil} e_{avg}, \tag{5.27}$$

where e_{avg} denotes the average edge length of a 3D mesh. Note that $\sigma(t)$ is the same for convolutions within an octave but changes across octaves. The DoG operator is then used to approximate the Laplacian operator. That is:

$$L_t = F_t - F_{t-1}. \tag{5.28}$$

Candidate keypoints are detected as these points having a local extrema value over one-ring neighborhoods in both the current and adjacent scales. Furthermore, a two-stage pruning approach is applied. **First**, all candidate keypoints are sorted according to their magnitudes in the DoG scale space, only some of the candidate keypoints with the top 5% magnitude are retained. **Second**, noncorner responses are eliminated using the eigenvalues of the Hessian matrix at each candidate keypoint. The symmetric Hessian matrix approximation is defined as:

$$\widetilde{H}_{a,b}(L_i(p_i)) = \frac{1}{2}(H_{a,b}(L_i(p_i)) + H_{b,a}(L_i(p_i))), \tag{5.29}$$

where

$$H_{a,b}(L_i(p_i)) = \begin{bmatrix} D_{aa}L_i(p_i) & D_{ab}L_i(p_i) \\ D_{ba}L_i(p_i) & D_{bb}L_i(p_i) \end{bmatrix}. \tag{5.30}$$

Here, D_{aa}, D_{ab}, D_{ba}, and D_{bb} are the second-order directional derivatives along the unit vectors a and b, which are two orthonormal vectors on the tangent plane to the mesh M at the point p_i.

Noncorner edge responses with $|\lambda_2/\lambda_1| < 10$ are finally removed, where λ_1 and λ_2 ($\lambda_1 < \lambda_2$) are the two eigenvalues of the Hessian matrix $\widetilde{H}_{a,b}$. An illustration of keypoints detected by MeshDoG on the Armadillo model is shown in Figure 5.8.

This method offers a canonical formula to detect both photometric and geometric keypoints on a mesh surface. It is capable to detect a sufficient number of repeatable keypoints. Note that, different from the 3D shape filtering method in Section 5.3.2.1, this method works on a photometric or geometric scalar function rather than on the coordinates of a surface.

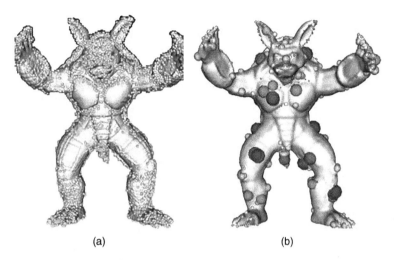

(a)   (b)

Figure 5.8 Keypoints detected on the Armadillo model by the MeshDoG [91] and IGSS [120] methods. Source: Panel (a) Reprinted with permission from Ref. [84]. ©2013, Springer. Panel (b) Reprinted with permission from Ref. [120]. ©2009, IEEE.

Other Work Inspired by the scale-invariant feature transform (SIFT) detector for 2D images, several other methods are also available. For example:

- **Geodesic scale-space** (GSS). Zou et al. [119] construct a GSS by convoluting a variable-scale geodesic Gaussian kernel with surface geometric attributes. Keypoints are then detected by searching the local extrema in both the spatial and scale dimensions of the GSS space, and are finally pruned according to their contrasts and anisotropies. Note that, the MeshDoG [91] and GSS [119] methods are quite similar in principle.
- **Hierarchical scale-space.** Hou and Qin [118] introduced a hierarchical scale-space using a pyramid representation and a geodesic metric. They first downsample the surface as the scale increases and build a scale-space of a specific scalar field on the surface using a Gaussian kernel. Keypoints are finally detected as local extrema in the scale-space. Simultaneous sampling in both the spatial and scale domains makes this method efficient and stable. This is because the hierarchical strategy enables the fast access of large geodesic neighborhoods in higher octaves of the pyramid. It is claimed that this method significantly reduces the processing time, as compared to GSS. It is also more stable under different mesh resolutions than MeshDoG [118].
- **Intrinsic geometric scale-space** (IGSS): Zou et al. [120] treat a surface as a Riemannian manifold and detect keypoints using an intrinsic geometric diffusion. Specifically, a scale-space of a 3D surface is generated by gradually smoothing the Gaussian curvatures via shape diffusion. Keypoints are identified as diffused curvature extrema in the IGSS space along both the spatial

and scale dimensions (as shown in Figure 5.8b). This method is based on the philosophy that the solution of the heat diffusion equation can equivalently derive the same scale-space as does a Gaussian smoothing.

5.3.2.2.3 Scale-Space on Transformed Domains
These methods consider a surface M as a Riemannian manifold and transform this manifold from the spatial domain to another domain (e.g. spectral domain). Subsequently, keypoints are detected in the transformed domain rather than the original spatial domain.

Heat Kernel Signature (HKS) Sun et al. [121] restrict the heat kernel to the temporal domain to detect keypoints. Let M be a compact Riemannian manifold. The heat diffusion process over M is governed by the heat equation:

$$\left(\Delta_M + \frac{\partial}{\partial t}\right) h(p, t) = 0, \tag{5.31}$$

where h is the heat, and Δ_M is the Laplace–Beltrami operator defined on M. The operator Δ is defined as:

$$\Delta u = \text{div}(\text{grad } u), \tag{5.32}$$

where div and grad denote the divergence and gradient operators, respectively. The fundamental solution $K_t(p, q)$ of the heat equation is called the heat kernel. $K_t(p, q)$ can be thought of as the amount of heat that is transferred from p to q in time t with a unit heat source located at p. They restrict the heat kernel $K_t(p, q)$ to the temporal domain $K_t(p, p)$ and use the local maxima of the function $K_t(p, p)$ to find keypoints. Here, the time parameter t is related to the neighborhood size of p, and therefore the time parameter provides the information of its inherent scale.

This method is invariant to isometric deformations (i.e. the deformations which preserve the geodesic distance). It usually captures the extremities of long protrusions on the surface [116]. It is able to detect highly repeatable keypoints and also robust to noise [98]. However, it is sensitive to varying mesh resolutions, and the number of detected keypoints is very small. Besides, it also requires a large computer memory [84].

Hu's Method Hu and Hua [122] extract keypoints in the Laplace–Beltrami spectral domain. Let f be a real function defined on a Riemannian manifold M. Using the Laplacian eigenvalue equation, f can be written as:

$$f = \sum_{i=1}^{n_v} \lambda_i \Lambda_i, \tag{5.33}$$

where Λ_i is the ith eigenvector, and λ_i is related to the ith eigenvalue. The geometry energy of a point p corresponding to the ith eigenvalue is defined as:

$$E_i(p) = \|\lambda_i \times \Lambda_i(p)\|_2. \tag{5.34}$$

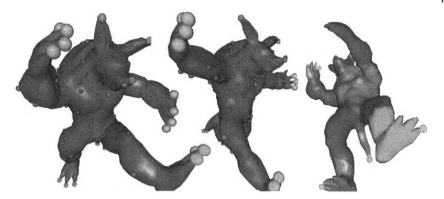

Figure 5.9 Keypoints detected on the Armadillo models with different poses by Hu and Hua [122]. Source: Reprinted with permission from Ref. [122]. ©2009, Springer.

The point whose geometry energy is larger than those of its neighboring points within several neighboring frequencies is picked up as a keypoint. Meanwhile, the spatial scale of a keypoint is provided by the "frequency" information in the spectral domain. An illustration of keypoints detected with this approach is given in Figure 5.9. Experimental results showed that the keypoints are very stable and invariant to rigid transformations, isometric deformations, and different mesh triangulations [122].

5.4 Local Feature Description

Given a local patch (surface) around a keypoint detected by a specific method (Section 5.3), the distinctive geometric information of the local surface is represented with a vector, called a *local feature descriptor*, which can then be used for various tasks. For example, local feature descriptors of two partial scans of a shape can be matched in order to register onto each other [87]. Local feature descriptors can also be aggregated to form a mid-level or high-level representation for shape retrieval [123]. Ideally, a feature descriptor should encapsulate the predominant local shape information of the underlying surface to provide sufficient descriptiveness. It should also be robust to a set of nuisances including noise, occlusion, clutter, varying point density, and viewpoint changes.

Existing methods can be classified into three categories according to their strategies used to generate feature descriptors, namely, (i) signature based, (ii) histogram based, and (iii) covariance-based methods. A comprehensive review of existing local feature descriptors is available in [85]. Guo et al. [86] introduce a set of metrics which can be used to evaluate the descriptiveness, robustness, scalability, efficiency, and performance of 3D local feature descriptors combined with 3D keypoint detectors. Ten popular descriptors are then evaluated on eight publicly available datasets. Several experimental comparisons between 3D local feature descriptors can also be found in [99, 124, 125].

5.4.1 Signature-Based Methods

The signature-based methods first resample the local surface around a keypoint with a set of points and then use the individual geometric attributes of these sample points to generate the local feature descriptor [86, 126]. Representative descriptors in this category include splash [127] and point signature [128].

5.4.1.1 Splash

Stein and Medioni [127] propose the splash descriptor, inspired by the picture of Prof. Edgerton (as shown in Figure 5.10a). Similar to the splashes generated by a drop falling into milk (Figure 5.10a), the *splash* descriptor encodes the local surface around a keypoint using the distribution (not histogram as used in 5.4.2) of surface normals (Figure 5.10b). Specifically, given a keypoint p, its surface normal $\tilde{\mathbf{n}}$ is used as the reference normal of a splash, and a circular slice around p is obtained using the geodesic radius r. Then, a set of sample points p_θ are sampled on the circular slice at regular angular intervals $\Delta\theta$. A local reference frame (LRF) is constructed using the normal $\tilde{\mathbf{n}}$ and the tangent plane at point p. For each sample point p_θ, two angles ϕ_θ and φ_θ can be generated as follows:

$$\phi_\theta = \arccos(\tilde{\mathbf{n}}, \tilde{\mathbf{n}}_\theta) \quad \text{and} \quad \varphi_\theta = \arccos(\mathbf{x}, \tilde{\mathbf{n}}_\theta^{z=0}), \tag{5.35}$$

where $\tilde{\mathbf{n}}_\theta$ is the surface normal at point p_θ, $\tilde{\mathbf{n}}_\theta^{z=0}$ is the projection of $\tilde{\mathbf{n}}_\theta$ on the XY plane, $\mathbf{x}$ is the X axis of the LRF. Each sample point p_θ can then be represented in the 3D space as (ϕ, φ, θ). Therefore, all sample points on the circular slice generate a set of 3D segments in the space defined by (ϕ, φ, θ). Then, several attributes are encoded into the 3D segments, including the curvature and torsion angles of a 3D segment, the maximum distance of the mapping from the θ

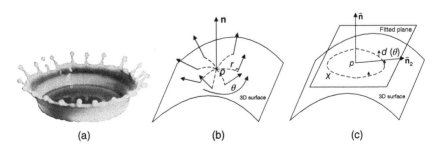

(a)   (b)   (c)

Figure 5.10 An illustration of signature-based methods. (a) Milk drop. (b) Splash [127]. (c) Point signature [128]. Source: Panels (b, c) Reprinted with permission from Ref. [85]. ©2014, IEEE.

axis, and the surface radius of the splash. The 3D segments encoded with these attributes work as the feature descriptor for the point p. Splash is one of the earliest well-known local feature descriptors for free-form 3D object recognition. It is shown to be robust to noise.

5.4.1.2 Point Signature

Chua and Jarvis [128] propose a 3D feature descriptor called point signature. For a keypoint p, a 3D space curve X is constructed by intersecting a sphere of radius r centered at p with the surface. Then, a plane Δ is fitted to the space curve and translated to p along a direction parallel to normal $\tilde{n}$ of that plane. The 3D space curve X is perpendicularly projected onto the plane Δ to generate a planar curve X'. The projection distance between each point on the curve X and its corresponding point on X' forms a signed distance d, which is negative if the projection has the same direction as $\tilde{n}$, and it is positive if the projection has an opposite direction as $\tilde{n}$. The projected point p_2 on the planar curve X' with the largest positive distance is then used to define a reference direction $\tilde{n}_2$. That is, $\tilde{n}_2$ is a unit vector from p to p_2, as shown in Figure 5.10c.

With this process, each point on the curve X can be characterized by two parameters: the signed distance d between the point and its corresponding projection on X', and the clockwise rotation angle θ about the axis $\tilde{n}$ from the reference direction $\tilde{n}_2$. The signed distance $d(\theta)$ along the rotation angle is used as the point signature descriptor at p.

In practice, an angular sampling is performed on the curve X from the reference direction $\tilde{n}_2$ to generate a discrete set of values $d(\theta_i)$ for $i = 1, \dots, n_\theta$, $0 \leq \theta_i \leq 2\pi$, where n_θ is the number of samples. The point signature descriptor is invariant to rotation and translation. Note that, the reference direction $\tilde{n}_2$ might not be unique since several points may have the same signed distance d. In this case, multiple signatures should be generated for a point p.

5.4.2 Histogram Based Methods

These methods represent the local surface around a keypoint with histograms of geometric or topological measurements (e.g. point count, mesh areas) in a specific domain (e.g. point coordinates, geometric attributes) [85, 126]. These methods can be classified into three different categories: (i) the histogram of spatial distributions, (ii) the histogram of geometric attributes, and (iii) the histogram of oriented gradients based methods [85].

5.4.2.1 Histogram of Spatial Distributions

These methods represent the local surface around a keypoint by generating histograms according to the spatial distributions of the local surface. Usually, a LRF or axis has to be defined for the keypoint to achieve invariance to rotation and translation. Consequently, 3D points on the local surface can be transformed to the LRF, and the dependence on the coordinate system is removed. Then, the 3D local space around the keypoint is divided into several bins. Each bin stores the occurrence frequency of some surface attributes.

Two issues are important for this type of descriptors: (i) the way the LRF is constructed and (ii) the strategy for the discretization of the neighborhood space. Different local reference construction approaches and neighborhood discretization strategies have been investigated, resulting in a number of descriptors in this category, including spin images [95], 3D Shape Context (3DSC) [96], ISS [103], and Rotational Projection Statistics (RoPS) [129].

5.4.2.1.1 Spin Images Johnson and Hebert [95] generate spin image descriptors using oriented points (i.e. 3D points with their associated surface normals). For a keypoint p with surface normal $\tilde{\mathbf{n}}$, an object-centered coordinate system can be defined using the point position and its surface normal. For each oriented point q on the local surface around p, two cylindrical coordinates can be defined with respect to p: the radial distance α and the signed distance β, which are defined as follows:

$$\alpha = \sqrt{||q - p||^2 - (\tilde{\mathbf{n}} \cdot (q - p))^2} \quad \text{and} \quad \beta = \tilde{\mathbf{n}} \cdot (q - p). \tag{5.36}$$

Here, α is the perpendicular distance between point q and the line through the surface normal at p. β is the signed perpendicular distance between the point q and the tangent plane defined by the keypoint p and its surface normal $\tilde{\mathbf{n}}$.

To generate a spin image, a 2D accumulator (histogram) indexed by (α, β) is first created. The (α, β) coordinates of all the points lying on the local surface around p are calculated. Once a point on the local surface falls into the bin of the 2D accumulator, the number in that bin is increased. Note that, a bilinear interpolation can be applied to smooth the contribution of each neighboring point. Figure 5.11a illustrates this method.

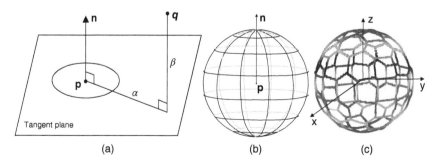

(a) (b) (c)

Figure 5.11 The local reference frame and neighborhood discretization for histogram of spatial distributions based methods. (a) Spin image [95]. (b) 3D shape context [96]. (c) Intrinsic shape signature [103]. Source: Reprinted with permission from Ref. [85]. ©2014, IEEE.

The 2D accumulator obtained with this method can be seen as an image where the dark areas in the image represent the bins containing a large number of projected points. The generation process of a spin image can also be visualized as spinning a sheet around the surface normal at the keypoint.

Spin images are invariant to rotation and translation, and are robust to clutter and occlusion. Besides, they are efficient and easy to implement. Spin images are one of the most popular local feature descriptors for 3D shapes. They have been widely used in various 3D vision-related applications such as surface registration and 3D object recognition. Also, in the literature, a large number of spin image variants have been proposed. Examples include face-based spin images [130], spin image signatures [131], multiresolution spin images [132], spherical spin images [133], and scale invariant spin images [134].

5.4.2.1.2   3D Shape Context   Inspired by the shape context descriptor for 2D images [135], Frome et al. [96] propose a 3DSC descriptor for 3D shapes, as shown in Figure 5.11b.

Given a keypoint p and its corresponding surface normal $\tilde{n}$, we first define a spherical support region using a sphere of radius r, centered at p, and whose north pole is aligned with the surface normal at p. The support region is then divided equally along the azimuth and elevation directions and logarithmically along the radial direction:

- The $L+1$ divisions along the azimuth direction are represented by $\Theta = \{\theta_0, \theta_1, \ldots, \theta_L\}$,
- The $J+1$ divisions along the elevation direction are represented by $\Phi = \{\phi_0, \phi_1, \ldots, \phi_J\}$,
- The $K+1$ radial divisions are represented by $R = \{r_0, r_1, \ldots, r_K\}$ such that

$$r_k = r_0 + \left(\frac{r_K}{r_0}\right)^{\frac{k}{K}}. \tag{5.37}$$

With this setup, the final feature descriptor is of dimension $L \times J \times K$. Each bin of indices (l, j, k) accumulates a weighted count $\omega(p_i)$ for each point p_i that falls inside the region of the bin. The contribution to the bin count for p_i is given by

$$\omega(p_i) = \frac{1}{\rho_{p_i} \sqrt[3]{V(l,j,k)}}, \tag{5.38}$$

where $V(l, j, k)$ is the volume of the (l, j, k)th bin, and ρ_{p_i} is the local point density, which is defined as the number of points in a sphere of radius δ around the point p_i. The normalization by the bin volume $V(l, j, k)$ accounts for the variation in bin size along the radius and elevation dimensions. The normalization by local point density is used to address the variation in sampling density.

Since only the surface normal is used as the reference axis to generate the 3DSC descriptor, there is a degree of freedom along the azimuth dimension.

Therefore, for the matching of feature descriptors of two surfaces, L descriptors, one for each initial azimuth values ϕ_0, should be generated for each keypoint in order to achieve the invariance to rotations.

5.4.2.1.3 Intrinsic Shape Signature (ISS)

To generate an ISS feature descriptor [103], an intrinsic reference frame F is first defined for the keypoint p with a support radius r using the eigenvalue analysis of its neighboring points. Specifically, a weight ω is calculated for each neighboring point p_j using the number of points within its spherical neighborhood of radius r_d. That is:

$$\omega_j = \frac{1}{|\{p_i : \|p_j - p_i\| < r_d\}|}, \tag{5.39}$$

where the denominator represents the number of points with distances to p_j less than a threshold r_d. Note that, the weight ω_j is used to improve the robustness of the descriptor with respect to an uneven point sampling. That is, a point from the sparsely sampled surface has a large contribution than the points from densely sampled surfaces.

Next, for the points within the neighborhood of a keypoint q, a weighted scatter matrix $C(p)$ can be calculated as follows:

$$C(p) = \frac{1}{\sum_{|p_j - p| < r} \omega_j} \sum_{|p_j - p| < r} \omega_j (p_j - p)(p_j - p)^\top. \tag{5.40}$$

By eigenvalue decomposition of the matrix $C(p)$, we obtain three eigenvectors $\{\Lambda_1, \Lambda_2, \Lambda_3\}$ sorted in the descending order of their respective eigenvalues. Then, an intrinsic reference frame F is generated using keypoint p as its origin, and $\Lambda_1, \Lambda_2, \Lambda_1 \times \Lambda_2$ as its X, Y and Z axes. Since there is an ambiguity in the direction of the Λ_1 and Λ_2, the intrinsic reference frame is not unique. Consequently, four reference frames should be calculated for a keypoint p.

Once a LRF is defined on the keypoint p, the neighboring space around p is aligned to it and then partitioned. Specifically, a discrete spherical grid is recursively generated from a base octahedron to partition the spherical space (r, θ, ϕ) into uniformly and homogeneously distributed bins, in the radial as well as in the angular directions, as shown in Figure 5.11c.

Finally, the ISS descriptor is computed by taking the sum of the weights (as defined in Eq. (5.39)) of all the points falling into each bin.

5.4.2.1.4 Rotational Projection Statistics (RoPS)

Guo et al. [129] construct a unique and unambiguous LRF by performing eigenvalue decomposition on the covariance matrix of all points lying on the local surface rather than just mesh vertices. For a keypoint p with a support radius r, a local surface containing m triangles and n vertices is extracted, as shown in Figure 5.12a,b. For the ith

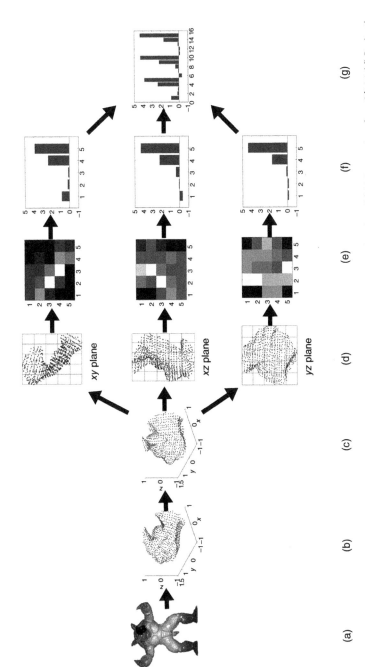

Figure 5.12 An illustration of the RoPS feature descriptor with one rotation. (a) Object (b) Local surface $\mathbf{Q}'$ (c) Rotated surface $\mathbf{Q}'(\theta_k)$ (d) Projection $\tilde{\mathbf{Q}}'(\theta_k)$ (e) Distrubution matrix $\mathbf{D}$ (f) Statistics $\{\mu_{mn}, e\}$ (g) Sub-feature $f_x(\theta_k)$ Source: Reprinted with permission from Ref. [129]. ©2013, Springer.

triangle with vertices p_{i1}, p_{i2}, and p_{i3} on the local surface, a point within the triangle can be represented using the three vertices of the triangle as follows:

$$p_i(a, b) = p_{i1} + a(p_{i2} - p_{i1}) + b(p_{i3} - p_{i1}), \tag{5.41}$$

where $0 \le a, b \le 1$, and $a + b \le 1$. Consequently, the scatter matrix $\mathbf{C}_i$ of points lying within the ith triangle is calculated as:

$$\mathbf{C}_i = \int_0^1 \int_0^{1-s} (p_i(s, t) - p)(p_i(s, t) - p)^\mathsf{T} dt\, ds. \tag{5.42}$$

Combining Eqs. (5.41) and (5.42), the scatter matrix $\mathbf{C}_i$ can be calculated as:

$$\mathbf{C}_i = \frac{1}{12} \sum_{j=1}^{3} \left(\sum_{k=1}^{3} (p_{ij} - p)(p_{ik} - p)^\mathsf{T} + (p_{ij} - p)(p_{ij} - p)^\mathsf{T} \right). \tag{5.43}$$

The scatter matrix $\mathbf{C}$ of the local surface is calculated as:

$$\mathbf{C} = \sum_{i=1}^{m} w_{i1} w_{i2} \mathbf{C}_i, \tag{5.44}$$

where

$$w_{i1} = \frac{|(p_{i2} - p_{i1}) \times (p_{i3} - p_{i1})|}{\sum_{i=1}^{m} |(p_{i2} - p_{i1}) \times (p_{i3} - p_{i1})|} \tag{5.45}$$

and

$$w_{i2} = \left(r - \left| p - \frac{p_{i1} + p_{i2} + p_{i3}}{3} \right| \right)^2. \tag{5.46}$$

The weight w_{i1} represents the ratio between the area of the ith triangle to the total area of the local surface. Since w_{i1} controls the contribution of each triangle according to its area, it can be used to improve the robustness of the LRF with respect to varying mesh resolutions. w_{i2} is related to the distance from the centroid of the ith triangle to the keypoint. That is, a point which is far from the keypoint has a smaller contribution to the overall scatter matrix than the point which is close to the keypoint. Therefore, the weight w_{i2} can be used to improve the robustness of LRF with respect to occlusion and clutter.

Next, an eigenvalue decomposition is performed on the scatter matrix $\mathbf{C}$ to obtain three eigenvectors $\{\Lambda_1, \Lambda_2, \Lambda_3\}$ in the descending order of magnitude of their associated eigenvalues. In order to eliminate the sign ambiguity of these eigenvectors, two unambiguous vectors $\tilde{\Lambda}_1$ and $\tilde{\Lambda}_3$ are obtained as follows:

$$\tilde{\Lambda}_1 = \Lambda_1 \cdot \text{sign} \left(\sum_{i=1}^{N} w_{i1} w_{i2} \left(\sum_{j=1}^{3} (p_{ij} - p) \Lambda_1 \right) \right), \tag{5.47}$$

$$\tilde{\Lambda}_3 = \Lambda_3 \cdot \text{sign} \left(\sum_{i=1}^{N} w_{i1} w_{i2} \left(\sum_{j=1}^{3} (p_{ij} - p) \Lambda_3 \right) \right). \tag{5.48}$$

Here, sign($\cdot$) represents the sign function. Consequently, a LRF is defined using the keypoint p as its origin, $\tilde{\Lambda}_1$, $\widetilde{\Lambda_3 \times \tilde{\Lambda}_1}$ and $\tilde{\Lambda}_3$ as its X, Y, and Z axes.

Once the LRF of the keypoint p is constructed, its neighboring points $Q = \{q_1, q_2, \ldots, q_n\}$ are aligned with the LRF to achieve rotation invariance. The RoPS descriptor is then generated using the following three steps:

- **Step 1 (Rotation).** The neighboring points Q are rotated around the X axis by a set of angles $\{\theta_k\}$, $k = 1, \ldots, T$. For each rotation (Figure 5.12c), the rotated points $Q(\theta_k)$ are projected onto the xy, xz, and yz coordinate planes to produce three projected point clouds $\tilde{Q}_i(\theta_k)$, $i = 1, 2, 3$. The projection process reduces the data dimensionality while preserving the 3D geometric information to some extent.

- **Step 2 (Projection).** A 2D distribution matrix $\mathbf{D}$ is produced for each projected point cloud $\tilde{Q}_i(\theta_k)$. Specifically, the bounding rectangle of the projected point cloud $\tilde{Q}_i(\theta_k)$ is divided into $L \times L$ bins (Figure 5.12d). The number of points within each bin is accumulated to generate the $L \times L$ distribution matrix $\mathbf{D}$ (Figure 5.12e). In order to achieve invariance to varying mesh resolutions, the matrix $\mathbf{D}$ is further normalized such that the sum of all bins is equal to one.

- **Step 3 (Statistics calculation).** The information in a distribution matrix $\mathbf{D}$ is encoded with the first four central moments $\{\mu_{11}, \mu_{21}, \mu_{12}, \mu_{22}\}$ and the Shannon entropy e (Figure 5.12f). The central moment μ_{mn} of order $m + n$ of a matrix $\mathbf{D}$ is calculated as:

$$\mu_{mn} = \sum_{i=1}^{L} \sum_{j=1}^{L} (i - \bar{i})^m (j - \bar{j})^n \mathbf{D}(i, j), \tag{5.49}$$

where

$$\bar{i} = \sum_{i=1}^{L} \sum_{j=1}^{L} i \mathbf{D}(i, j) \quad \text{and} \quad \bar{j} = \sum_{i=1}^{L} \sum_{j=1}^{L} j \mathbf{D}(i, j). \tag{5.50}$$

The Shannon entropy e is defined as:

$$e = - \sum_{i=1}^{L} \sum_{j=1}^{L} \mathbf{D}(i, j) \log(\mathbf{D}(i, j)). \tag{5.51}$$

The subfeature $f_x(\theta_k)$ generated from the kth rotation around the X axis is obtained by concatenating the statistics obtained from the xy, xz, and yz planes (Figure 5.12g). The subfeatures generated from all rotations around the X axis are further concatenated to produce $\{\mathbf{x}_x(\theta_k)\}$, $k = 1, \ldots, T$. Similarly, the point cloud Q is also rotated around the Y and Z axes to generate $\{\mathbf{x}_y(\theta_k)\}$, $k = 1, \ldots, T$ and $\{\mathbf{x}_z(\theta_k)\}$, $k = 1, \ldots, T$, respectively. The RoPS descriptor is finally obtained by concatenating the subfeatures from all rotations:

$$\mathbf{x} = \{\mathbf{x}_x(\theta_k), \mathbf{x}_y(\theta_k), \mathbf{x}_z(\theta_k)\}, \quad k = 1, \ldots, T. \tag{5.52}$$

It is demonstrated by extensive experiments that the RoPS descriptor is more discriminative than Signature of Histograms of OrienTations (SHOT) and spin image. It is also more robust with respect to noise and varying mesh resolutions than SHOT and spin image [86, 129].

5.4.2.2 Histogram of Geometric Attributes

These methods represent the local surface around a keypoint with histograms generated according to the geometric attributes of the local surface. Examples of descriptors in this category include point feature histograms (PFH) [136], Fast Point Feature Histograms (FPFH) [137], and SHOTs [126].

5.4.2.2.1 Point Feature Histograms (PFH)

PFH descriptor is proposed by Rusu et al. [136]. Given a keypoint p and its neighboring points within the support radius r, the surface normal $\tilde{n}$ is calculated. For each pair of points q_i and q_j within the neighborhood of keypoint p, their associated surface normals $\tilde{n}_i$ and $\tilde{n}_j$ are estimated. A source point q_s is selected as the point having the smallest angle between its normal and the line connecting these two points, while another point is considered as target point q_t. That is, if $\arccos(\tilde{n}_i(q_j - q_i)) \leq \arccos(\tilde{n}_j(q_i - q_j))$, then $q_s = q_i, q_t = q_j$. Otherwise, $q_s = q_j, q_t = q_i$. A Darboux frame is then constructed using the source point q_s as its origin. The coordinate axes of the Darboux frame are defined as:

$$u = \tilde{n}_s, \qquad v = (q_t - q_s) \times u, \qquad w = u \times v. \tag{5.53}$$

Note that, in differential geometry, a Darboux frame is a moving frame constructed on a surface. It is named after French mathematician Jean Gaston Darboux.

Then, four features are defined as:

$$f_1 = v\tilde{n}_t, \tag{5.54}$$

$$f_2 = \|q_t - q_s\|, \tag{5.55}$$

$$f_3 = \frac{u(q_t - q_s)}{\|q_t - q_s\|}, \tag{5.56}$$

$$f_4 = \arctan\left(\frac{w\tilde{n}_t}{u\tilde{n}_t}\right). \tag{5.57}$$

Next, the range of each feature is divided into N_{div} bins. Consequently, these four features can be accumulated into a N_{div}^4 histogram, which forms the PFH descriptor. Each bin indexed by idx represents the percentage of source points in the neighborhood of a keypoint p having the corresponding idx value. Here, idx is calculated as:

$$idx = \sum_{i=1}^{4} step(s_i, f_i) N_{div}^{i-1}. \tag{5.58}$$

In [136], N_{div} is set to 2. $step(s_i, f_i)$ is 0 if $f_i < s_i$ and 1 otherwise, where s_i is set to be the center of the value range of f_i. Consequently, the dimensionality of the PFH descriptor is 2^4.

5.4.2.2.2 Fast Point Feature Histograms (FPFH)
The computational complexity of PFH descriptor is $O(n^2)$, where n is the number of neighboring points of the keypoint p. In order to reduce the computational complexity to $O(n)$, Rusu et al. [137] propose the FPFH descriptor. For each point p_i within the neighborhood of keypoint p, the features f_1, f_3, and f_4 between p_i and its neighboring points are calculated using Eqs. (5.54)–(5.57). These features are then accumulated into a simplified point feature histogram (SPFH) by dividing the range of each feature into five bins. Consequently, the dimensionality of a SPFH is $5^3 = 125$. The final FPFH descriptor of a keypoint p is calculated as the weighted sum of the SPFH values of its neighboring points, i.e.

$$\text{FPFH}(p) = \text{SPFH}(p) + \sum_{i=1}^{n} \frac{1}{\omega_i} \text{SPFH}(p_i), \tag{5.59}$$

where the weight ω_i represents the distance between the keypoint p and its neighboring point p_i.

Note that, the FPFH descriptor at a keypoint is generated by calculating the relationship between the keypoint and its neighboring points, while the PFH is obtained by calculating the relationship between all pairs of neighboring points. Consequently, the computational complexity of the FPFH is $O(n)$ instead of $O(n^2)$. Meanwhile, the majority of descriptiveness of PFH is preserved in FPFH.

5.4.2.2.3 Signature of Histograms of Orientations (SHOT)
The SHOT descriptor is introduced by Tombari et al. [126]. Given a keypoint p and its support radius r, the weighted covariance matrix of its neighboring points is calculated as:

$$\mathbf{C} = \frac{1}{\sum_{i=1}^{n}(r - d_i)} \sum_{i=1}^{N}(r - d_i)(p_i - p)(p_i - p)^{\top}, \tag{5.60}$$

where p_i is a neighboring point of the keypoint p, and d_i is the distance between p and p_i. The weighting strategy used in Eq. (5.60) improves the repeatability and robustness in the presence of clutter.

Eigenvalue decomposition is then performed on the covariance matrix $\mathbf{C}$ to obtain three eigenvectors $\{\Lambda_1, \Lambda_2, \Lambda_3\}$ sorted in the descending order of their associated eigenvalues. In order to remove the sign disambiguation of these eigenvectors, the X axis of the LRF is defined as:

$$x = \begin{cases} \Lambda_1, & |\{i : (p_i - p)\Lambda_1 \geq 0\}| \geq |\{i : (p_i - p)\Lambda_1 < 0\}|, \\ -\Lambda_1, & \text{Otherwise.} \end{cases} \tag{5.61}$$

The same approach is then used to remove the sign disambiguation for the Z axis, and the Y axis is finally obtained as $z \times x$. Consequently, a unique and unambiguous LRF is constructed using the keypoint p as its origin, and X, Y, and Z axes as its axes.

Next, the neighboring points are transformed to the coordinate system defined by the LRF, and an isotropic spherical grid is superimposed on the support region of the keypoint p. The support region is then partitioned into 32 volumes along the radial, azimuth, and elevation axes (eight azimuth divisions, two elevation divisions, and two radial divisions). For each volume, a local histogram is generated according to the dot product between the surface normal $\tilde{n}_p$ at the keypoint p and the surface normal $\tilde{n}_{p_i}$ at each neighboring point p_i in this volume. To avoid boundary effects, for each point being accumulated into a specific bin of the local histogram, a quadrilinear interpolation is applied to its neighbors, including the neighboring bins in the local histogram and the bin with the same index in the local histograms of the neighboring volumes. The local histograms from all volumes are concatenated to generate the SHOT descriptor. To improve the robustness to point density variations, the SHOT descriptor is further normalized such that the sum of its elements is equal to 1.

5.4.2.3 Histogram of Oriented Gradients

Inspired by SIFT and HOG descriptors for 2D images, these methods represent the local surface around a keypoint with histograms generated according to the oriented gradients of coordinates/geometric attributes of the local surface. A typical descriptor in this category is MeshHOG [91]. Other descriptors, which follow a similar approach, include [111, 118].

The MeshHOG descriptor was introduced by Zaharescu et al. [91]. They define a scalar function f on each point of a 3D shape, and then calculate the gradients ∇f at all points. Note that, the function f can be defined in terms of surface curvatures, texture information, or any other geometric or photometric attributes. Given a keypoint p, a LRF is generated. Then, the gradient vectors of the points on the local surface around the keypoint are projected onto three planes associated with the LRF. Each plane is further divided into four polar slices. Consequently, each point on the local surface will fall within one of these polar slices during projection. For each polar slice, its space is further divided into eight orientation slices. A histogram of dimension 8 is then generated for each polar slice according to the orientations of the gradients ∇f of points falling within that polar slice. The MeshHOG descriptor is then obtained by concatenating the histograms obtained from all the polar slices on the three planes.

5.4.3 Covariance-Based Methods

Signature-based (Section 5.4.1) and histogram-based (Section 5.4.2) methods can only encode the shape of a local surface using attributes of the same type.

In practice, however, different heterogeneous descriptors capture different aspects of shape and thus combining them in an efficient way may lead to better performance. Along this line, Tabia et al. [138, 139] proposed to use the covariance of geometric features instead of the features themselves. The advantage is that covariance matrices can encode multiple and heterogeneous attributes with different dimensions into a single compact representation without any normalization or joint probability estimation. They also measure the relationship between different attributes of a 3D local surface.

Let $\{p_i, i = 1, \ldots, n\}$ be a set of points in the neighborhood of a keypoint p. Tabia et al. [138, 139] starts by computing, for each point p_i, various types of attributes, which are then concatenated to form one attribute vector $\mathbf{x}_i = (x_1^i, \ldots, x_d^i)$ of dimension d. Then, the local surface around the keypoint p is represented with the symmetric positive definite (SPD) covariance matrix:

$$K = \frac{1}{n-1} \sum_{i=1}^{n} (\mathbf{x}_i - \overline{\mathbf{x}})(\mathbf{x}_i - \overline{\mathbf{x}})^{\top}. \tag{5.62}$$

The diagonal elements of the covariance matrix K represent the variance of each attribute. Its off-diagonal elements represent how attributes of different types covary. Note that, different types of attributes can be encoded into the covariance descriptor. Tabia et al. [138, 139], for example, used the magnitude of the gradient, the mesh Laplacian value, the location of each point with respect to the center of the local surface around that point, the distance between neighboring points, and the volume of the parallelepiped formed by the coordinates of each point. For 3D scene recognition, Cirujeda et al. [140], used a set of six attributes $(R_i, G_i, B_i, \alpha_i, \beta_i, \gamma_i)$ where R_i, G_i, and B_i are the RGB color space values for a point p_i, α_i is the angle between the surface normal at the patch center p and the segment from p to p_i, β_i is the angle between the segment from p to p_i and the surface normal at p_i, and γ_i is the angle between both surface normals at p and p_i.

Another important property of covariance descriptors is that they can be used to capture multiscale shape properties without increasing its size. In fact, a multiscale covariance (MCOV) descriptor [140] can be obtained by generating a set of covariance matrices using several patch sizes (i.e. different radius values around the keypoint) as follows:

$$K_M = \{K(r)\}, \quad r = 1, \ldots, n_r, \tag{5.63}$$

where $K(r)$ is the covariance descriptor generated with a radius r, and n_r is the number of scales. This mutiscale covariance descriptor has been successfully used to match 3D color scenes under challenging conditions [140].

Finally, since covariance matrices lie on a nonlinear manifold, we need to use geodesic distances on the Riemannian manifold for comparing them since the Euclidean distance is extrinsic to the space of covariance matrices.

5.5 Feature Aggregation Using Bag of Feature Techniques

In general, a local descriptor captures only a few information about a 3D object because of its local support. However, aggregating many local descriptors, computed at different locations of a 3D object, can result in powerful descriptors, which outperform the global descriptors described in Chapter 4 and evaluated in Chapter 12. This section describes one popular technique, called the bag of features (BoF) [123, 141, 142], which originated from the text analysis community but has been widely adopted in image and 3D shape analysis.

The approach builds a global signature from local features in three steps. In an off-line training step, the approach (see Figure 5.13) takes a set of 3D models (the training set), computes local descriptors on each model, and puts all these local descriptors in a single bag. The approach then builds, using some clustering techniques, a dictionary (also called codebook) by clustering the descriptors in the bag and taking the centers (or centroids) of the clusters as key shapes (called codewords or vocabulary). This is the *dictionary construction* step.

At run time, given a 3D model, local features are first extracted. Each feature is then encoded into the dictionary. This is the *coding* step. Finally, the codes of the 3D model are pooled (i.e. aggregated) to form one global signature, which describes the shape of the 3D model. This is the *pooling* step.

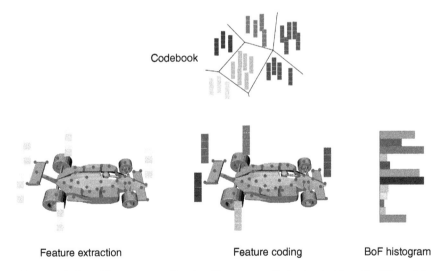

Codebook

Feature extraction   Feature coding   BoF histogram

Figure 5.13 Bag of features encoding; local features are first extracted from the 3D models in the training set. A codebook of size k is then constructed by grouping these features, using some clustering techniques, into k clusters. The centroids (or centers) of the clusters, called also key shapes, form the codewords. A 3D model is represented using the histogram of occurrences of the key shapes in that 3D model.

This approach, which is largely inspired by the Bag of Words concept used in the analysis and retrieval of text documents, is somehow similar to the shape distributions or distribution-based descriptors presented in Chapter 4. The main conceptual difference is the way the histograms are constructed. In the case of shape distributions, some shape properties (e.g. local descriptors) are first measured at various locations on the 3D shape. Then, a signature is constructed by taking the histogram of the measured properties. Each bin of the histogram can be seen as a codeword learned only from a single shape. In contrast, the BoF approach uses a set of training models to learn the vocabulary that best represents the population of the 3D shapes being analyzed.

Existing techniques differ in the way the dictionary is constructed and the coding and pooling techniques used to build the shape signature.

5.5.1 Dictionary Construction

5.5.1.1 Feature Extraction

There are two types of features that can be used in the BoF approach: low-level features and semantic features. Low-level features can be any of the local descriptors described in this chapter, some of them can be computed directly on the 3D mesh while others can be extracted from the 2D views of the 3D models. The most important point to consider is that the features have to be of the same type. For instance, Ohbuchi et al. [143] and Lian et al. [141] use 2D SIFT features extracted from a set of 2D views of the 3D models. Liu et al. [144] and Li and Godil [145] use Spin Image descriptors computed on a dense set of feature points uniformly sampled on the surface of the 3D model. Bronstein et al. [123] proposed a similar approach but use HKSs on a dense set of points in order to make the signatures invariant to nonrigid deformations.

Instead of using low-level features, Tabia et al. [146] represent each 3D object as a vector of weighted occurrences of patches extracted around some key feature points. Toldo et al. [147], on the other hand, segment the shape into regions (or subparts) and describe each region with several descriptors. A 3D shape is then modeled as a histogram of subpart occurrences. Segmentation-based BoF have the advantage of capturing some semantics of the 3D objects.

Note that the properties of BoF signatures depend heavily on the properties of the individual features used to build the dictionary. For instance, the invariance of the signature to shape-preserving transformations (translation, scale, rotation, and eventually bending) is inherited directly from the invariance of the features to these transformations.

5.5.1.2 Codebook Construction

Let us now assume that N features have been extracted from the training set. The next step is to create a codebook of k codewords. One approach for building the codebook is by clustering the feature vectors of the training set into

k different clusters using, for example, the k-means clustering algorithm. The codewords, which form the codebook, are defined as the centers (or centroids) of the learned clusters. In what follows, we denote by $\mathbf{B} = \{\mathbf{b}_i \in \mathbb{R}^d\}_{i=1}^k$ the list of codewords, or dictionary, obtained with this approach, and d is the dimension of the feature vectors.

Another approach for building the codebook is by fitting a Gaussian Mixture Model (GMM) to the feature vectors of the training set. Let $\Theta = (\mu_j, \Sigma_j, \pi_j | j = 1, \ldots, k)$ be the parameters of the GMM. Here, μ_j and Σ_j are, respectively, the mean and covariance matrix of the jth mode. The parameters π_j are the mixing weights of the mixture. One can safely assume that Σ_j is a diagonal covariance matrix whose diagonal elements are denoted by $\sigma_{lj}, l = 1, \ldots, d$. With this approach, the codewords $\mathbf{B}$ are the means of the modes of the GMM, i.e. $\mathbf{b}_j = \mu_j, j = 1, \ldots, k$.

Note that the codebook construction is an offline process. To achieve a high performance, the training set used for constructing the codebook should capture all the possible shape variations.

5.5.2 Coding and Pooling Schemes

Let M be a 3D model, and $\mathbf{X} = \{\mathbf{x}_1, \ldots, \mathbf{x}_N\}$ a set of N local descriptors computed at N different locations on M. Each descriptor $\mathbf{x}_i$ is a d-dimensional vector of the form $\mathbf{x}_i = (x_1^i, \ldots, x_d^i)$. Coding is the process of representing each descriptor $\mathbf{x}_i$ using the codebook $\mathbf{B}$. In other words, each descriptor $\mathbf{x}_i$ will be encoded using a k-dimensional code vector $\mathbf{u}_i = (u_1^i, \ldots, u_k^i)^{\mathsf{T}}$. Pooling is the process of aggregating all the code vectors $\mathbf{u}_i, i = 1, \ldots, N$, into a single shape signature $\mathbf{x}$ of dimension k. Below, we present a few pooling and coding techniques, which have been extensively used in the literature.

5.5.2.1 Sparse Coding

In this approach, each local descriptor $\mathbf{x}_i \in \mathbf{X}$ votes for one codeword $\mathbf{b}_j \in \mathbf{B}$, its nearest neighbor in the codebook. That is, the code vector $\mathbf{u}_i$ is a vector whose elements are all set to zero except the jth entry, which takes the value of one. In other words, each descriptor $\mathbf{x}_i$ is assigned to its nearest codeword. This is referred to as a hard assignment. The final signature $\mathbf{x}$ is given as:

$$\mathbf{x} = \sum_{i=1}^{N} \mathbf{u}_i. \tag{5.64}$$

The jth element of $\mathbf{x}$ encodes the number of occurrences of the jth codeword in the shape M. Thus, $\mathbf{x}$ is a vector of occurrence counts, which can be also seen as a histogram over the vocabulary. It is a sparse vector (or a sparse histogram) since many of its entries are zero.

5.5.2.2 Fisher Vectors

Instead of assigning each local descriptor $\mathbf{x}_i$ to its nearest codeword, one can perform a soft assignment. That is, each descriptor $\mathbf{x}_i$ will be assigned to one codeword $\mathbf{b}_j$ with a certain probability w_{ij}, which indicates how likely that the descriptor $\mathbf{x}_i$ belongs to the codeword $\mathbf{b}_j$.

Let $\Theta = (\mu_j, \Sigma_j, \pi_j | j = 1, \ldots, k)$ be the parameters of the GMM used to build the dictionary. The GMM associates each vector $\mathbf{x}_i$ to a mode j in the mixture (or a codeword $\mathbf{b}_j$ in the vocabulary) with a strength w_{ij} given by the posterior probability:

$$w_{ij} = \frac{1}{A} \exp\left[-\frac{1}{2}(\mathbf{x}_i - \mu_j)^\top \Sigma_j^{-1}(\mathbf{x}_i - \mu_j) \right],$$

where

$$A = \sum_{j=1}^{k} \exp\left[-\frac{1}{2}(\mathbf{x}_i - \mu_j)^\top \Sigma_j^{-1}(\mathbf{x}_i - \mu_j) \right]. \tag{5.65}$$

Note that $w_{ij} \geq 0$ and $\sum_{j=1}^{k} w_{ij} = 1$. Now, for each mode j, consider the mean and covariance deviation vectors:

$$u_{lj} = \frac{1}{N\sqrt{\pi_j}} \sum_{i=1}^{N} w_{ij} \frac{x_{li} - \mu_{lj}}{\sigma_{lj}}, l = 1, \ldots, d$$

and

$$v_{lj} = \frac{1}{N\sqrt{2\pi_j}} \sum_{i=1}^{N} w_{ij} \left[\left(\frac{x_{li} - \mu_{lj}}{\sigma_{lj}} \right)^2 - 1 \right], \tag{5.66}$$

where $l = 1, \ldots, d$ spans the dimension of the local descriptors. The Fisher vector $\Phi(M)$ of the 3D model M is the stacking of the vectors $\mathbf{u}_k = [u_{1k}, \ldots, u_{dk}]^\top$ followed by the vectors $\mathbf{v}_k = [v_{1k}, \ldots, v_{dk}]^\top$ for each of the k modes in the Gaussian mixtures:

$$\Phi(M) = [\mathbf{u}_1, \ldots \mathbf{u}_K, \mathbf{v}_1 \ldots \mathbf{v}_k]^\top. \tag{5.67}$$

5.5.3 Vector of Locally Aggregated Descriptors (VLAD)

Similar to Fisher vectors, let w_{ij} be the strength of the association of the data vector $\mathbf{x}_i$ with the jth cluster such that $w_{ij} \geq 0$ and $\sum_{j=1}^{k} w_{ij} = 1$. This association may be either soft (e.g. obtained using the posterior probabilities of the GMM distribution fitted to the training set, see Section 5.5.2.2) or hard (e.g. obtained using vector quantization with K-means, see Section 5.5.1). To build the VLAD descriptor, we take each of the k clusters and compute the sum of residuals (i.e. the d-dimensional vector differences between each descriptor and the cluster center) as follows:

$$\mathbf{v}_j = \sum_{i=1}^{N} w_{ik}(\mathbf{x}_i - \mu_j), \quad j = 1, \dots, k. \tag{5.68}$$

Then, the k sums of residuals are stacked together to form a $d \times k$ matrix of the form:

$$\hat{\Phi}(M) = [\mathbf{v}_1, \dots, \mathbf{v}_j, \dots, \mathbf{v}_k]. \tag{5.69}$$

This is referred to as the unnormalized VLAD descriptor, which usually needs to be normalized before its usage. The normalization can be done in four different ways:

- **Component-wise mass normalization.** Each vector $\mathbf{v}_j$ is divided by the total mass of the features associated to it, i.e. $\sum_{i=1}^{N} w_{ij}$.
- **Square-rooting.** The function $\text{sign}(z)\sqrt{\|z\|}$ is applied to all the scalar components of the VLAD descriptor.
- **Component-wise $\mathbb{L}^2$ normalization.** The vectors $\mathbf{v}_j$ are divided by their $\mathbb{L}^2$ norm $\|\mathbf{v}_j\|$.
- **Global $\mathbb{L}^2$ normalization.** The VLAD descriptor $\hat{\Phi}(M)$ is divided by its $\mathbb{L}^2$ norm $\|\hat{\Phi}(\mathbf{v})\|_2$.

5.5.4 Vector of Locally Aggregated Tensors (VLAT)

The vector of locally aggregated tensors (VLAT) [148, 149], see Figure 5.14, measures the deviation of the covariance of the local features in each cluster from the covariance of the local features in a 3D shape. Le μ_j and K_j be the mean descriptor and mean tensor (covariance) of all the descriptors which fall within the jth cluster. Then, for every 3D model M_i, we compute one tensor K_i per cluster j by aggregating the centered tensors of the centered descriptors as follows:

$$K_i = \frac{1}{n_i} \sum_{r=1}^{n_i} (\mathbf{x}_{ir} - \mu_j)(\mathbf{x}_{ir} - \mu_j)^\top. \tag{5.70}$$

Here, μ_j is the center of the jth cluster. The sum in Eq. (5.71) is over the local descriptors of M_i that belong to the jth cluster. n_i is the number of such local descriptors.

Next, compute the deviation K_{ij} of the tensor K_i of the 3D model M_i from the tensor of the jth cluster:

$$K_{ij} = \frac{1}{n_{ij}} \sum_r (\mathbf{x}_{ir} - \mu_j)(\mathbf{x}_{ir} - \mu_j)^\top - K_j. \tag{5.71}$$

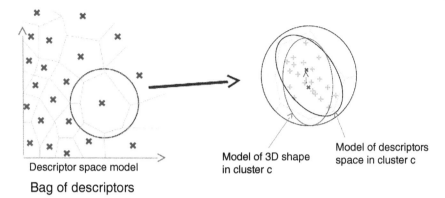

Bag of descriptors

Figure 5.14 Illustration of the VLAT encoding. The features extracted from a 3D model are first assigned to their closest key-shapes (cluster) in the codebook. Then, we measure the deviation of the covariance of the features in each cluster from the covariance of the local features of the 3D model. The measured deviation forms the descriptor of the 3D shape with respect to the cluster.

The tensor matrix K_{ij} is then flattened into a feature vector $\mathbf{T}_{ij}$, which is used as the descriptor of M_i with respect to the jth cluster. Finally, all the vectors $\mathbf{x}_{ij}$ of M_i are concatenated to form one single descriptor $\Phi(M_i) = \{\mathbf{T}_{ij}\}, j = 1, \ldots, k$ where k is the size of the dictionary.

5.6 Summary and Further Reading

5.6.1 Summary of 3D Keypoint Detection

To comprehensively evaluate the performance of a keypoint detection method, Tombari et al. [84] introduced a set of evaluation metrics including absolute repeatability, relative repeatability, scale repeatability, quantity bias, descriptor matching, and efficiency. Eight keypoint detectors including LSP [101], ISS [103], keypoint quality (KPQ) [93], MeshDoG [91], LBSS [109], KPQ-AS KPQ [93], and SP [106] have been tested on five datasets. Several conclusions can be briefly summarized as follows:

- For 3D shape retrieval, KPQ provides the best performance in terms of repeatability, distinctiveness, and robustness to noise. Meanwhile, ISS is very efficient, and achieves a good tradeoff between absolute repeatability and relative repeatability.
- For 3D object recognition, adaptive-scale keypoint detectors have a higher repeatability compared to fixed-scale keypoint detectors. MeshDoG and KPQ-AS achieve the best overall performance among the eight detectors. KPQ-AS is highly distinctive and also robust to noise. ISS is very efficient

and also achieves an acceptable repeatability. It is highly suitable for time-crucial applications.
- LBSS can achieve a very high scale repeatability.

Despite the extensive research on local feature detection, there are still avenues for further research. Below, we summarize a few of them.

- Most existing 3D keypoint detectors are hand-engineered. They are unable to extract global optimal keypoints for a particular application. Learning a detector to obtain keypoints that are optimized for a specific task can be an interesting future work.
- The computational and memory cost of existing methods are still very high. This limits their applications in time-crucial scenarios, e.g. on mobile computing platforms. Research on 3D keypoint detection with high efficiency and distinctiveness is of great importance.

5.6.2 Summary of Local Feature Description

Guo et al. [86] performed a comprehensive performance evaluation of many of the 3D local feature descriptors presented in this chapter. Guo et al. also proposed a set of evaluation metrics, which measure the descriptiveness, compactness, robustness to different nuisances, scalability, and efficiency of local descriptors. The metrics also evaluate the performance of different combinations of local descriptors with 3D keypoint detectors. The local descriptors have been tested on eight datasets, including the Retrieval [84], Random Views [84], Laser Scanner [97], Space Time [84], Kinect [84], LIDAR [150], Dense Stereo [150], and 2.5D Views [97] datasets. The conclusions of this comprehensive evaluation can be summarized as follows:

- FPFH and SHOT are appropriate for time-crucial applications. FPFH is compact and highly efficient for feature generation and matching. Its descriptiveness is also reasonably high. Meanwhile, SHOT achieves a good balance between descriptiveness and computational efficiency.
- FPFH and RoPS are appropriate for space-crucial applications with limited memory and storage resources. FPFH is the most compact 1 among the 10 descriptors, with the feature dimensionality of only 33. Its memory and storage requirement is very low. Besides, RoPS achieves a higher descriptiveness at the cost of a slight increase in storage requirement.
- RoPS achieves a high feature matching accuracy on datasets acquired by different techniques. Its variation in performance across different datasets is very small, with a good performance achieved on various types of datasets. Its high descriptiveness is attributed to the fact that the descriptor encodes the "complete" information by observing the local surface from different

viewpoints through rotation. Its strong robustness can be explained by at least two factors. First, RoPS uses the low-order moments of 2D distribution matrices to generate the overall descriptor. It is therefore less sensitive to noise and varying mesh resolutions. Second, the LRF of RoPS is obtained by calculating the scatter matrix of all points on a local surface rather than the discretely sampled vertices. Consequently, LRF is more robust to the change of mesh resolutions.

- The feature matching performance of a specific descriptor can be enhanced when combined with 3D keypoint detection methods rather than uniform sampling or random sampling. This implies that 3D keypoint detection plays an important role in local feature matching. It is also observed that for each local feature descriptor, there is a specific feature with which it can be combined to achieve the optimal feature matching performance [151].

Despite the extensive research in this area, there are still several avenues for future research. Examples include

- **Extraction of local features from low resolution and noisy data**. Most of the existing 3D local features are designed for high-quality 3D shape data. They achieve a very poor performance on low-quality data with low resolution and high-level of noise. With further developments of low-cost 3D cameras, 3D shape data will be available in commodity devices such as smartphones. Thus, designing advanced local features for low-quality 3D shape data acquired by these devices is necessary.
- **Learning deep features for 3D shape representation**. Deep neural networks have been widely used for image representation in numerous applications, e.g. face recognition and object detection. Learning 3D shape features with deep neural networks has become a new trend in this area, with many methods already available [83, 152, 153]. Current methods in this direction include the deep learning (DL) architectures exploiting descriptors extracted from 3D data, the DL architectures exploiting directly 3D data, and the DL architectures exploiting 2D projections/views of 3D objects [83]. DL-based features remain an open research area.

5.6.3 Summary of Feature Aggregation

One limitation of the BoF approach for feature aggregation stems from the fact that it ignores the spatial relationships between the features. As such, several extensions have been proposed in the literature to take into account the spatial relations between features. Examples include the work of Lavoué [154] and Bronstein et al. [123]. Another limitation is that many of the previous works assume that the local features are elements of the Euclidean space and use the $\mathbb{L}^2$ metric for clustering the features during the codebook construction. However, there are many types of features that are not elements of the Euclidean space, e.g. the covariance matrices of Tabia et al. [138, 139]. To overcome this problem, Tabia et al. [138, 139] used geodesic distances in the space of covariance matrices to build the bag of covariance (BoC) signatures and showed that geodesic distances outperform their Euclidean counterpart. This approach has been later extended to spatially sensitive BoC signatures [139], which capture the spatial relations between covariance features.

Part III

3D Correspondence and Registration

6

Rigid Registration

6.1 Introduction

Rigid surface registration is a fundamental problem in computer vision. The goal is to find a rigid transformation (i.e. a combination of translations, rotations, and scaling) which aligns two or multiple 3D surfaces to a common reference frame. Rigid surface registration works as a crucial intermediate step for various vision systems. For example, in a 3D modeling system, point clouds of an object acquired from different viewpoints are transformed into a common reference frame to produce a complete 3D model. In 3D object recognition, a candidate 3D model is registered to a scene with occlusion and clutter to determine if an object is present in the scene. Registration has numerous applications in various areas including robotics, reverse engineering, mould fabrication, industrial inspection, artefact reproduction, virtual museums, games, remote sensing, and medical imaging [155–157].

Existing surface registration methods can broadly be divided into two categories: (i) **coarse registration** and (2) **fine registration** [155]. In general, coarse registration is first applied to estimate a rough initial transformation between two surfaces without any prior alignment. The estimated transformation is then refined by fine registration to produce a more accurate alignment between two surfaces [87]. For coarse registration, either manual or automatic methods can be used. A manual method usually uses a turntable, robot arms, positioning devices, or attached markers on objects to obtain the initial transformation between two surfaces of an object. This method relies on human intervention, and the object has to be scanned in a fully controlled scenario [87]. In contrast, an automatic method estimates the initial transformation between two surfaces through correspondences generated by feature matching. The features used in this application can be points, lines, or surface patches. Automatic coarse registration method is more appropriate for practical applications than a manual method. We therefore focus in this chapter on automatic coarse registration

3D Shape Analysis: Fundamentals, Theory, and Applications, First Edition.
Hamid Laga, Yulan Guo, Hedi Tabia, Robert B. Fisher, and Mohammed Bennamoun.
© 2019 John Wiley & Sons, Inc. Published 2019 by John Wiley & Sons, Inc.

methods. Fine registration is usually formulated as minimizing a specific error metric calculated by an iterative process, which starts from the initial state specified by coarse registration. Typical fine registration methods include the iterative closest point (ICP) [158] algorithm, its various variants, and the coherent point drift (CPD) [159] algorithm.

In this chapter, we first present, in Section 6.2, the basic concepts and the representative methods for coarse registration. We then describe, in Section 6.3, the methods for fine registration. We finally summarize this chapter in Section 6.4.

6.2 Coarse Registration

The task of coarse registration is to estimate an initial rigid transformation between two surfaces of an object using feature correspondences between these two surfaces.

Given two 3D point clouds $\mathbf{P}$ and $\mathbf{Q}$, the registration problem can be formulated as the minimization of the least squares error e:

$$e = \sum_{i=1}^{n_c} \|p_i - (Oq_i + t)\|^2, \tag{6.1}$$

where $O \in SO(3)$ is a rotation matrix, t is a translation vector, n_c is the number of points in the point cloud $\mathbf{P}$, p_i is the ith point in $\mathbf{P}$, and q_i is the corresponding point of p_i in the point cloud $\mathbf{Q}$. If the pointwise correspondences are known, then the optimal transformation $[O, t]$ can be obtained in a closed form using singular value decomposition (SVD) [160] or eigen-system analysis [161]. In practice, true feature correspondences between two 3D surfaces are hard to know or to automatically generate. The feature correspondences generated by feature matching usually contain a large number of false positive correspondences. In this case, several approaches have been proposed to generate an accurate transformation from correspondences obtained by feature matching.

Note that, correspondences can be generated using a number of different features, including points, lines, or surfaces. Therefore, the existing coarse registration methods can be divided into point correspondence-based, line correspondence-based, and surface correspondence-based methods.

6.2.1 Point Correspondence-Based Registration

Point correspondence-based surface registration methods have been extensively investigated in the literature and have been frequently used in many practical systems. In this section, we will first describe the typical pipeline of this approach and then present several representative methods of this category.

6.2.1.1 The Typical Pipeline

The typical pipeline for a point correspondence-based coarse registration system usually consists of three major modules: (i) keypoint detection, (ii) feature description, and (iii) correspondence generation.

Given two 3D surfaces (e.g. in the forms of point clouds) for registration, the first step is to detect a set of keypoints (Section 5.3) on each surface to reduce the number of points and subsequently the computational time for feature extraction and matching. Next, a local surface can be extracted around each keypoint using the location and the scale of the keypoint. Then, the shape of the local surface is encoded using a feature descriptor (or local descriptor). For 3D rigid registration applications, the local feature descriptor should be highly distinctive, invariant to rotation and translation, and robust to different nuisances including noise and point density variations. We refer the reader to Section 5.4 for more details on local descriptors.

Once the keypoints and their associated descriptors have been extracted from the two surfaces under registration, hereinafter denoted as M_a and M_b, respectively, then the features of M_a are matched against all the features of M_b to generate feature correspondences. Several different strategies can be used to determine a successful match between two features:

- **Threshold-based matching.** Two features are considered as a match if the distance between their associated feature descriptors is below a threshold.
- **Nearest neighbor (NN)-based matching.** The distances between a feature x_a of M_a and all the features of M_b are calculated. If the smallest distance is below a threshold, then a feature correspondence is established between x_a and the feature of M_b with the smallest distance to x_a.
- **Nearest neighbor distance ratio-based matching.** The ratio between the smallest distance and the second smallest distance is calculated. If this ratio is below a threshold, then x_a and the feature on M_b with the smallest distance to x_a are considered to be in correspondence.

Note that, searching for the nearest feature to a given feature from a large set is very time-consuming. Therefore, it is necessary to reduce the computational time by minimizing the search space or using a feature indexing approach.

Once the feature correspondences are established, a possible rigid transformation between the two surfaces can be obtained using a few pairs of feature correspondences. The registration can then be further verified to obtain a rough transformation between the two surfaces.

6.2.1.2 Transformation Estimation from a Group of Correspondences

Johnson and Hebert [95] use a group of geometrically consistent feature correspondences, generated by the matching of spin images, to estimate the rigid transformation that aligns two surfaces. Specifically, the spin images of all points on M_a are first generated and stored in a library. A point is then randomly selected from M_b to generate its associated spin image descriptor. A feature correspondence is established between the randomly selected point on M_b and the point on M_a if they have the highest correlation coefficient of their spin images. The above procedure is repeated for several times by randomly selecting a point on M_b each time, resulting in about 100 feature correspondences. These feature correspondences are further grouped using geometric consistency (Section 11.2.2.1) and the outliers are gradually removed. The groups of geometrically consistent feature correspondences are finally used to estimate the rigid transformations for the registration of the two surfaces M_a and M_b.

6.2.1.3 Transformation Estimation from Three Correspondences

Example 6.1 To estimate the rigid transformation between two surfaces, Yamany and Farag [162] use every three feature correspondences, generated by the matching of surface signatures. Specifically, keypoints are first detected on M_a and M_b using the simplex angles of points. Point signature descriptors (Section 5.3.1) are then used to generate feature correspondences. Given a signature image $\mathbf{x}_p$ for a point on M_a and a signature image $\mathbf{x}_q$ for a point on M_b, a similarity between these two descriptors is defined as

$$s = \frac{1}{n_d^2} \sum_{(i,j)} |\mathbf{x}_p(i,j) - \mathbf{x}_q(i,j)|^2, \tag{6.2}$$

where n_d is the number of pixels in a signature image. The similarity s is further augmented by adding an overlap ratio to handle partial matching. The signature images on M_a are matched against all signature images on M_b, resulting in a number of feature correspondences. Next, several groups of feature correspondences with three correspondences satisfying the geometric consistency constraint are obtained. These groups are ranked based on the distances between feature correspondences. Finally, a rigid transformation can be estimated for each group of feature correspondences for a coarse registration between M_a and M_b.

Example 6.2 Masuda [163] generate feature correspondences by matching of log-polar height maps (LPHM) features. Given a point p on M_a and its LPHM feature $\mathbf{x}_p$, and a point q on M_b and its LPHM feature $\mathbf{x}_q$, a feature correspondence is established between p and q if $\mathbf{x}_q$ is the nearest feature on M_b to $\mathbf{x}_p$ and $\mathbf{x}_p$ is also the nearest feature on M_a to $\mathbf{x}_q$. Once a set of feature correspondences C is obtained, the random sample and consensus (RANSAC) algorithm [164] is used to remove incorrect feature correspondences and to estimate the rigid transformation. The RANSAC procedure performs as follows:

- **Step 1–Initialization.** Set the iteration time $t \leftarrow 0$, the set of inlier feature correspondences $C_{inlier} \leftarrow$ null, and the maximum number of iterations n_{iter}.

- **Step 2–Sampling.** Multiple (e.g. three) feature correspondences are randomly selected from the set C to estimate a candidate transformation T_c by minimizing the least squared error of the coordinates of these corresponding points.
- **Step 3–Updating.** The feature correspondences that are consistent with the transformation T_c are extracted. Specifically, the transformation T_c is applied to each pair of corresponding points p and q, and the correspondence is considered as an inlier if $\|T_c p - q\| < \delta$ and $(T_c \tilde{n}_p) \tilde{n}_q > \tau_n$, where $\tilde{n}_p$ and $\tilde{n}_q$ are the surface normals at points p and q, respectively. Otherwise, the correspondence is considered as an outlier. If the number of inlier feature correspondences is larger than $|C_{inlier}|$, the set of inlier feature correspondences C_{inlier} and the transformation T are updated using the inliers. Otherwise, the process goes to Step 5.
- **Step 4–Refinement.** Refinement is performed by alternatively extracting inlier feature correspondences and updating the transformation. Specifically, the inner iteration process is repeated until the RMS error difference of the registered inlier corresponding points between two adjacent iterations is smaller than a threshold or a maximum number of iterations is reached.
- **Step 5–Loop Test.** $t \leftarrow t + 1$. If t reaches the maximum number of iterations n_{iter}, the iteration process is stopped, and the last transformation T is used to register the surfaces M_a and M_b. Otherwise, the process goes back to Step 1.

6.2.1.4 Transformation Estimation from Two Correspondences

Malassiotis and Strintzis [165] use two feature correspondences, generated by the matching of snapshot features, to estimate the transformation between two surfaces. Given two points p_1 and p_2 on M_a, and their associated normals $\tilde{n}_1$ and $\tilde{n}_2$, their corresponding points q_1 and q_2 on M_b are obtained as follows:

- For p_1, its candidate corresponding point q_1 is obtained by calculating the normalized correlation coefficient over overlapping pixels of two snapshot features. Note that, several candidate corresponding points might be obtained during this process.
- For each candidate corresponding point q_1, the corresponding point q_2 on M_b is obtained such that $\|p_1 - p_2\| \approx \|q_1 - q_2\|$ and $\angle(\tilde{n}_1, \tilde{n}_2) \approx \angle(\tilde{n}'_1, \tilde{n}'_2)$, where $\tilde{n}'_1$ and $\tilde{n}'_2$ are the surface normals of q_1 and q_2, respectively.
- A candidate rotation matrix O is estimated by first aligning the surface normals $\tilde{n}_1$ and $\tilde{n}'_1$ and then aligning the surface normals $\tilde{n}_2$ and $\tilde{n}'_2$.
- A candidate translation is estimated by aligning the points p_1 and p_2 with their corresponding points q_1 and q_2.

Once the feature correspondences between M_a and M_b are obtained, the following procedure is used to obtain the final transformation between two surfaces.

- **First**, keypoints are randomly selected on M_a. Then n_s pairs of points p_1 and p_2 are selected by discarding the pairs of points that are too close or too far.
- **Second**, for each pair of points p_1 and p_2, their corresponding points q_1 and q_2 are obtained. From these, a candidate transformation is generated.
- **Third**, the two surfaces M_a and M_b are aligned using the candidate transformation, and the average distance between points on the overlapping areas of these two surfaces is calculated. If the average distance or the overlap does not satisfy a constraint, the candidate transformation is discarded.
- **Fourth**, the final transformation is estimated from the remaining candidate transformations. Specifically, if the number of remaining transformations is small, the one with the smallest registration error is selected. Otherwise, a group of feature correspondences is selected to minimize the median of the registration error using the RANSAC algorithm. These selected feature correspondences are used to derive the final transformation.

6.2.1.5 Transformation Estimation from One Correspondence

Example 6.3 Mian et al. [97] use only one feature correspondence, generated by the matching of 3D tensors, to estimate the transformation between two surfaces. Specifically, given two surfaces M_a and M_b, a tensor M_a is matched with the tensors of M_b that satisfy two position constraints: (i) their overlap ratio is higher than a threshold (e.g. 0.5) and (ii) their linear correlation coefficient is also higher than a threshold (e.g. 0.5). Then, a transformation can be calculated using the local reference frames of these two corresponding 3D tensors. That is, the rotation matrix O and translation vector t that align M_a with M_b can be calculated as follows:

$$O = B_b^\top B_a, \text{ and } t = p_1 - Op_2. \tag{6.3}$$

Here, B_a and B_b are 3×3 matrices consisting of the x, y, and z axes of the local reference frame, for these two corresponding points, respectively. p_1 and p_2 are the coordinates of the two corresponding points, respectively. Note that, in this method, only one feature correspondence is sufficient to estimate the rigid transformation between two surfaces.

Example 6.4 Given two surfaces M_a and M_b, RoPS features [87] are first extracted from both surfaces and then matched using the NN distance ratio-based approach, resulting in a set of feature correspondences $C = \{c_1, c_1, \ldots, c_{n_c}\}$. A candidate transformation T_k can be estimated from each feature correspondence c_k using the local reference frames of the corresponding points p and q. Then, all feature correspondences, whose estimated transformations are similar to T_k, are selected. Specifically, a candidate transformation can be estimated from each feature correspondence. The rotation of the transformation is converted into three Euler angles, and the difference between the two transformations can be measured by the distance between their Euler angles and the distance between their translation vectors. Consequently, for a feature correspondence c_k, all feature correspondences with their associated transformations that are close to T_k are used to form a group of consistent feature correspondences C_k. Then, a refined transformation $\tilde{T}_k$ is obtained from these consistent feature correspondences C_k. Therefore, N_c refined transformations $\mathcal{T} = \{\tilde{T}_1, \tilde{T}_1, \dots, \tilde{T}_{N_c}\}$ can be obtained from the feature correspondences. In order to select the optimal transformation estimation from $\mathcal{T}$, the two surfaces M_a and M_b are sampled to two low-resolution meshes $\tilde{M}_a$ and $\tilde{M}_b$, respectively. These meshes $\tilde{M}_a$ and $\tilde{M}_b$ are then aligned using each of these transformations in $\mathcal{T}$. The transformation that results in the largest number of inlier points is considered as the final transformation for the registration between M_a and M_b. Since simplified meshes are used to select the optimal transformation, the algorithm is computationally efficient.

6.2.2 Line-Based Registration

Besides points, lines can be extracted from each surface to establish line correspondences between two surfaces. The transformation can then be estimated from the line correspondences.

6.2.2.1 Line Matching Method

Several line matching methods are available in the literature. In this section, we describe the method proposed by Stamos and Leordeanu [166]. That is, lines are first extracted from each 3D surface. Pairs of lines are then matched to estimate the optimal rotation and translation between two 3D surfaces. The overall procedure is described as follows:

1) **Line extraction.** Planar regions and linear features are first extracted at the internal and external boundaries of planar surfaces, using a segmentation algorithm [167]. Then, the planar regions P, the inner and outer borders of planar regions B_{in} and B_{out}, the inner and outer 3D border lines L_{in} and L_{out} are generated. Each border line is represented by $(p_{start}, p_{end}, p_{id}, \tilde{n}, p_{size})$, where p_{start} and p_{end} are the two endpoints of the line, and p_{id}, $\tilde{n}$, and p_{size} are the identity, the normal vector, and the size of the underlying plane Π of the line. The size p_{size} of a plane is estimated using the number of surface points on the plane, the distance from the origin of the coordinate system to the plane, and the normal of the plane. The plane information associated with each line can be used to speedup the registration process.

2) **Surface registration.** Given two surfaces M_a and M_a, their line features, $\{(p_{start}, p_{end}, p_{id}, \tilde{n}, p_{size})\}$ and $\{(p'_{start}, p'_{end}, p'_{id}, \tilde{n}', p'_{size})\}$, respectively, are matched and verified to generate the optimal transformation between the two surfaces. This is done as follows:

- **Preprocessing**: Each line feature on M_a is paired with each line feature on M_b. For a pair of lines, if both the length ratio of the two lines and the plane size ratio of their associated planes are smaller than a threshold, the pair of lines is discarded. The remaining pairs of lines (l, r) are then considered for the following procedures.
- **Initialization**: The current maximum number of line matches n_{max} is set to 0. The best transformation (O_{best}, t_{best}) so far is initialized to $O_{best} = I_{3 \times 3}$, $t_{best} = [0, 0, 0]$, where $I_{3 \times 3}$ is a 3×3 identity matrix.
- **Step 1**: The next pair of lines (l_1, r_1) is selected to calculate a rotation matrix O and a translation t_{est}. The set of pairs of matched planes is initialized as $M_{planes} = \emptyset$. The rotation matrix O is then applied to all pairs (l, r) with $l > l_1$. If the directions of two lines in a pair, or the normals of two associated planes of the lines in a pair, do not match after rotation, the pair is discarded. If the number of remaining pairs of lines is larger than n_{max}, the match between lines (l_1, r_1) is accepted, and the two associated planes of (l_1, r_1) are added to M_{planes}. Otherwise, the process returns to Step 1.
- **Step 2**: The next pair (l_2, r_2) is selected from the remaining pairs of lines. If this pair is inconsistent with the estimated translation t_{est}, the pair of lines is rejected and the process goes to Step 2. Otherwise, an exact translation t is estimated from these two pairs of lines (l_1, r_1) and (l_2, r_2). The exact translation t is further applied to these two pairs of lines. If these lines are in correspondence, the match between the lines l_2 and r_2 is accepted, and the two associated planes of (l_2, r_2) are added to M_{planes}.
- **Step 3**: All remaining pairs of lines are transformed to a common coordinate system using the transformation (O, t). The number of valid pairs of lines (which are consistent with the transformation) is used to rate the transformation. If a pair of lines is valid, its associated planes are added to M_{planes}. If the number of line matches is larger than n_{max}, then n_{max} is updated with the current number of line matches, and (O_{best}, t_{best}) is updated with (O, t). If there are more pairs of lines to be considered, the process returns to Step 2. Otherwise, the process goes to Step 1.
- **Step 4**: All lines on surface M_a are transformed to the coordinate system of the surface M_b using the transformation (O_{best}, t_{best}). Then, all pairs of line matches are determined and further employed to refine the transformation (O_{best}, t_{best}) using a weighted least square algorithm.

The final transformation (O_{best}, t_{best}) is used to register the surfaces M_a and M_b. This method is suitable for the registration of large-scale urban scenes.

6.2.2.2 Line Clustering Method

Chao and Stamos [168] first extract linear segments and plane normals from a 3D surface using a segmentation algorithm [167]. They then extract three major directions for each surface by clustering these linear segments. A local coordinate system can be assigned to each 3D surface using these major three directions. The rotation between pairs of surfaces is then estimated by matching the coordinate systems of the two surfaces. The translation, on the other hand, is estimated by matching linear segments. Below, we describe in detail the procedure.

6.2.2.2.1 Rotation Estimation It is assumed that man-made urban scenes are usually characterized by a large number of linear features. These linear features have a major vertical direction and several horizontal directions. Once two surfaces M_a and M_b are given for registration, linear segmentation and plane surfaces are extracted from each surface, using a segmentation algorithm [167], and further clustered into three major directions. Specifically, if the angle between a line vector and the cluster centroid is smaller than a threshold, the line vector is considered as being part of that cluster. Consequently, these line vectors can be grouped into several clusters. Usually, three major directions that are perpendicular to each others can be extracted. In the case where only two perpendicular major directions are obtained, the third major direction is obtained as the cross product of the two initial directions. Note that it is rare that only one major direction can be extracted from a 3D surface of a man-made urban scene. Once the coordinate axes $\{x_a, y_a, z_a\}$ for surface M_a and the coordinate axes $\{x_b, y_b, z_b\}$ for surface M_b are obtained, 24 candidate rotation matrices can be derived to align these two coordinate systems due to the sign ambiguity of the estimated major directions. In practice, the number of candidate rotation matrices can significantly be reduced using several heuristics.

Each candidate rotation O_c can be represented as

$$O_c = \begin{bmatrix} r_{00} & r_{01} & r_{02} \\ r_{10} & r_{11} & r_{12} \\ r_{20} & r_{21} & r_{22} \end{bmatrix}. \tag{6.4}$$

Since the Y axis (i.e. the vertical axis) is almost the same for different 3D surfaces of an urban scene, r_{11} of a valid rotation is assumed to be larger than a threshold of 0.7. Besides, surfaces acquired from a scene at successive viewpoints have similar coordinate systems. Therefore, r_{00} and r_{22} are required to be positive. Consequently, several candidate rotations can be removed using these restrictions. As reported in [168], usually less than 5 candidate rotations are left for further processing.

For the remaining candidate rotations, the sum of the diagonal elements $(r_{00} + r_{11} + r_{22})$ of O_c is used to rank these rotations. The candidate rotation O with the largest sum is used as the final rotation, and the other remaining rotations are saved for user interface-based operations.

6.2.2.2.2 Translation Estimation Once the rotation O is estimated, the two surfaces M_a and M_b can be aligned in orientation. Then, the distance between each pair of parallel linear segments can be calculated as the vector connecting the midpoints of these two segments. Given two pairs of lines (l_1, r_1) and (l_2, r_2), a translation can be estimated as follows:

- If the four lines l_1, r_1, l_2, and r_2 are parallel to each other, and the distance between l_1 and r_1 is the same as the distance between l_2 and r_2, the candidate translation is calculated as the average of these two distances.
- If l_1 is parallel to r_1, l_2 is parallel to r_2, but l_1 is not parallel to l_2, and the distance between l_1 and r_1 is the same as the distance between l_2 and r_2, the candidate translation is calculated as the solution of an over-constrained linear system.

Consequently, a large number of candidate translations can be estimated from these line pairs, which are further clustered into several groups of translations. Then, the n_{ct} largest clusters of candidate translations are selected, and their centroids are calculated as the final candidate translations. The final translation is considered as the candidate translation (i.e. the centroid) with the maximum number of line matches between two surfaces.

6.2.3 Surface-Based Registration

These methods use the information of a whole surface to estimate the coarse registration between two surfaces. Typical methods include principal component analysis (PCA), RANSAC based data aligned rigidity constrained exhaustive search (DARCES), and 4 points congruent sets (4PCS).

6.2.3.1 Principal Component Analysis (PCA)

This method uses the direction of the principal axes of the surfaces to register two 3D surfaces. It is assumed that the principal axes of two surfaces will almost coincide if their overlap is sufficient. Therefore, the rigid transformation can be estimated by aligning the axes of these two surfaces. A typical PCA-based surface registration method is proposed in [169].

Given two surfaces M_a and M_b, the initial transformation between these two surfaces is estimated by the eigenvectors of the covariance matrices of the points on these two surfaces. For a surface M_a with 3D points $\{p_i\}$, the weighted covariance matrix C is defined as:

$$C = \frac{1}{W} \sum_{i=1}^{n} \{\omega_i (p_i - \bar{p})(p_i - \bar{p})^\top\}, \tag{6.5}$$

where $W = \sum_{i=1}^{n} \omega_i$, $\bar{p} = \frac{1}{W} \sum_{i=1}^{n} \omega_i p_i$, and n is the number of points on the surface. The weight ω_i is estimated by:

$$\omega_i = \frac{\|v\| \|\tilde{n}_i\|}{v^\top \tilde{n}_i}, \tag{6.6}$$

where v represents the viewing direction vector for the range finder, and $\tilde{n}_i$ is the unit normal vector to the surface at p_i. Consequently, three eigenvectors of the covariance matrix C can be generated to form a matrix H_a for surface M_a. Similarly, a matrix H_b can also be generated for surface M_b. Then, an approximate rotation between the two surfaces M_a and M_b can be estimated by

$$O = H_a^T H_b. \tag{6.7}$$

Similarly, an approximate translation between M_a and M_b is calculated as:

$$t = c_b - Oc_a, \tag{6.8}$$

where c_a and c_b are the center of mass of the points on surfaces M_a and M_b. Finally, the two surfaces can be registered using the approximate rotation O and translation t.

This method is computationally more efficient compared to point correspondence-based and line-based algorithms (Sections 6.2.1 and 6.2.2). However, this method still has several limitations. **First**, it can achieve a good coarse registration only when there is a large overlap between the input surfaces [170]. **Second**, registering surfaces that contain symmetries is very hard with this method. That is because, for a symmetric surface, there are two eigenvalues that are quite similar. In this case, the order of the eigenvectors may change and the resulting rotation is far from the true solution. **Finally**, the registration achieved by this method is relatively approximate, but it provides a good initial guess for subsequent fine registration.

6.2.3.2 RANSAC-Based DARCES

RANSAC [164] can obtain the appropriate parameters of a fitting model from a set of samples containing outliers. This idea can also be used to perform 3D surface registration. A typical method is proposed in [171].

Given two surfaces M_a and M_b, several reference points are selected from each surface by performing uniform sampling or using all the points. Then, three control points are selected from these reference points on each surface to generate a transformation between two surfaces. The estimated transformation is used to align these reference points. The transformation producing the largest number of corresponding points is considered as the surface registration solution. A naive approach is to try all combinations of control point triplets on the two surfaces. This would result in $C_{n_a}^{n_b}$ combinations, where n_a and n_b are the numbers of points on M_a and M_b, which is computationally very expensive. Instead, the DARCES approach [171] can be used to speedup the search process.

Specifically, a primary control point p_p is randomly selected from M_a. Two other control points on M_a are then selected such that their distances to other vertices in the triangle are larger than d_{min}. For the primary control point p_p, each reference point q_p on M_b can be assumed to be its corresponding point.

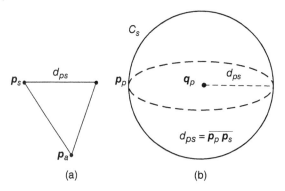

Figure 6.1 The search for the corresponding point of the secondary point p_s. The corresponding point q_s of point p_s should lie on a spherical surface C_s with a center at q_p and a radius of d_{ps} (b), where d_{ps} is the distance between points p_p and p_s (a).

Then, the candidate points corresponding to the secondary point p_s can be obtained using the rigidity constraint. That is, the corresponding point q_s of point p_s should lie on a spherical surface C_s with a radius of d_{ps} and centered at q_p, as shown in Figure 6.1b. Here, d_{ps} is the distance between p_p and p_s, as shown in Figure 6.1a. That is, $C_s = \{q \mid \|q - q_p\| = d_{ps}\}$. Consequently, once the corresponding point for the primary control point p_p is already given, the search space for the corresponding point of the secondary control point is highly reduced, which is limited to a spherical surface.

Once the corresponding point q_s for the secondary control point p_s is hypothesized, the candidate corresponding point q_a for the auxiliary point p_a can be searched using q_p, q_s, and d_{qa}, where d_{qa} is the distance between p_a and the line $\overline{p_p p_s}$ connecting points p_p and p_s, as shown in Figure 6.2a. Suppose p_c is the orthogonal projection of point p_a onto the line $\overline{p_p p_s}$, the candidate corresponding point q_a should lie on a circle C_a which is perpendicular to $\overline{q_p q_s}$ and centered at point q_c with a radius of d_{qa}, where q_c is the point corresponding to p_c, as shown in Figure 6.2. That is, $C_a = \{q \mid \|q - q_c\| = d_{qa} \text{ and } \overline{qq_c} \perp \overline{q_p q_s}\}$.

Next, a rigid transformation can be derived from the three pairs of corresponding points (p_p, q_p), (p_s, q_s), and (p_a, q_a) using the Arun method [160]. All the reference points on M_a and M_b are then aligned using the estimated transformation, and the number of successfully aligned reference points are counted (i.e. the overlapping number of points). For each plausible combination of feature correspondences, a rigid transformation and its overlapping number can be obtained. The transformation resulting in the largest overlapping number is used as the candidate solution.

If the transformation derived from the current primary control point p_p has an overlapping number larger than a threshold, the transformation is considered as the final solution for surface registration. Otherwise, another primary control point p_p is randomly selected from the reference points on M_a. The above procedure is repeated until a rigid transformation with an overlapping number that is larger than the threshold is found.

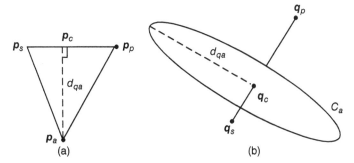

Figure 6.2 The search for the corresponding point of the auxiliary control point. (a) Suppose p_c is the orthogonal projection of point p_a onto the line $\overline{p_p p_s}$. (b) The candidate corresponding point q_a of p_a should lie on a circle C_a which is perpendicular to $\overline{q_p q_s}$ and centered at point q_c with a radius of d_{qa}, where q_c is the point corresponding to p_c, d_{qa} is distance between p_a and the line $\overline{p_p p_s}$ connecting points p_p and p_s.

This method is robust to outliers. However, it is computationally very expensive [170]. There are several ways to improve the efficiency of this method. For example, if both the position and the normal vector of each point is used, only two pairs of oriented points are sufficient to estimate a rigid transformation between two surfaces. In this case, the search space of the RANSAC-based DARCES method can further be reduced.

6.2.3.3 Four-Points Congruent Sets (4PCS)

Given two surfaces M_a and M_b, the basic idea of the 4PCS method [172] is to use a set of four coplanar points from M_a as base $\mathfrak{B}$, and then find all the sets of 4-points from M_b that are approximately congruent to the base $\mathfrak{B}$. Here, *congruent* means that the two sets of 4-points can be aligned using a rigid transformation, within an error tolerance δ.

6.2.3.3.1 *Affine Invariants of 4-Points Set* Given four coplanar points that are not all colinear $X = \{a, b, c, d\}$, the lines $\overline{ab}$ and $\overline{cd}$ can intersect at an intermediate point e. Let

$$r_1 = \|a-e\|/\|a-b\| \quad \text{and} \quad r_2 = \|c-e\|/\|c-d\|. \tag{6.9}$$

These two ratios are invariant to rigid transformations. As shown in Figure 6.3, for 4-points $\{a, b, c, d\}$ on M_a, if M_b matches M_a with the corresponding 4-points $\{a', b', c', d'\}$ lying on the overlapping area, then $\|a-e\|/\|a-b\| = \|a'-e'\|/\|a'-b'\|$, and $\|c-e\|/\|c-d\| = \|c'-e'\|/\|c'-d'\|$.

Thus, given 4-points $\{a, b, c, d\}$ on M_a and two ratios r_1 and r_2 (as shown in Figure 6.4a), every 4-points on M_b defined by r_1 and r_2 can be extracted. That is, for each pair of points q_1 and q_2 on surface M_b, there are two assignments

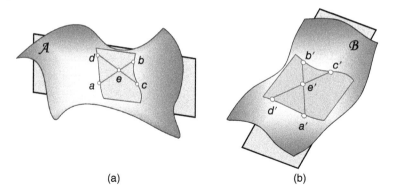

Figure 6.3 An illustration of congruent 4-points. For 4-points (a) $\{a, b, c, d\}$ on surface M_a, if surface M_b matches surface M_a with the corresponding 4-points (b) $\{a', b', c', d'\}$ lying on the overlapped area, then $\|a-e\|/\|a-b\| = \|a'-e'\|/\|a'-b'\|$, and $\|c-e\|/\|c-d\| = \|c'-e'\|/\|c'-d'\|$.

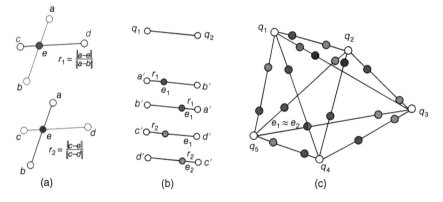

Figure 6.4 Congruent 4-points extraction. (a) Given 4-points $\{a, b, c, d\}$ on surface M_a, two ratios r_1 and r_2 are obtained. (b) Four possible intermediate points can be produced for each pair of points q_1 and q_2 using different assignments of r_1 and r_2. (c) If points e_1 and e_2 are coincident, these 4-points $\{q_5, q_3, q_4, q_1\}$ are probably a rigidly transformed copy of $\{a, b, c, d\}$.

corresponding to points $\{a, b\}$:

$$e_1 = q_1 + r_1(q_2 - q_1), \tag{6.10}$$

$$e_1 = q_2 + r_1(q_1 - q_2). \tag{6.11}$$

There are another two assignments corresponding to points $\{c, d\}$:

$$e_2 = q_1 + r_2(q_2 - q_1), \tag{6.12}$$

$$e_2 = q_2 + r_2(q_1 - q_2). \tag{6.13}$$

Consequently, there are four possible intermediate points for each pair of points q_1 and q_2, as shown in Figure 6.4b.

For any two pairs of points $\{q_1, q_2\}$ and $\{q_3, q_4\}$, four intermediate points can be calculated for each pair of points using Eqs. (6.11)–(6.13). If points e_1 and e_2 coincide, these 4-points $\{q_1, q_2, q_3, q_4\}$ are probably a rigid transformed copy of $\{a, b, c, d\}$. Here, coincidence means that the distance between two points is sufficiently small. As shown in Figure 6.4c, the 4-points $\{a, b, c, d\}$ are approximately congruent to the 4-points $\{q_5, q_3, q_4, q_1\}$. Note that, in Figure 6.4c, only two intermediate points are shown for each pair of points for the sake of simplicity.

6.2.3.3.2 Congruent 4-Points Extraction

Given a base $\mathcal{B}$ of 4 coplanar points on surface M_a, the set of all 4-points on M_b that are approximately congruent to the 4-points base $\mathcal{B}$ (up to an approximation tolerance δ) have to be extracted as follows.

- **First**, the two ratios r_1 and r_2 are computed for the set of 4-points $\mathcal{B}$.
- **Then**, all sets of 4-points on M_b that can be related to the 4-points base $\mathcal{B}$ by rigid transforms are extracted using the method described in Section 6.2.3.3.1
- **Next**, using the 4-points base $\mathcal{B}$ and each candidate congruent 4-points on M_b, a rigid transformation can be estimated using a closed form solution [173].

This process requires the computation and storage of $O(n_b^2)$ intermediate points, where n_b is the number of points on M_b. The process is, therefore, expensive in terms of computation and storage requirement. Since the Euclidean distance between points can be preserved by rigid transformation, the extraction process can be accelerated. That is, for a 4-points base $\{a, b, c, d\}$, the distances $d_1 = \|a - b\|$ and $d_2 = \|c - d\|$ can be calculated. Then, we only consider the point pairs whose distance is approximately equal to d_1 or d_2. Consequently, the search space can significantly be reduced.

6.2.3.3.3 The 4PCS Algorithm

Given a surface M_a, three points are picked randomly, and then the fourth point is selected such that the 4-points can form a wide and approximately coplanar base $\mathcal{B}_i$. A wide base means that the points are far from each others, which is helpful to achieve a more stable surface registration. However, a wide base also means that the sensitivity to overlap is increased since the 4-points may not be in the overlapping area of the two surfaces. In practice, the knowledge of the overlap fraction can be used to guide the selection of the maximum distance between these 4-points.

For the selected coplanar base $\mathcal{B}_i$, the invariant ratios r_1 and r_2 can be calculated using Eq. (6.9). Then, the method described in Section 6.2.3.3.2 is used

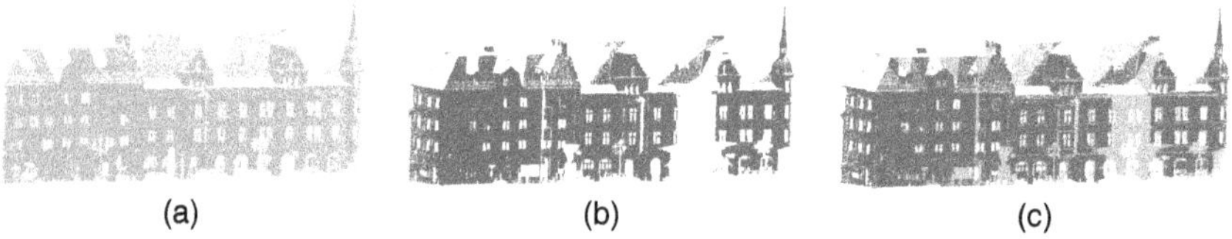

(a)   (b)   (c)

Figure 6.5 Point cloud registration achieved by the 4PCS method. (a,b) Two point clouds of building facades acquired from two different viewpoints. (c) The registration result achieved by the 4PCS method.

to extract sets of 4-points $U = \{U_1, U_2, \ldots, U_s\}$ on M_b, that are congruent to the coplanar base $\mathfrak{B}_i$ within an approximation tolerance δ. For each 4-points U_j on M_b, a rigid transformation T_j is estimated to align U_j and $\mathfrak{B}_i$. To further verify the transformation T_j, T_j is used to align the two surfaces M_a and M_b, and the number of corresponding points on these two surfaces under transformation T_j is calculated as the score. The transformation with the highest score is considered as the transformation generated by the coplanar base $\mathfrak{B}_i$. For the M_a, several coplanar bases $\{\mathfrak{B}_i\}$ are tested, resulting in a set of candidate transformations. The transformation with the highest score among all these bases is used to register surfaces M_a and M_b.

Figure 6.5a,b shows two point clouds of building facades acquired from two different viewpoints. The registration result achieved by the 4PCS method is shown in Figure 6.5c. It can be observed that the two surfaces of the building facades can be successfully aligned. Note that, a major advantage of the 4PCS method is that it can be used to register 3D point clouds with only a few geometric features. Besides, the 4PCS method is robust to noise and outliers.

6.3 Fine Registration

The task of fine registration is to obtain a highly accurate alignment of two or more surfaces (e.g. in the form of point clouds), starting with a coarse registration as an initial guess, followed by an iterative refinement process. The general process is as follows: given the initial estimate of the rigid transformation between two surfaces M_a and M_b, these two surfaces are first coarsely registered using one of the methods described in the previous section. Then, the transformation is refined by minimizing a distance defined between the two surfaces. Different distance measures have been proposed in the literature. Examples include the point-to-point [158] and the point-to-plane distances [174] used in the Iterated Closest Point (ICP) algorithms [158, 174].

6.3.1 Iterative Closest Point (ICP)

The ICP algorithm, proposed concurrently by Besl and McKay [158] and Chen and Medioni [174], became a standard registration method due to its simplicity, robustness, and reliability. Given two coarsely registered surfaces M_a and M_b, the points on these two surfaces are matched by searching for each point on M_a its closest point on surface M_b. This is referred to as *the closest point search*. The average distance between these associated closest points is then minimized to estimate the transformation between the two surfaces. This is referred to as *the transformation estimation*. The transformation is then applied to these surfaces to achieve a more accurate registration. This process is iterated until its convergence.

Below we discuss these two steps and then summarize the basic algorithm. Note that the ICP algorithm operates on surfaces represented as point sets. This, however, is not a restriction since any surface representation can trans-formed into a point set in a straightforward manner. For example, polygonal meshes can be converted into points by only considering the vertices and dis-carding the triangulation. Another way is to sample a dense point set at different locations on the surface of the input 3D shapes. In what follows, let M_a and M_b be two input surfaces. We refer to M_a as the data surface and M_b as the model surface. We assume that the number of points on M_a is n_p, and the number of points on M_b is n_q.

6.3.1.1 Closest Point Search

For each data point p_i on M_a, its closest point q_i on M_b is defined as:

$$q_i = \underset{q \in M_b}{\operatorname{argmin}} \|q - p_i\|. \tag{6.14}$$

By doing this for each point on M_a, we obtain a set of closest points $Q = \{q_i\}$. That is,

$$Q = C(M_a, M_b), \tag{6.15}$$

where C represents the closest point operator.

6.3.1.2 Transformation Estimation

Given a set of closest point pairs $\{p_i, q_i\}$, the next step is to estimate a rigid transformation $T = [O, t]$, composed of a rotation matrix O and a translation vector t, which brings the surface M_a as close as possible to M_b. This is done by minimizing the least square error cost function $\varphi(T)$:

$$\varphi(T) = \frac{1}{n_p} \sum_{i=1}^{n_p} \|q_i - Op_i - t\|^2, \tag{6.16}$$

where n_p is the number of points in p. Let

$$\mu_p = \frac{1}{n_p}\sum_{i=1}^{n_p} p_i \quad \text{and} \quad \mu_q = \frac{1}{n_p}\sum_{i=1}^{n_p} q_i \tag{6.17}$$

be the centers of mass of point sets p and q, respectively,

$$C = \frac{1}{n_p}\sum_{i=1}^{n_p}(p_i - \mu_p)(q_i - \mu_q)^{\mathsf{T}} = \frac{1}{n_p}\sum_{i=1}^{n_p} p_i q_i^{\mathsf{T}} - \mu_p \mu_q \tag{6.18}$$

the cross-covariance matrix between the point sets p and q, and

$$\widehat{C}_{ij} = (C - C^{\mathsf{T}})_{ij} \tag{6.19}$$

the anti-symmetric matrix whose elements are denoted by $\widehat{C}_{ij}$. The matrix $\widehat{C}$ is used to form the column vector $\triangle = [\widehat{C}_{23}, \widehat{C}_{31}, \widehat{C}_{12}]^{\mathsf{T}}$. The eigen decomposition of the 4×4 matrix

$$W = \begin{bmatrix} \text{tr}(C) & \triangle^{\mathsf{T}} \\ \triangle & C + C^{\mathsf{T}} - \text{tr}(C)I_3 \end{bmatrix}, \tag{6.20}$$

where I_3 is a 3×3 identity matrix, results in four eigenvalues r_0, r_1, r_2, and r_3. These can be used to form a unit quaternion $T_R = [r_0, r_1, r_2, r_3]$ from which we obtained the optimal rotation matrix

$$O = \begin{bmatrix} r_0^2 + r_1^2 - r_2^2 - r_3^2 & 2(r_1 r_2 - r_0 r_3) & 2(r_1 r_3 + r_0 r_2) \\ 2(r_1 r_2 + r_0 r_3) & r_0^2 + r_2^2 - r_1^2 - r_3^2 & 2(r_2 r_3 - r_0 r_1) \\ 2(r_1 r_3 - r_0 r_2) & 2(r_2 r_3 + r_0 r_1) & r_0^2 + r_3^2 - r_1^2 - r_2^2. \end{bmatrix}. \tag{6.21}$$

Once the rotation matrix O is estimated, the optimal translation vector $\mathbf{t}$ is given as:

$$t = \mu_q - O\mu_p. \tag{6.22}$$

Plugging the optimal transformation $T = [O, t]$ into Equation (6.16) provides the registration error.

6.3.1.3 Summary of the ICP Method
Given the two surfaces M_a and M_b, the procedure of the ICP method can be summarized as follows:

- **Initialization.** Set the rigid transformation T to $[1, 0, 0, 0, 0, 0, 0]$, i.e. $O \leftarrow I_3$, and $\mathbf{t} \leftarrow [0, 0, 0]^{\mathsf{T}}$. Set the iteration number k to 0, and p_0 to the points on M_a.
- **Step 1**: Compute, for each point on M_a, its closest points on M_b. The computational complexity of this process is of order $O(n_p n_q)$, in the worst case, and of $O(n_p \log n_q)$ on average. Here, n_p and n_q are the numbers of points on M_a and M_b, respectively.

- **Step 2**: Compute the transformation T_k, which aligns p^0 onto q^k. This process has a computational complexity of $O(n_p)$.
- **Step 3**: Compute the registration error using Eq. (6.16).
- **Step 4**: Apply the transformation T_k to p^0, i.e. $p^{k+1} \leftarrow T_k(p^0)$. This process has a computational complexity of $O(n_p)$.

If the change of the registration error in two iterations is smaller than a threshold τ_d (i.e. $|d_k - d_{k+1}| < \tau_d$), the iteration is stopped, and the transformation T_k is used to register the two surfaces M_a and M_b. Otherwise, steps 1–4 are repeated to produce a more accurate registration.

6.3.2 ICP Variants

Although the basic ICP algorithm usually achieves a satisfactory performance, it has several limitations in terms of accuracy, robustness, and speed of convergence. For instance, the computation of the basic ICP algorithm is relatively high since the algorithm has to search for the closest points during each iteration. A large number of variants have been proposed to improve different aspects of the basic ICP algorithm [175].

6.3.2.1 Point Selection

The process of finding closest points between the two surfaces, which is very time consuming, can be accelerated using special data structures such as k-d trees. By doing so, the computational complexity remains still as high as $O(\log n_q)$. An alternative approach for accelerating the registration process further is to only use a subset of points from the point clouds. Such points can be selected by uniform or random sampling. Other more sophisticated selection techniques include:

- *Normal space sampling [175]*: The motivation is that, for some scenes such as incised planes, some small features are very important for the registration. Both uniform sampling and random sampling can select only a small number of points from these small features, which is insufficient for a reliable registration. In contrast, normal space sampling selects points such that the normals of the selected points are uniformly distributed. Thus, more points can be selected from areas which have tiny features. Consequently, the registration performance can be improved.
- *Intensity Gradient based Sampling [176]*: One can use, when available, the color or intensity information of each point to guide the selection process. For instance, Weik [176] select the points that have a strong intensity gradient.
- *Multi-resolution sampling*: Another approach is to use a different number of points at different iterations of the ICP algorithm. For instance, one can use a small number of points at the start, and then increase the number of points

with the number of iterations [87]. The former speeds up the process while the latter improves the registration accuracy.

Note that these point selection approaches can be performed on either the model point cloud or the data point cloud, or both of them.

6.3.2.2 Point Matching

Several approaches can be used to find the corresponding points for a set of sampling points. Examples include:

- *Closest point*: The closest point on M_b is obtained for each sampling point on M_a. A brute-force search for the closest point to a sampling point takes $O(n_q)$ time, while a k-d tree search takes $O(\log n_q)$ time.
- *Normal shooting*: A ray is defined to originate from the sampling point on M_a and along the direction of the normal at the sampling point. The intersection of this ray with model M_b is considered to be the corresponding point to the sampling point.
- *Reverse calibration*: The sampling point on M_a is projected onto the model surface M_b from the viewpoint of the range camera for M_b. The projected point on M_b is considered as the corresponding point of that sampling point.

An additional compatibility based on color or angle between normals can further be enforced to obtain more accurate corresponding points.

6.3.2.3 Point Pair Weighting

In the basic ICP method, each corresponding point pair is given an equal weight. The registration performance can be improved by assigning different weights to different pairs. Examples of weighting schemes include:

- *Distance-based weighting*: The point pairs with a larger distance are given smaller weights. For example, the weight can be defined as $\omega = 1 - \frac{\text{Dist}(p_i, q_i)}{\max_j(\text{Dist}(p_j, q_j))}$ [177].
- *Normal-based weighting*: The weight can be calculated using the compatibility of surface normals, e.g. $\omega = \tilde{\mathbf{n}}_p \cdot \tilde{\mathbf{n}}_q$.
- *Color-based Weighting*: The color consistency between two points can also be integrated into the weight.

6.3.2.4 Point Pair Rejection

Among the point pairs obtained by point matching, there are usually several outliers, which will decrease the accuracy and stability of the final registration result. Consequently, several point pair rejection strategies have been proposed. Examples include:

- *Distance-based rejection*: The corresponding point pairs with a distance larger than a threshold are rejected.

- *Ranking-based rejection*: The corresponding point pairs are first ranked according to their distances. The worst $\alpha\%$ of pairs are then rejected [178].
- *Distance deviation-based rejection*: The standard deviation σ_d of the distances of all point pairs is first calculated, and the point pairs with a distance larger than $\alpha\sigma_d$, $0 < \alpha < 1$, are rejected.
- *Consistency-based rejection*: One can assume that point pairs should be consistent with their neighboring pairs. For example, two corresponding point pairs (p_1, q_1) and (p_2, q_2), they are consistent only if $|\text{dist}(p_1, p_2) - \text{dist}(q_1, q_2)|$ is smaller than a threshold. Otherwise, they are considered as inconsistent and are rejected. In practice, each point pair is checked against 10 point pairs, if more than 5 point pairs are inconsistent with that point pair, it is rejected [179].
- *Boundary-based rejection*: the corresponding point pairs containing points on the boundary of a surface patch are rejected [180].

6.3.2.5 Error Metrics

The basic ICP algorithm presented above minimizes the error metric of Eq. (6.16). This is called the *point-to-point distance*. As seen above, this metric leads to a closed-form solution for the estimation of the rigid transformation which aligns the two surfaces. There are, however, several other error metrics that one can use. Examples include:

- *Point-to-plane distance.* Chen and Medioni [181] define the distance between a point on M_a and the surface M_b as the shortest distance between that point and all the planes that are tangent to M_b. The error metric is then defined as the average of this distance over all the points in M_a. In theory, the point-to-plane distance requires less iterations for convergence than the point-to-point distance. However, under this distance measure, there is no closed-form solution. Thus, the optimal transformation needs to be obtained using a generic nonlinear or linear (after linearizing the problem) optimization method.
- *Point-to-point distance with color.* In addition to the geometric information, one can also use other attributes, e.g. color or texture, to compute the error metric. The error metric then becomes a weighted sum of multiple terms, each one measures a distance based on a specific type of attributes.

6.3.3 Coherent Point Drift

The CPD method [159] considers the registration of two point clouds as a probability density estimation problem. Specifically, the set of n_q points $\mathbf{Q}$ on M_b are treated as centroids of a Gaussian Mixture Model (GMM), while the other set of n_p points $\mathbf{P}$ on M_a, which represents the data points, are generated from the GMM. The correspondence between the two point sets is obtained using the posterior probabilities of the GMM components. Here, the probability density function prob (p) of the GMM is defined as:

$$
\text{prob}\,(p) = \sum_{j=1}^{n_q+1} \text{Prob}\,(j)\text{prob}\,(p|j). \tag{6.23}
$$

For $j = 1, 2, \ldots, n_q$, we have

$$
\text{prob}\,(p|j) = \frac{1}{(2\pi\sigma^2)^{3/2}} \exp\left(-\frac{\|p - q_j\|^2}{2\sigma^2}\right). \tag{6.24}
$$

For $j = n_q + 1$, we have

$$
\text{prob}\,(p|n_q + 1) = \frac{1}{n_p}. \tag{6.25}
$$

Equation (6.25) defines a uniform distribution, which is used to model noise and outliers. For simplicity, the same standard deviation σ and membership probability $\text{Prob}\,(j) = \frac{1}{n_q}$ are used for all the components of the GMM. If we assume that the weight for the uniform distribution is ω, then Eq. (6.23) can be rewritten as:

$$
\text{prob}\,(p) = \omega\frac{1}{n_p} + (1 - \omega) \sum_{j=1}^{n_q} \frac{1}{n_q}\text{prob}\,(p|j). \tag{6.26}
$$

Note that in this method, ω, which gives a flexible control over the robustness and accuracy, is the only parameter which needs to be tuned in this method.

The correspondence probability between two points $p_k \in M_a$ and $q_j \in M_b$ is then defined as the posterior probability of the GMM centroid given the data point. That is:

$$
\text{Prob}\,(j|p_k) = \frac{\text{Prob}\,(j)\text{prob}\,(p_k|j)}{\text{prob}\,(p_k)}. \tag{6.27}
$$

The expectation–maximization (EM) method is then used to find the rotation O and translation t between two point clouds, and the standard deviation σ of GMM. The process is as follows;

First, the old parameters O^{old}, t^{old}, and σ^{old} are used to obtain the posteriori probability distributions of mixture components $\text{Prob}^{old}(j|p_k)$. That is:

$$
\text{Prob}^{old}(j|p_k) = \frac{\exp^{-\frac{1}{2}\left\|\frac{p_k - O^{old}q_j - t^{old}}{\sigma^{old}}\right\|^2}}{\sum_{i=1}^{n_q} \exp^{-\frac{1}{2}\left\|\frac{p_k - O^{old}q_i - t^{old}}{\sigma^{old}}\right\|^2} + c}, \tag{6.28}
$$

where $c = (2\pi\sigma^2)^{3/2}\frac{\omega}{1-\omega}\frac{n_q}{n_p}$.

Then, the new parameters are obtained by minimizing the objective function E, which is defined as the expectation of the negative log-likelihood function:

$$E = -\sum_{k=1}^{n_p} \sum_{j=1}^{n_q+1} \text{Prob}^{old}(j|p_k) \log\left(\text{Prob}^{new}(j)\, \text{prob}^{new}(p_k|j)\right). \qquad (6.29)$$

By removing the constants that are independent of O, $\mathbf{t}$, and σ, Eq. (6.29) can be rewritten as:

$$E(O, \mathbf{t}, \sigma) = \frac{1}{2\sigma^2} \sum_{k=1}^{n_p} \sum_{j=1}^{n_q} \text{Prob}^{old}(j|p_k)\|p_k - Oq_j - \mathbf{t}\|^2 + \frac{3N_{\mathbf{P}}}{2}\log\sigma^2, \qquad (6.30)$$

where $N_{\mathbf{P}} = \sum_{k=1}^{n_p}\sum_{j=1}^{n_q}\text{Prob}^{old}(j|p_k)$. To obtain the optimal rotation O, we first calculate $\hat{\mathbf{P}}$ and $\hat{\mathbf{Q}}$ as follows:

$$\hat{\mathbf{P}} = \mathbf{P} - \mathbf{1}\mu_p^{\mathrm{T}} \quad \text{and} \quad \hat{\mathbf{Q}} = \mathbf{Q} - \mathbf{1}\mu_q^{\mathrm{T}}, \qquad (6.31)$$

where $\mathbf{1}$ is a column vector with all of its elements set to one, and

$$\mu_p = \frac{1}{n_p}\mathbf{P}^{\mathrm{T}}\mathbf{P}_r^{\mathrm{T}}\mathbf{1} \quad \text{and} \quad \mu_q = \frac{1}{n_p}\mathbf{Q}^{\mathrm{T}}\mathbf{P}_r\mathbf{1}. \qquad (6.32)$$

Here, each element $\text{Prob}(j,k)$ of $\mathbf{P}_r$ is defined by $\text{Prob}^{old}(j|p_k)$.

Next, the minimization of E with respect to O is equivalent to the maximization of $\text{tr}(\mathbf{A}^{\mathrm{T}}O)$ subject to $O^{\mathrm{T}}O = \mathbf{I}$ and $\det(O) = 1$. Here, $\mathbf{I}$ is an identity matrix, and $\mathbf{A}$ is defined as:

$$\mathbf{A} = \hat{\mathbf{P}}^{\mathrm{T}}\mathbf{P}_r^{\mathrm{T}}\hat{\mathbf{Q}}. \qquad (6.33)$$

Two matrices $\mathbf{U}$ and $\mathbf{V}$ can then be obtained by SVD decomposition of the matrix $\mathbf{A}$. That is, $\mathbf{USV} = svd(\mathbf{A})$. The optimal rotation O is calculated as

$$O = \mathbf{UCV}, \text{ where } \mathbf{C} = \mathrm{d}(1, \dots, 1, \det(\mathbf{UV}^{\mathrm{T}})). \qquad (6.34)$$

Here, $\mathrm{d}(\mathbf{v})$ represents the diagonal matrix formed from vector $\mathbf{v}$. Once the rotation O is obtained, the translation $\mathbf{t}$ is calculated as:

$$\mathbf{t} = \mu_p - O\mu_q. \qquad (6.35)$$

In summary, the CPD procedure performs as follows:

- **Initialization:** The rotation O is set as a 3×3 identity matrix and $\mathbf{t} = [0, 0, 0]^{\mathrm{T}}$. Set ω to a value between 0 and 1, and $\sigma^2 \leftarrow \frac{1}{3n_p n_q}\sum_{k=1}^{n_p}\sum_{j=1}^{n_q}\|p_k - q_j\|^2$.
- **EM optimization:** Iterate between the following E-step and M-step until convergence.
  - E-Step: $\mathbf{P}_r$ is obtained by assigning each element $\text{Prob}(j,k)$ of $\mathbf{P}_r$ as $\text{Prob}^{old}(j|p_k)$ (defined in Eq. (6.28)).

- M-Step: O is calculated using Eq. (6.34), t is calculated using Eq. (6.35), σ^2 is estimated as $\frac{1}{3n_p}(\text{tr}(\hat{\mathbf{P}}^T\text{d}(\mathbf{P}_r^T\mathbf{1})\hat{\mathbf{P}}) - \text{tr}(\mathbf{A}^TO))$.
- **Transformation**: Apply the computed rotation and translation to the point cloud Q.
- **Correspondence**: the probability of correspondence is defined by $\mathbf{P}_r$.

Myronenko and Song [159] showed that the CPD method is insensitive to the initial pose of the two point clouds [159]. Specifically, the CPD method works well if the initial rotation between two point clouds is about 70°, while the ICP method is trapped into a local minimum when the rotation is larger then 40°. Myronenko and Song [159] also showed that the CPD method is more robust to noise and outliers than the ICP method. To improve the computational efficiency, Myronenko and Song [159] used the Fast Gauss Transform (FGT) to calculate the matrix-vector products, such as $\mathbf{P}_r\mathbf{1}$, $\mathbf{P}_r^T\mathbf{1}$, and $\mathbf{P}_r\mathbf{P}$. Further, Lu et al. [182] presented an accelerated coherent point drift (ACPD) algorithm by integrating the global squared iterative expectation–maximization (gSQUAREM) technique and the Dual-Tree Improved Fast Gauss Transform (DT-IFGT) technique into the CPD method.

6.4 Summary and Further Reading

Rigid registration of 3D surfaces is a fundamental problem for various 3D vision tasks such as 3D modeling, 3D object recognition, and SLAM (simultaneous localization and mapping). It has been studied for more than two decades, with many problems being properly solved. Coarse registration methods usually consist of feature extraction, feature matching, and transformation estimation. A key component of a successful registration is feature extraction, a sufficient number of reliable and robust features provide the basis for accurate feature matching and ultimately transformation estimation. Consequently, different types of features have been investigated for surface registration, such as keypoints, lines, and surfaces. Besides, various methods have also been proposed to remove false feature correspondences generated by feature matching. For fine registration, ICP is the most widely used method. A number of variants have also been introduced to improve the ICP method in terms of accuracy, robustness and efficiency.

For 3D surfaces with rich geometric features, the problem of rigid registration of surfaces is almost solved. However, several open problems remain for challenging scenarios. For example, accurately registering point clouds of featureless scenes (such as flat and white walls), and scenes with repetitive structures (such as building facade) is still challenging. More efforts should be given to those specific registration problems in the future.

7

Nonrigid Registration

7.1 Introduction

In the previous chapter, we discussed techniques for rigid registration, i.e. finding correspondences between and aligning to each other 3D objects or partial 3D scans that are equivalent up to a rigid transform. In this chapter, we shift our focus to the more complex problem where the objects under consideration undergo elastic deformations, which involve bending and stretching. Such 3D objects appear in many applications. For instance, anatomical organs deform in an elastic way during their growth and during their alteration as a result of disease progression. Similarly, human body shapes, as shown in Figure 7.1a–c, significantly bend and stretch during their motion and their growth. 3D reconstruction, growth and motion analysis, motion interpolation, and deformation transfer are examples of applications that require matching features across such 3D objects.

This chapter describes a set of techniques for finding correspondences and computing registrations between 3D objects, which deform in a nonrigid way. While the literature is very rich, see for example [183] for a detailed survey of this topic, we restrict our attention to the methods that formulate correspondences as (a subclass of) diffeomorphisms. We first provide, in Section 7.2, a general formulation and lay down the mathematical framework. Section 7.3 describes some important mathematical concepts and tools on which the methods presented in this chapter are based upon. We then focus on methods which assume that the type of deformations 3D objects undergo are limited to bending, hereinafter referred to as *isometric* deformations (Section 7.4). Finally, we turn our attention to the general problem where deformations include bending and stretching (Section 7.5).

For the remaining parts of this chapter, we represent the boundary of a 3D object S as a function $f : D \to \mathbb{R}^3$, which assigns to every domain point $s \in D$ a three-dimensional point $f(s) = (x(s), y(s), z(s))$. The choice of the domain D depends on the application. For instance, open surfaces such as 3D human faces can be easily embedded on a planar domain $D \subset \mathbb{R}^2$. For closed genus-0

3D Shape Analysis: Fundamentals, Theory, and Applications, First Edition.
Hamid Laga, Yulan Guo, Hedi Tabia, Robert B. Fisher, and Mohammed Bennamoun.

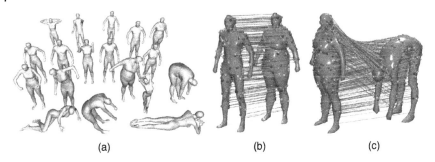

Figure 7.1 Examples of (a) 3D models that undergo nonrigid deformations, (b) correspondences when the 3D models undergo only stretching, and (c) correspondences when the 3D models stretch and bend at the same time. Source: Reproduced with permission of Morgan & Claypool Publishers.

surfaces, the unit sphere $\mathbb{S}^2$ is the most natural choice. The space of such surfaces is denoted by $\mathcal{F}$. As discussed in Chapter 2, translation, scale, rotation, and reparameterization are shape-preserving transformations. Translation and scale can easily be dealt with by normalizing the objects, i.e. moving their center of mass to the origin and rescaling them to have a unit surface area. Let C_f be the space of origin-centered and scale-normalized surfaces. This space is called *the preshape space*. Unless explicitly indicated, we assume that the input surfaces are elements of C_f. The remaining transformations, i.e. rotation and reparameterization, will be dealt with algebraically as discussed in the following sections.

7.2 Problem Formulation

We are given two origin-centered and scale-normalized surfaces $S_1, S_2 \in C_f$, represented with their functions f_1 and f_2, respectively. We seek to:

- find a one-to-one correspondence between the two surfaces, and
- optimally and smoothly deform S_2 to align it onto S_1.

The former is known as *the correspondence* problem while the latter is known as *the registration* problem.

As discussed in Section 2.1.2, both rotations and reparameterizations are shape-preserving transformations, i.e. $\forall O \in SO(3)$ and $\forall \gamma \in \Gamma$, f and $Of \circ \gamma$ are equivalent. Furthermore, diffeomorphisms provide correspondence. Thus, the problem of finding correspondences between two surfaces f_1 and $f_2 \in C_f$ can be formulated as the problem of finding the optimal rotation O^* and the optimal reparameterization γ^* such that $O^* f_2 \circ \gamma^*$ is as close as possible to f_1.

The closeness is measured using some metric or dissimilarity measure. In other words, we seek to solve the following optimization problem:

$$(O^*, \gamma^*) = \underset{O \in SO(3), \gamma \in \Gamma}{\operatorname{argmin}} \, d_C(f_1, Of_2 \circ \gamma). \tag{7.1}$$

The quantity:

$$d(f_1, f_2) = \underset{O, \gamma}{\min} \, d_C(f_1, Of_2 \circ \gamma) \tag{7.2}$$

is the dissimilarity between the shapes of f_1 and f_2. Here, $d_C(\cdot, \cdot)$ is a certain measure of closeness or dissimilarity between surfaces in C_f. It can be defined as the minimal deformation energy that one needs to apply to f_2 in order to align it onto f_1. In other terms, it should quantify the amount of bending and stretching one needs to apply to one surface in order to align it onto the other. Consider the examples of Figure 7.2 where f_1 and f_2 are the surfaces rendered in light gray. Deforming f_1 onto f_2 results in a sequence of m intermediate shapes $F_i : \mathbb{S}^2 \to \mathbb{R}^3, i = 1, \ldots, m$ such that $F_0 = f_1, F_m = f_2$, and $F_i = F_{i-1} + V_{i-1}$ (see the intermediate shapes in Figure 7.2a–c). Here, $V_{i-1} : D \to \mathbb{R}^3$ is a (deformation) vector field such that $F_i(s) = F_{i-1}(s) + V_{i-1}(s)$. The sequence $F = \{F_i, i = 1, \ldots, m\}$ can be seen as a path or a curve in the preshape space C_f. When $m \to \infty$, F becomes a parameterized path $F : [0, 1] \to C_f$ such that $F(0) = f_1, F(1) = f_2$. Its length is given by:

$$L(F) = \int_0^1 \left\langle\!\!\left\langle \frac{dF}{d\tau}(\tau), \frac{dF}{d\tau}(\tau) \right\rangle\!\!\right\rangle^{\frac{1}{2}} d\tau. \tag{7.3}$$

Here, $\frac{dF}{d\tau}(\tau)$ is in fact an infinitesimal vector field, which deforms the surface $F(\tau)$. It can also be interpreted as a small perturbation of the surface. The inner product, $\langle\!\langle \cdot, \cdot \rangle\!\rangle$, also known as *the metric*, measures the norm of this vector field. Thus, integrating over the time parameter τ provides the length of the path, which is a measure of the strength of the deformation that one needs to apply to f_2 in order to align it onto f_1. There are many paths that can deform f_2 onto f_1. Figure 7.2 shows three examples of these paths. For instance:

- The deformation path of Figure 7.2a is obtained with an arbitrary parameterization of the two surfaces. Since the correspondences are incorrect, the intermediate shapes along the deformation path contain many artifacts.
- The deformation path of Figure 7.2b is computed after putting in correspondence the two surfaces but the path is not the shortest (optimal) one under a physically motivated metric. As one can see, many parts in the intermediate shapes are shrinking.
- The deformation path of Figure 7.2c is obtained after putting the two surfaces in correct correspondence. It is also the shortest path under a physically motivated metric.

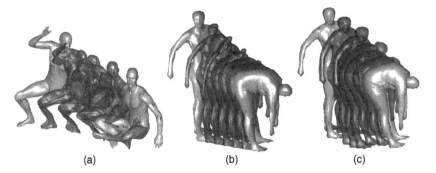

(a)   (b)   (c)

Figure 7.2 Examples of deformation paths between nonrigid shapes. In each example, the source and target shapes are rendered in light gray. The intermediate shapes, along the deformation path, are rendered in a semi transparent dark gray. (a) A deformation path when the correspondences are not optimal, (b) a deformation path, which is not optimal with respect to a physically motivated metric, and (c) the shortest deformation path, or a geodesic, under a physically motivated metric.

We are interested in the third, i.e. in the shortest one under a physically motivated metric. This path is called *the geodesic* in the preshape space C_f and is given by:

$$F^* = \arg \min_F L(F). \tag{7.4}$$

The length of the geodesic path is the dissimilarity between the two surfaces:

$$d_C(f_1, f_2) = \inf_{\substack{F:[0,1] \to C_f, \\ F(0)=f_1, F(1)=f_2}} L(F) = L(F^*). \tag{7.5}$$

Putting the Eqs. (7.2) and (7.5) together, the joint elastic correspondence and registration problem reduces to the solution of the following optimization problem:

$$F^* = \min_{(O, \gamma) \in SO(3) \times \Gamma} \left\{ \min_{\substack{F:[0,1] \to C_f, \\ F(0)=f_1, F(1)=O(f_2 \circ \gamma)}} L(F) \right\}, \tag{7.6}$$

where $L(F)$ is given by Eq. (7.3) and depends on the choice of the representation and the metric. Equation (7.6) is composed of two nested optimization problems;

- The inner minimization is the alignment step. It assumes that the surfaces are in correspondence, i.e. fixes the rotation O and the reparameterization γ, and finds the shortest path in C_f which connects f_1 to $O(f_2 \circ \gamma)$.
- The outer optimization is the correspondence computation step. It finds the optimal rotation O and reparameterization γ that bring f_2 as close as possible to f_1 without changing its shape.

Clearly, these two problems, i.e. correspondence and alignment, are interrelated and should be solved jointly using the same metric (or optimality criteria). This is, however, not straightforward since the search space is large as the solution involves a search over all possible rotations, diffeomorphisms, and deformation paths. The following sections discuss two families of solutions. The first one (Section 7.4) restricts the type of deformations that the two surfaces undergo to those which preserve the intrinsic properties of the shapes. In other words, the surface f_2 is a bent version of f_1. Section 7.5 treats the more general case of surfaces that bend and stretch.

7.3 Mathematical Tools

Before describing in detail the methods used for elastic correspondence and registration, we will first review some mathematical tools and concepts that will be used throughout the chapter.

7.3.1 The Space of Diffeomorphisms

We have seen in Section 2.1.2 that diffeomorphisms provide registration. Since we are dealing with surfaces, which are spherically parameterized, we will look at special cases and properties of the diffeomorphisms $\gamma \in \Gamma$, which map the sphere onto itself, i.e. $\gamma: \mathbb{S}^2 \to \mathbb{S}^2$. There are three special cases of such diffeomorphisms; the identity diffeomorphism, the rotations of the spherical grid, and the Möbius transformations.

Example 7.1 (The identity diffeomorphism). The identity diffeomorphism is a diffeomorphism $\gamma_{id}: \mathbb{S}^2 \to \mathbb{S}^2$, which maps every point $s \in \mathbb{S}^2$ to itself, i.e. $\forall s \in \mathbb{S}^2, \gamma_{id}(s) = s$. In other words, for every surface $f, f \circ \gamma_{id} = f$.

Example 7.2 (Rotations). A special case of diffeomorphisms is the rigid rotations of the parameterization domain $\mathbb{S}^2$. Let $s = (u, v)$ be the spherical coordinates such that $u \in [0, \pi]$ and $v \in [0, 2\pi)$. A rotation of the spherical grid with an angle (θ, ϕ) is a diffeomorphism $\gamma_{(\theta,\phi)}$, which transforms s into $s' = \gamma_{(\theta,\phi)}(s) = (u', v')$. The new spherical coordinates (u', v') are computed as follows;

- $u' \leftarrow u + \theta, v' \leftarrow v + \phi$.
- if $u' > \pi$ or $u' < 0$ then,
  - $u' \leftarrow 2\pi - u'$.
  - $v' \leftarrow v' + \pi$.
- $u' \leftarrow u' \pmod{2\pi}$.
- $v' \leftarrow v' \pmod{2\pi}$.

Here, (mod) refers to the modulus operator.

Example 7.3 (Möbius transformations). Another special case of diffeomorphisms is the Möbius transformation. A Möbius transformation is a one-to-one and onto map Ψ of the extended complex plane $\mathbb{C}$ (or equivalently the sphere $\mathbb{S}^2$) to itself. It is defined as:

$$\Psi(z) = \frac{az + b}{cz + d}, \quad z \in \mathbb{C}, \tag{7.7}$$

where a, b, c, and d are complex numbers such that $ad - bc \neq 0$. (Recall that there is a one-to-one correspondence between the points on the surface of $\mathbb{S}^2$ minus the north pole and the points in the complex plane through stereographic projection.)

A Möbius transformation can be represented using the 2×2 matrix $\begin{pmatrix} a & b \\ c & d \end{pmatrix}$, which has a nonvanishing determinant. An interesting property of the Möbius transformation is that for two given triplets $z_1, z_2, z_3 \in \mathbb{C}$ and $y_1, y_2, y_3 \in \mathbb{C}$, the coefficients a, b, c, and d of the Möbius transformation Ψ, which interpolates them, can be found analytically by taking:

$$\begin{pmatrix} a & b \\ c & d \end{pmatrix} = \begin{pmatrix} y_2 - y_3 & y_1 y_3 - y_1 y_2 \\ y_2 - y_1 & y_1 y_3 - y_3 y_2 \end{pmatrix}^{-1} \begin{pmatrix} z_2 - z_3 & z_1 z_3 - z_1 z_2 \\ z_2 - z_1 & z_1 z_3 - z_3 z_2 \end{pmatrix}. \tag{7.8}$$

Recall that z_i and y_i are complex numbers and so are the coefficients of the Möbius transformation matrix.

Definition 7.1 (The tangent space $T_\gamma(\Gamma)$). Let $\gamma \in \Gamma$ be a diffeomorphism. Then, $T_\gamma(\Gamma)$ is the tangent space of Γ at γ. Geometrically, an element $\partial\gamma \in T_\gamma(\Gamma)$ is a smooth field of vectors that are tangential to $\mathbb{S}^2$. They can be also seen as incremental changes of γ.

7.3.2 Parameterizing Spaces

Many problems in shape analysis are formulated as the optimization of an energy function over the space of some parameters. Often, these spaces are very large and of an infinite dimension. Take as example Eq. (7.6), which is a joint optimization over the space of rotations, diffeomorphisms, and deformation paths. A naive search over the space of solutions is usually computationally prohibitive. One approach for overcoming this problem is to define a set of orthonormal basis $\mathbf{B} = \{\mathbf{b}_j, j = 1, \ldots\}$ so that every element p in the space of solutions can be represented as a linear combination of the elements of $\mathbf{B}$. In other words, one can write

$$p = \sum_i \alpha_i \mathbf{b}_i, \quad \text{such that } \alpha_i \in \mathbb{R} \text{ and } \mathbf{b}_i \in \mathbf{B}. \tag{7.9}$$

By doing so, when searching for the optimal p, one only needs to find the coefficients α_i. In theory, some spaces would require an infinite number of bases.

In practice, however, one can approximate the elements of the space under consideration using a finite, eventually large, number of bases. Now, the choice of the basis (and their number) depends on the space under consideration. Here, we will look at two specific cases, which are of interest to the elastic correspondence and registration of 3D shapes, namely, how to define the basis for the space of closed genus-0 surfaces and for the tangent space $T_{\gamma_{id}}(\Gamma)$. These will be used when solving the optimization problem of Eq. (7.6).

Example 7.4 (Basis for the space of surfaces). Since the elements of $\mathcal{F}$ are smooth mappings of the form $f : \mathbb{S}^2 \to \mathbb{R}^3$, one can form an orthonormal basis, **B**, for the space of surfaces using spherical harmonics (SHs), in each coordinate. Let $\{Y_i(s)\}$ be the real-valued SH functions. Then, a basis for representing surfaces $f : \mathbb{S}^2 \to \mathbb{R}^3$ can be constructed as $\{Y_i e_1\} \cup \{Y_j e_2\} \cup \{Y_k e_3\}$, where e_1, e_2, e_3 is the standard basis for $\mathbb{R}^3$. We will refer to the resulting basis as $\mathbf{B} = \{\mathbf{b}_j\}$, for j in some index set.

The set **B** is a generic basis. It can be used to represent arbitrary spherically parameterized surfaces. However, complex surfaces will require a large number of basis elements in order to capture all the surface details. Alternatively, depending on the context and on the availability of training data, it may be computationally more efficient to use a principal component analysis (PCA) basis instead of a generic basis. For instance, in the case of human shapes, there are several publicly available data sets, which can be used to build a PCA basis. Thus, given a set of training samples, we can compute its PCA in $\mathcal{F}$ and use the first principal components to construct the basis. In practice, we found that the number of PCA basis elements required for representing the space of human body shapes is of the order of 100. This is significantly smaller than the number of generic basis elements, e.g. harmonic basis, which would be required (more than 3000).

Example 7.5 (Basis for the tangent space $T_{\gamma_{id}}(\Gamma)$). One approach to construct an orthonormal basis for the tangent space of Γ at the identity element γ_{id} is by using the gradient of the SH basis. Let $Y_l^m : \mathbb{S}^2 \to \mathbb{C}$ denote the complex SH function of degree l and order m, $m = 0, \ldots, l$. The real and imaginary parts of Y_l^m, for all l and m, labeled as ψ_i (indexed by i in an arbitrary order), form an orthonormal basis of the Hilbert space $\mathbb{L}^2(\mathbb{S}^2, \mathbb{R})$. Since the imaginary part of Y_l^0 is always zero, we have $(l + 1)^2$ distinct functions on $\mathbb{S}^2$ for the first l harmonics. By definition, the gradient of the SH basis ψ_i, which is given by:

$$\nabla_{(\theta,\phi)}\psi_i = \left[\frac{\partial \psi_i}{\partial \theta}, \frac{1}{\sin(\theta)} \frac{\partial \psi_i}{\partial \phi} \right],$$

is a tangent vector field on $\mathbb{S}^2$. These vector fields, after excluding $\nabla_{(\theta,\phi)}\psi_0$, which is a zero vector, provide half of the orthonormal basis elements of $T_{\gamma_{id}}(\Gamma)$. The remaining elements are obtained by rotating each vector in the vector field counterclockwise, when seen from outside of the sphere, by $\pi/2$ radians. Thus, for $\nabla_{(\theta,\phi)}\psi_i$, its rotation is simply the vector

$$* \nabla_{(\theta,\phi)}\psi = \left[\frac{1}{\sin(\theta)} \frac{\partial \psi}{\partial \phi}, -\frac{\partial \psi}{\partial \theta} \right]. \tag{7.10}$$

We can then define two sets

$$\mathbf{B}_1 = \left\{ \tilde{\mathbf{b}}_i = \frac{\nabla \psi_i}{\|\nabla \psi_i\|} \right\} \quad \text{and}$$

$$\mathbf{B}_2 = \left\{ * \tilde{\mathbf{b}}_i = \frac{* \nabla \psi_i}{\| * \nabla \psi_i\|} \right\}, \qquad i = 1, 2, \dots . \tag{7.11}$$

It turns out that the $\mathbf{B}_l = \mathbf{B}_1 \cup \mathbf{B}_2$ provides an orthonormal basis of $T_{\gamma_{id}}(\Gamma)$, the set of all smooth tangent vector fields on $\mathbb{S}^2$. A complete proof of this proposition is given in [184]. We will use $\mathbf{B} = \{\mathbf{b}_i, \quad i = 1, ..., 2(l+1)^2 - 2\}$ to denote individual elements of $\mathbf{B}_l$.

7.4 Isometric Correspondence and Registration

In this section, we restrict ourselves to the case where the surfaces S_1 and S_2, represented, respectively, with functions f_1 and f_2, are isometric, i.e. they are equivalent up to a rigid transformation and/or isometric motion. In other words, S_2 is a bended version of S_1. Although this may sound restrictive, there are many practical problems which fall in this category. For instance, the human body motion is nearly isometric, and thus this class of methods can be used to put in correspondence and register surfaces of the same 3D human body but in different poses. They can also be used to track and model the motion of 3D human bodies.

The main observation is that isometries are a subset of the Möbius group, which has a low-dimensionality. In other words, if the input surfaces are conformally parameterized then instead of optimizing Eq. (7.1) (and equivalently Eq. (7.2)) over the entire space of diffeomorphisms, one only needs to do it over the space of Möbius transformations. Moreover, the Möbius transformation that aligns one surface onto another can be computed in a closed-form if the correspondence between a few pairs of points (three in the case of spherically parameterized surfaces) is given. Thus, under the isometry assumption, the correspondence problem becomes a problem of finding a few pairs of points which are in correct correspondence. This can be solved using random sampling and voting, a well-known technique which has proven to be very effective in solving many computer vision and graphics problems.

7.4.1 Möbius Voting

Let us assume that the input surfaces S_1 and S_2, which undergo nearly isometric deformations, are conformally parameterized onto the unit sphere $\mathbb{S}^2$, or equivalently the extended complex plane $\mathbb{C}$ (see Section 3.3.2). Thus, they are represented with spherical functions $f_1, f_2 \in C_f$. From each surface f_i, $i \in \{0, 1\}$, we sample m landmarks. Let s_k^i, $k = 1, \ldots, m$ be the location of these landmarks on $\mathbb{S}^2$ and let z_k^i be their corresponding coordinates in the complex plane $\mathbb{C}$.

To find the Möbius transformation that maps f_2 onto f_1, one only needs to find three points on f_2 and their correspondences on f_1. One way of doing this is by computing local shape descriptors at each landmark point and use the similarity in the descriptor space for matching the landmarks across the surfaces (see Chapter 5 for examples of local descriptors). This is, however, not reliable since complex surfaces contain many self similar and symmetric parts. Another approach is to use a voting procedure in which we randomly sample three landmarks on one surface and three others on the other surface, compute the Möbius map induced by these three pairs, measure the matching error, and repeat the process for a large number of iterations. The triplet of pairs, which has the smallest error, can be used to compute the best Möbius map which aligns the two surfaces. Lipman et al. [185] used this idea but instead of voting for a Möbius transformation, they vote for landmark correspondences. The procedure can be summarized as follows:

- Sample one random triplet (s_1^1, s_2^1, s_3^1) from the landmarks of f_1 and another random triplet (s_1^2, s_2^2, s_3^2) from the landmarks of f_2.
- The sampled pair of triplets defines a Möbius transform ψ, which transforms s_k^2 into $\psi(s_k^2) = s_k^1, k \in \{1, 2, 3\}$. It also transforms each landmark on f_2 onto some points (not necessarily one of the sampled landmarks) on f_1 (see Section 7.3.1).
- Evaluate the quality of ψ by computing an alignment error. The inverse of the alignment error can be interpreted as the confidence of the k-th point on f_1 being in correspondence with the l-th point on f_2.
- This process is repeated a certain number of iterations and the correspondence confidences are accumulated into a fuzzy correspondence matrix $C = (C_{k,l})$. Here, $C_{k,l}$ is the confidence of having the k-th landmark on f_1 in correspondence with the l-th landmark on f_2.
- Finally, the fuzzy correspondence matrix C is processed to produce a discrete set of correspondences with high confidence. This step is a standard assignment problem, which can be efficiently solved using the Hungarian algorithm [186].

Let $\tilde{s}_k^2 = \psi(s_k^2)$ and let s_l^1 be the closest point to $\tilde{s}_k^2$ in terms of arc-length distance on $\mathbb{S}^2$. The alignment error ξ_{kl} can be measured as the length of the shortest arc on $\mathbb{S}^2$ which connects $\tilde{s}_k^2$ to s_l^1. For each Möbius transformation, the confidence matrix is updated by adding the quantity $(\epsilon + \xi_{kl}/n)^{-1}$ to its (k, l)-th entry $C_{k,l}$. Here, n is the number of computed pairs of correspondences, and ϵ is a small number.

In theory, this procedure is guaranteed to find a correct solution if (i) the surfaces differ only by an isometric transformation, (ii) a large number of

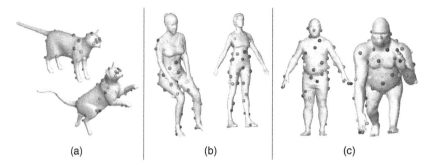

(a)	(b)	(c)

Figure 7.3 Examples of correspondences computed using the Möbius voting approach. (a) 3D shapes undergoing isometric motion, (b) 3D shapes undergoing nearly isometric motion, and (c) 3D shapes undergoing large elastic deformations. Source: The figure is adapted from Reference [187] ©2009 ACM, Inc. Reprinted by permission. http://doi.acm.org/10.1145/1531326.1531378.

landmarks are used, and (iii) a sufficiently large number of iterations is used. For more details about this procedure, we refer the reader to [187]. In practice, the algorithm is able to find a sparse set of correspondences even under large elastic motion, see Figure 7.3.

Note that many other papers extended this approach to relax the isometry condition. For example, Kim et al. [188] observed that efficient methods exist to search for nearly isometric maps (e.g. the Möbius Voting method described above), but no single solution found with these methods provides a low-distortion everywhere for pairs of surfaces differing by large deformations. As such, Kim et al. [188] suggested using a weighted combination of these maps to produce a *blended map*.

7.4.2 Examples

Figure 7.3a shows an example of matching the surfaces of a 3D model of the same cat but in two different poses. Despite the large isometric motion, the algorithm is able to match a few points on the two surfaces without using any manually specified landmarks to guide the search for correspondences. Figure 7.3b shows an example of two human body shapes, which are nearly isometric, i.e. they bend and slightly stretch. Figure 7.3c shows a result for the particularly hard case of matching a man to a gorilla. Even though the two surfaces are very far from being isometric, and no landmarks are used to guide the search for correspondences, the algorithm is able to correctly match many semantic features.

Figure 7.3 illustrates one of the limitations of this approach: it only finds correspondences between a few landmarks. In particular, the approach does not find many correspondences in highly extruded regions such as the tail of the cat (Figure 7.3a), the arms of the human body shapes (Figure 7.3b), and the legs of the human body shape and gorilla (Figure 7.3c). The main reason for that is the fact that the approach requires 3D surfaces that are conformally parameterized onto the sphere. Conformal parameterization, however, results in high distortions near elongated parts. In fact, the distortion increases exponentially with the size of the elongated parts. As such, these regions will be undersampled during the conformal parameterization process, and subsequently, any analysis method will be unable to match these parts.

7.5 Nonisometric (Elastic) Correspondence and Registration

In this section, we turn our attention to the general case where the 3D shapes being analyzed undergo complex elastic deformations, which include bending and stretching. Unlike bending, which is an isometric deformation, stretching does not preserve the intrinsic properties of surfaces. As such, solving the double optimization problem of Eq. (7.6) becomes very challenging since one has to simultaneously solve for the optimal rotation O, the optimal diffeomorphism γ, and the optimal deformation path. Before discussing how this is done, we first need to explicitly define the inner term of Eq. (7.6). In fact, this term measures the strength of the deformation one needs to apply to one surface in order to align it onto another. Section 7.5.1 will discuss different deformation models and how the optimal deformation path (i.e. registration) can be computed under fixed rotation and parameterization.

Section 7.5.2 describes one representation called square-root normal fields (SRNFs), which brings the complex elastic deformation models into a more simple and computationally tractable representation. Sections 7.5.3–7.5.6 discuss how elastic correspondence and registration are performed under the SRNF representation.

7.5.1 Surface Deformation Models

7.5.1.1 Linear Deformation Model

A surface f_2 can be optimally deformed to another surface f_1 by simply adding to each point $f_2(s)$ a displacement vector $\mathbf{d}(s)$ such that $f_1(s) = f_2(s) + \mathbf{d}(s)$. The optimal path F^* that connects f_1 to f_2 is then the straight line between the two points f_1 and $f_2 \in C_f$. Subsequently, the difference, or dissimilarity, between f_1 and f_2 is given by

$$d(f_1, f_2) = \int_{\mathbb{S}^2} \|\mathbf{d}\| ds = \int_{\mathbb{S}^2} \|f_1(s) - f_2(s)\| ds. \tag{7.12}$$

Figure 7.4 Interpolation of a straight cylinder bending to a curved one. (a) Linear interpolation in $\mathcal{F}$ $((1 - \tau)f_1 + tf_2)$, (b) Geodesic path by SRNF inversion.

(a)   (b)

Under this deformation model, the metric of Eq. (7.3) becomes the $\mathbb{L}^2$ metric, i.e. the standard inner product. This is equivalent to treating C_f as a Euclidean space, which is equipped with the $\mathbb{L}^2$ metric.

The Euclidean metric in C_f is easy to compute but has several limitations. First, it does not capture natural deformations. This is best illustrated with the example of Figure 7.4, which shows optimal deformation paths between a straight cylinder and a bent one. Figure 7.4a shows the linear path between the two surfaces, after full registration. The path is obtained using the $\mathbb{L}^2$ metric, i.e. by connecting each pair of corresponding points with a straight line. As one can see, the intermediate shapes along this path shrink unnaturally.

The second limitation of the $\mathbb{L}^2$ metric on C_f is that the action of the elements of Γ on C_f is not by isometry, i.e. in general, $\|f_1 - f_2\| \neq \|f_1 \circ \gamma - f_2 \circ \gamma\|$ for $\gamma \in \Gamma$ (unless γ is area-preserving). Thus, the $\mathbb{L}^2$ metric in C_f is not suitable for re-parameterization invariant shape analysis.

7.5.1.2 Elastic Deformation Models

Instead of using the $\mathbb{L}^2$ metric in C_f, one would like to explicitly capture the type of deformations that one needs to apply to f_2 in order to align it onto f_1. Such deformations can be of two types; bending and stretching. This requires redefining Eq. (7.3) in terms of an energy function that penalizes these deformations:

$$d_C^2(f_1, f_2) = \int_0^1 \{\alpha_s E_s(F(\tau)) + \alpha_b E_b(F(\tau))\} d\tau, \tag{7.13}$$

where E_s and E_b are, respectively, the stretching and bending energies. The parameters α_s and $\beta_s \in \mathbb{R}_{>0}$ control the contribution of each of the terms to the energy function. One can compose this deformation model by combining different bending and stretching models as defined previously in Section 2.2.2. For instance, Jermyn et al. [18] showed that by combining the bending model of Eq. (2.43) and the stretching model of Eq. (2.45), the elastic deformation model leads to what is called the *full elastic model*. It can be written as follows;

$$\langle\langle(\delta g, \delta \hat{n}), (\delta g, \delta \hat{n})\rangle\rangle = \int_{\mathbb{S}^2} ds \sqrt{|g|} \{\alpha \text{trace}((g^{-1}\delta g)^2) +$$
$$\beta \text{trace}(g^{-1}\delta g)^2 + \mu\langle\delta\hat{n}(s), \delta\hat{n}(s)\rangle\}. \tag{7.14}$$

Here, g is the first fundamental form (or the metric), $\hat{n}$ is the unit normal vector, δg and $\delta \hat{n}$ are perturbations of these quantities, and $\langle \cdot, \cdot \rangle$ is the standard inner product. Equation (7.14) measures the strength of these perturbations. The first term accounts for changes in the shape of a local patch that preserve patch area. The second term quantifies changes in the area of the patches. Thus, the first two terms capture the full stretch. The last term measures the changes in the direction of the surface normal vectors and thus captures bending. The parameters α, β, and μ are positive weights, which one can use to penalize the different types of deformations, e.g. if one wants to favor stretching over bending then the third term should be given a higher weight.

Next, we describe one simplification of this elastic deformation model, discuss its properties, and show how it can be efficiently implemented.

7.5.2 Square-Root Normal Fields (SRNF) Representation

Jermyn et al. [18] defined a new representation of surfaces called the *SRNFs*, which are essentially surface normals scaled by the square root of the local surface area. Mathematically, the SRNF $Q(f)$ of a surface f is defined at each point $s = (u, v) \in \mathbb{S}^2$ as follows:

$$Q(f)(s) = q(s) = \frac{n(s)}{\sqrt{|n(s)|}}, \tag{7.15}$$

where $n(s) = \frac{\partial f}{\partial u} \times \frac{\partial f}{\partial v}$. The space of SRNFs, hereinafter denoted by C_q, has very nice properties that are relevant to shape analysis. In particular, Jermyn et al. [18] showed that the $\mathbb{L}^2$ metric in C_q is a special case of the full elastic metric of Eq. (7.14). It corresponds to the case where $\alpha = 0$, $\beta = \frac{1}{4}$, and $\mu = 1$:

$$\langle \delta q, \delta q \rangle = \int_{\mathbb{S}^2} \sqrt{|g|} \left\{ \frac{1}{4} \text{trace}(g^{-1}\delta g)^2 + \langle \delta \hat{n}(s), \delta \hat{n}(s) \rangle \right\} ds, \tag{7.16}$$

where $\langle \cdot, \cdot \rangle$ is the standard inner product. Since $\sqrt{|g|}$ is just the norm of the normal vector, i.e. $\sqrt{|g|} = |n|$, Eq. (7.16) is equivalent to

$$\langle \delta q, \delta q \rangle = \frac{1}{4} \int_{\mathbb{S}^2} \left\{ \frac{\delta a(s)\delta a(s)}{a(s)} + \langle \delta \hat{n}(s), \delta \hat{n}(s) \rangle a(s) \right\} ds. \tag{7.17}$$

This metric is called the *partial elastic metric* since it only quantifies two components of the elastic deformations, namely, the changes in the surface normals and the change in the local surface area.

The SRNF representation has many important properties which are suitable for 3D shape analysis. For instance:

- The $\mathbb{L}^2$ metric in the space of SRNFs is equivalent to a weighted sum of the change in the surface area and in the surface bending. This has many positive implications on elastic shape analysis. In fact, computational tools in Euclidean spaces are not only straightforward but they are computationally very efficient. For example, shortest paths between two points are just straight lines. Means and modes of variations can be easily computed using standard PCA. Thus, one can perform all of these tasks in the space of SRNFs, C_q, under the $\mathbb{L}^2$ metric, and then map back the results to the space of surfaces $\mathcal{F}$ or C_f for visualization. This brings large gains in computational efficiency compared to other metrics such as those defined on the space of thin shells [13].

- If a surface is reparameterized according to $f \rightarrow f \circ \gamma, \gamma \in \Gamma$, its corresponding SRNF map is given by $(q, \gamma) = (q \circ \gamma)\sqrt{J_\gamma}$, where J_γ is the determinant of the Jacobian of γ.
- The action of γ is by isometries under the $\mathbb{L}^2$ metric, i.e. $\|q_1 - q_2\| = \|(q_1, \gamma) - (q_2, \gamma)\|, \forall \gamma \in \Gamma$. This is not the case for the space of surfaces C_f when equipped with the $\mathbb{L}^2$ metric [18, 184, 189–192].

The SRNF representation has been used for optimal re-parameterization, and thus registration, of spherical and quadrilateral surfaces [18, 184, 190, 193] which undergo isometric as well as elastic deformations, and surfaces which contain missing parts. They have been also used to compute geodesic paths between such surfaces [192, 193]. The strength of this representation is that shape comparison, registration, and geodesic computation can be performed under the same Riemannian elastic metric.

The main limitation of the SRNF representation is the fact that there is no analytical expression for Q^{-1} (the inverse SRNF map) for arbitrary points in C_q. In other words, if we are given a point $q \in C_q$, one cannot compute analytically a surface f such that $Q(f) = q$. Moreover, the injectivity and surjectivity of Q remain to be determined, meaning that for a given $q \in C_q$, there may be no $f \in C_f$ such that $Q(f) = q$, and if such an f does exist, it may not be unique. If one cannot invert the representation, one can always pull the $\mathbb{L}^2$ metric back to C_f under Q and perform computations there, as in [194]. This is, however, computationally expensive, and rather defeats the purpose of having an $\mathbb{L}^2$ metric in the first place. Fortunately, there are a few numerical strategies that one can use to estimate the best f whose SRNF map $Q(f)$ is as close as possible to a given SRNF map q. These will be discussed in Section 7.5.3. This means that one can use the standard $\mathbb{L}^2$ metric in the space of SRNFs for the analysis of shapes instead of the complex and expensive pullback elastic metrics.

7.5.3 Numerical Inversion of SRNF Maps

We describe in this section a numerical procedure, which finds for a given $q \in C_q$, a surface $f \in C_f$ such that $Q(f)$ is as close as possible to q. This can be formulated as an optimization problem. Let us first define an energy function $E_0 : \mathcal{F} \rightarrow \mathbb{R}_{\geq 0}$ by

$$E_0(f; q) = \inf_{(O, \gamma) \in SO(3) \times \Gamma} \|Q(f) - O(q, \gamma)\|_2^2, \tag{7.18}$$

where $(q, \gamma) = (q \circ \gamma)\sqrt{J_\gamma}$, and J_γ is the determinant of the Jacobian of γ. Finding an $f \in \mathcal{F}$ such that $Q(f) = q$ is then equivalent to seeking zeros of E_0. We denote by f^* an element of $\mathcal{F}$ satisfying $E_0(f^*; q) = \inf_{f \in \mathcal{F}} E_0(f; q)$. From a computational point of view, it will be easier to deal with deformations of a surface, rather than the surface itself. We, therefore, set $f = f_0 + w$, where f_0 denotes the current estimate of f^*, and w is a deformation of f_0. Then, we minimize

$$E(w; q) = \inf_{(O,\gamma) \in SO(3) \times \Gamma} \|Q(f_0 + w) - O(q, \gamma)\|_2^2, \tag{7.19}$$

with respect to w. One can view f_0 as an initial guess of the solution or a known surface with a shape similar to the one being estimated. If no initial guess is possible, one can initialize f_0 to a unit sphere. Equation (7.19) can then be solved by iterating over the following two steps:

- Given f_0 and w, find a rotation O and a diffeomorphism γ such that the energy E is minimized.
- Given a fixed O and γ, optimize E with respect to w using gradient descent.

The first step finds the best rotation O using singular value decomposition (SVD), given a fixed γ. Then, it finds the best γ by a search over the space of all diffeomorphisms using a gradient descent approach. In practice, restricting the search over Γ to the space of rigid re-parameterizations, $\Gamma_0 \equiv SO(3)$, is sufficient to achieve a good reconstruction accuracy, and is also computationally very efficient since it can be solved using the Procrustes method.

To minimize E with respect to w using a gradient descent approach, we need the derivative of E. Since $\mathcal{F}$ is an infinite-dimensional vector space, we will approximate the derivative using a finite orthonomal basis **B** for $\mathcal{F}$ and set $w = \sum_{\mathbf{b} \in \mathbf{B}} c_b \mathbf{b}$, with $c_b \in \mathbb{R}$. As discussed in Example 7.4, these bases can be generic, by using SHs, or domain specific (e.g. PCA basis) if training data are available. One can also combine the PCA and SH bases in order to account for the shape deformations which have not been observed in the training dataset.

The directional derivative of E at $f_0 + w$ in the direction of b, $c_b \equiv \nabla_b E(w; q)$, is given by:

$$\nabla_b E(w; q, f_0) = \frac{d}{d\epsilon}\big|_{\epsilon=0} \|Q(f_0 + w + \epsilon\mathbf{b}) - q\|_2^2$$

$$= 2\langle Q(f_0 + w) - q, Q_{*, f_0+w}(\mathbf{b})\rangle. \tag{7.20}$$

(Note that to simplify the notation, we have dropped O and γ.) Here $Q_{*,f}$ denotes the differential of Q at f. It is defined as:

$$Q_{*,f}(\mathbf{b})(s) = \frac{n_b(s)}{\sqrt{|n(s)|}} - \frac{n(s) \cdot n_b(s)}{2|n(s)|^{5/2}} n(s), \tag{7.21}$$

where $n_b(s) = (f_u(s) \times \mathbf{b}_v(s)) + (\mathbf{b}_u(s) \times f_v(s))$. Finally, the update is determined by the gradient

$$\nabla E(f_0; q) = \sum_{b \in B} (\nabla_b E(\mathbf{b}; q, f_0)) \, \mathbf{b}. \tag{7.22}$$

A naive implementation of this procedure results in the algorithm becoming trapped in local minima, see Figure 7.6b. This, however, can be efficiently overcome using a multiscale and a multiresolution representation of the elements of $\mathcal{F}$ and C_q.

7.5.3.1 SRNF Inversion Algorithm

The gradient descent method at a fixed resolution finds it difficult to reconstruct complex surfaces which contain elongated parts and highly nonconvex regions, such as the octopus surface shown in Figure 7.6. Such parts, however, correspond to the high frequency components of the surface, which can be easily captured using a multiscale analysis. More specifically, we redefine the surface representation as follows; Each surface S will be represented with a multiscale and multiresolution function $\mathbf{f} = (f^1, \ldots, f^m = f)$ where the surface $f^i \in C_f$ at level i is a smoothed, sub-sampled, and thus coarse version of f^{i+1}. With a slight abuse of notation, we also define the multiresolution SRNF map sending $\mathbf{f} \mapsto Q(\mathbf{f}) = \mathbf{q} = (q^i)$, where $q^i = Q(f^i)$.

Figure 7.5 shows an example of a multiresolution surface obtained using spherical wavelet decomposition [193]. Observe that at low resolutions, the surface is blobby and very similar to a sphere since all the extruding parts, which undergo complex deformations, disappear.

With this representation, the SRNF inversion problem can be reformulated as follows. Given a multiscale SRNF $\mathbf{q} = (q^1, \ldots, q^m)$, the goal is to find a multiscale surface $\mathbf{f} = (f^1, \ldots, f^m)$ such that $Q(f^i)$ is as close as possible to q^i, i.e. $E_0(f^i; q^i)$ is minimized. Gradient descent procedures are known to provide good results if the initialization is close to the solution. Under this setup, the following iterative procedure can be used to solve the optimization problem:

- **First**, find a surface $\tilde{f}^1$ that minimizes $E_0(f; q^1)$, initializing the gradient descent with a sphere f^0.

Figure 7.5 A spherical wavelet decomposition of a complex surface. Observe that at low resolutions, high frequency components disappear and the surface looks very similar to a sphere.

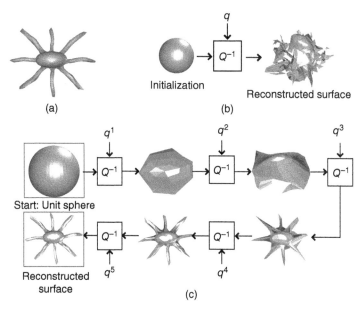

Figure 7.6 Comparison between the single resolution vs. multiresolution-based SRNF inversion procedure. (a) Ground truth surface, (b) SRNF inversion using single-resolution-based gradient descent, and (c) multiresolution SRNF inversion.

- **Second,** set $i \leftarrow i + 1$.
- **Third,** find the surface $\tilde{f}^i$ which minimizes $E_0(f; q^i)$, initializing the gradient descent with $\tilde{f}^{(i-1)}$.
- **Fourth,** repeat the second and third steps until $i > m$.

This procedure is illustrated in Figure 7.6. Figure 7.6a shows a ground truth surface. Figure 7.6b displays the inversion result when Eq. (7.19) is solved using a single-resolution representation. The result is a degenerate surface since the initialization is too far from the correct solution. The multiresoltion procedure is able to recover the surface as shown in Figure 7.6c. Note also that the multiresolution procedure significantly reduces the computational cost: although the optimizations at coarser scales require a large number of iterations, they operate on spherical grids of low resolution (e.g. 16×16) and are thus very fast. At the finest scale, although the resolution is high (128×128), the gradient descent requires only a few iterations since the initialization is already very close to the solution.

7.5.4 Correspondence

Now that we have reviewed various deformation models and laid down a representation which reduces the complex partial elastic metric to an $\mathbb{L}^2$ metric in the space of SRNFs, we turn our attention to the problem of computing one-to-one correspondences by solving the outer optimization of Eq. (7.6), or equivalently Eq. (7.1). The inner optimization of Eq. (7.6) will be dealt with in Section 7.5.5. We will then show in Section 7.5.6, how the same formulation can be used for the coregistration of multiple shapes. Let f_1 and f_2 be two elements in C_f and q_1 and q_2 their respective SRNFs. Let C_q be the image of C_f using the SRNF map. The correspondence problem of Eq. (7.1) can be reformulated as follows;

- Find the optimal rotation and reparameterization which align q_2 onto q_1. This is equivalent to solving the following optimization problem

$$(O^*, \gamma^*) = \underset{O \in SO(3), \gamma \in \Gamma}{\text{argmin}} \ \|q_1 - O(q_2, \gamma)\|, \tag{7.23}$$

  where $\| \cdot \|$ is the $\mathbb{L}^2$ norm.
- Apply O^* and γ^* to f_2, i.e. $f_2 \leftarrow O^* f_2 \circ \gamma^*$.

The optimization of Eq. (7.23) can be solved by iterating between the optimization over rotations, assuming a fixed parameterization, i.e. the correspondences are given, and optimization over diffeomorphisms assuming a fixed rotation.

7.5.4.1 Optimization Over SO(3)

If the correspondences are given, finding the best rotation O^* which aligns q_2 onto q_1 is straightforward. Under the $\mathbb{L}^2$ metric, the best rotation O^* can be analytically found using SVD. The procedure can be summarized as follows;

- Compute the 3×3 matrix $A = \int_{\mathbb{S}^2} q_1(s) q_2(s)^\top ds$.
- Compute the SVD decomposition of A such that $A = U\Sigma V^\top$.
- If the determinant of A is negative then the best transformation which aligns f_2 onto f_1 is a reflection. In that case, change the sign of $V^\top$.
- Finally, we can define the optimal rotation as $O^* = UV^\top$.

7.5.4.2 Optimization Over Γ

One option for optimizing over the space of diffeomorphisms Γ is by using a gradient descent approach, which, although converges to a local solution, is still general enough to be applicable to general surfaces. Gradient descent is an iterative algorithm, which starts with an initial guess of the solution and then iteratively updates it with a fraction of the gradient of the energy function being optized. Let us first focus on the current iteration, and define the reduced cost function $E_{reg} : \Gamma \to \mathbb{R}_{\geq 0}$:

$$E_{reg}(\gamma) = \|q_1 - (\tilde{q}_2, \gamma)\|^2 = \|q_1 - \phi(\gamma)\|^2, \tag{7.24}$$

where $\tilde{q}_2 = (q_2, \gamma_0)$ and $\phi(\gamma) = (\tilde{q}_2, \gamma)$. Here, γ_0 and γ denote, respectively, the current and the incremental reparameterizations. The gradient descent algorithm optimizes the reduced cost function by iterating over the following two steps:

- Update the current estimate of the incremental diffeomorphism γ as follows:

$$\gamma_{new} \leftarrow \gamma_{(id)} + \epsilon \frac{\partial E_{reg}}{\partial \gamma}. \tag{7.25}$$

- Reparameterize q_2 according to $q_2 \leftarrow (q_2, \gamma_{new})$.

Until $\|d\gamma\|$ is sufficiently small. The key step of the gradient descent is the computation of the gradient of the reduced cost function $dE_{reg} = \frac{\partial E_{reg}}{\partial \gamma}$. Note that dE_{reg} is an element of $T_{\gamma_{id}}(\Gamma)$ (see Section 7.3.1) since it corresponds to a perturbation of the current estimate of the optimal reparameterization. Thus, if one has an orthonormal basis $\mathbf{B} = \{\mathbf{b}_i\}$ for $T_{\gamma_{id}}(\Gamma)$, we can approximate the full gradient of E_{reg} with respect to γ by

$$d\gamma \simeq \sum_{\mathbf{b}_i \in \mathbf{B}_l} \langle q_1 - \tilde{q}_2, \phi_*(\mathbf{b}_i) \rangle_2 \mathbf{b}_i, \tag{7.26}$$

where ϕ_* is the differential of ϕ. The term inside the sum is the directional derivative of E_{reg} in the direction of the basis element $\mathbf{b}_i$. This linear combination of the orthonormal basis elements of $T_{\gamma_{id}}(\Gamma)$ provides the incremental reparameterization of $\tilde{q}_2$. This leaves three remaining issues: (i) the specification of an orthonormal basis of $T_{\gamma_{id}}(\Gamma)$, which is already covered in Example 7.5, (ii) an expression for ϕ_*, and (iii) the initialization of the gradient.

7.5.4.3 Differential of ϕ [184]

We need to determine the differential $\phi_*(\mathbf{b})$ for each element $\mathbf{b}$ of the orthonormal basis $\mathbf{B}_l$. Let $\tilde{q}_2 = (\tilde{q}_2^1, \tilde{q}_2^2, \tilde{q}_2^3)$, where each $\tilde{q}_2^j$ is a real-valued function on $\mathbb{S}^2$. Then, $\phi_*(\mathbf{b})$ also has three components $(\phi_*^1, \phi_*^2, \phi_*^3)$. Let ∇b be the divergence of $\mathbf{b}$ and $\nabla \tilde{q}_2^j$ the gradient of $\tilde{q}_2^j$. Since the Laplacian of a SH Y_l^m is simply $-l(l+1)$, then:

$$\phi_*^j \left(\frac{\nabla \psi_i}{\|\nabla \psi_i\|} \right) = \frac{1}{\|\nabla \psi_i\|} (-l_i(l_i + 1)\psi_i + (\nabla \tilde{q}_2^j \cdot \nabla \psi_i)), \tag{7.27}$$

and since the divergence of $* \Delta \psi_i$ is zero then:

$$\phi_*^j \left(\frac{* \nabla \psi_i}{\| * \nabla \psi_i\|} \right) = \frac{1}{\| * \nabla \psi_i\|} (\nabla \tilde{q}_2^j \cdot * \nabla \psi_i). \tag{7.28}$$

7.5.4.4 Initialization of the Gradient [184]

Gradient-based approaches require a good initialization to avoid the convergence to a local minima. A simple, but efficient in practice, approach is to start with an exhaustive search in order to bring q_2 as close as possible to q_1. The solution of the exhaustive search is then used to initialize the gradient descent procedure, which finds the accurate solution. Since Γ is large, we perform the exhaustive search over the subset of Γ which is formed by rigid rotations of $\mathbb{S}^2$.

To build this subset of the rotations of the sphere, Kurtek et al. [184] used the group of symmetries of the regular dodecahedron. A dodecahedron has 20 vertices, 12 flat facets (each facet has five vertices), and 30 edges. It has 60 rotational symmetries:

- Four (4) rotations (by multiples of $2\pi/5$) about the centres of six pairs of opposite faces. This makes a total of 24 rotational symmetries.
- One (1) rotation (by π) about the centres of the 15 pairs of opposite edges. This makes a total of 15 rotational symmetries.
- Two (2) rotations (by $\pm 2\pi/3$) about 10 pairs of opposite vertices. This makes a total of 20 rotational symmetries.

Together with the identity, this accounts for a total of 60 rotational symmetries. Let us denote these rotations by $\gamma_1, \gamma_2, \ldots, \gamma_{60}$ (recall that those are also elements of Γ). The exhaustive search is performed by solving for

$$i^* = \underset{i=1,\ldots,60}{\mathrm{argmin}} \| q_1 - (O_i^* q_2 \circ \gamma_i) \|^2. \tag{7.29}$$

Here, O_i^* is the optimal rotation of $q_2 \circ \gamma_i$ to best match q_1, obtained through Procrustes analysis, as described in Section 7.5.1.1.

Figure 7.7 shows examples of correspondences computed using this framework. In Figure 7.7a the source and target shapes differ only with isometric motion. In Figure 7.7b, the shape undergoes complex elastic motion, which

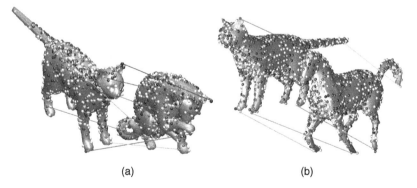

(a)   (b)

Figure 7.7 Examples of correspondence between complex 3D shapes which undergo (a) isometric (bending only) and (b) elastic deformations (bending and stretching). The correspondences have been computed using the SRNF framework, which finds a one-to-one mapping between the source and target surfaces. For clarity, only a few correspondences are shown in this figure.

includes bending and stretching. In both cases, the SRNF framework is able to find a one-to-one mapping between the input surfaces. Note that in Figure 7.7, for visualization clarity, we only illustrate a few correspondences.

7.5.5 Elastic Registration and Geodesics

Now we have all the ingredients to solve the inner optimization problem of Eq. (7.6), i.e. to find a smooth deformation path, or geodesic, $F^* : [0, 1] \to C_f$ which minimizes Eq. (7.3). Recall that when using the $\mathbb{L}^2$ on C_f, the solution is the straight line which connects f_1 to f_2. It can be expressed analytically as

$$F^*(\tau) = f_1 + \tau(f_2 - f_1), \text{with } \tau \in [0, 1]. \tag{7.30}$$

As shown in Figure 7.4, using the $\mathbb{L}^2$ metric in C_f leads to undesirable distortions even when the correspondences are correct. Instead, we take linear paths in the SRNF space and map them back to the space of surfaces using the SRNF inversion procedure described in Section 7.5.3. That is, given two surfaces f_1 and f_2, such that f_2 is in full correspondence with f_1, and their corresponding SRNF maps q_1 and q_2, one wants to construct a geodesic path $\alpha(t)$ such that $\alpha(0) = f_1$ and $\alpha(1) = f_2$. Let $\beta : [0, 1] \to C_q$ denote the straight line which connects q_1 and q_2. Then, for any arbitrary point $\beta(\tau), \tau \in [0, 1]$, we want to find a surface $\alpha(\tau) \in \mathcal{F}$ such that $\|Q(\alpha(\tau)) - \beta(\tau)\|$ is minimized. We will accomplish this using the following multiresolution analysis.

- Let $\mathbf{f}_1$ and $\mathbf{f}_2$ be the multiresolution representations of f_1 and f_2, respectively.
- We compute the multiresolution SRNFs of $\mathbf{f}_1$ and $\mathbf{f}_2$, which we denote by $\mathbf{q}_1$ and $\mathbf{q}_2$, respectively. Recall that $\mathbf{q}_i = (q_i^1, \dots, q_i^m)$ where $q_i^j = Q(f_i^j)$.
- To compute the point $\alpha(\tau)$ on the geodesic path α, we compute $\alpha^j(\tau)$, for $j = 1$ to m, by SRNF inversion of $\beta^j(\tau) = (1 - \tau)q_1^j + \tau q_2^j$, using $\alpha^{j-1}(\tau)$ as an initialization for the optimization procedure.
- The final solution $\alpha(\tau)$ is set to be $\alpha^m(\tau)$.

For the first iteration, one can set $\alpha^0(\tau)$ to be either a unit sphere or f_1^1.

Figure 7.8 shows three examples of geodesics computed with this procedure. In each example, we compute the one-to-one correspondences and the geodesic between the most left and the most right shapes. The first row shows a geodesic between two carpal bones, which usually do not bend a lot like the human shapes shown in the middle and last rows of Figure 7.8.

7.5.6 Coregistration

In the previous sections, we have considered the pairwise correspondence and registration problem. In other words, we are given two surfaces and seek to find

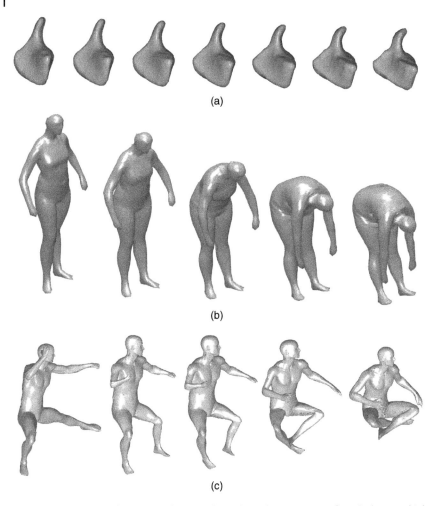

(a)

(b)

(c)

Figure 7.8 Examples of correspondence and geodesics between complex 3D shapes which undergo elastic deformations. In each row, the most left shape is the source and the most right one is the target. Both correspondences and geodesics are computed using the SRNF framework presented in this chapter. (a) A geodesic between two carpal bones which stretch. (b) A geodesic between two human body shapes, which bend and stretch. (c) A geodesic between two human body shapes with large isometric (bending) deformation.

the optimal rotation, diffeomorphism, and geodesic which align one surface onto the other. In shape analysis, there are many situations where one deals with a collection of multiple objects $\{f_i \in C_f, i = 1, \dots, n\}$. A typical example is the statistical analysis of 3D shapes where we seek to compute the mean shape and the modes of variations. In these situations, one needs to put in correspondence and align simultaneously all the shapes in the collection. We refer to this problem as *co-registration*.

One way to extend the correspondence and registration procedure described in the previous sections to the case of multiple objects is by selecting one reference object and aligning all the remaining ones to it. This, however, is suboptimal and the solution will often depend on the choice of the reference object. A more efficient solution is to compute the *mean* or *average* shape and align all the models to it.

Let us assume that the input objects have been normalized for translation and scale. Their mean shape can be defined as the object whose shape is as close (similar) as possible to all the shapes of the other objects in the collection. This is known as the *Karcher* mean. Mathematically, we seek to find a shape $\bar{f}$ and a set of optimal rotations O_i^* and diffeomorphisms $\gamma_i^*, i = 1, \dots, n$, which align every surface f_i onto the computed mean $\bar{f}$:

$$(\bar{f}, O_i^*, \gamma_i^*) = \underset{f, O_i, \gamma_i}{\operatorname{argmin}} \sum_{i=1}^{n} d_C^2(f, O_i f_i \circ \gamma_i), \tag{7.31}$$

where $d_C(\cdot, \cdot)$ is a certain measure of dissimilarity (or distance) that quantifies shape differences.

There is a generic algorithm that computes the Karcher mean under any arbitrary metric d_C, see for example [19]. When d_C is nonlinear (an example is the physically motivated elastic metrics presented in this chapter), the generic algorithm is computationally very expensive since it involves the computation of at least $n \times k$ geodesics, where n is the number of shapes in the collection and k is the number of iterations needed for the convergence of the algorithm. SRNF representations offer an elegant and computationally efficient alternative solution. Instead of using a complex elastic metric, one can simply map all the surfaces to the SRNF space, C_q, resulting in $\{q_1, \dots, q_n\}$. Then, the mean shape in the SRNF space, denoted by $\bar{q}$, is computed by iterating between coregistering all the q_is to $\bar{q}$, and subsequently updating $\bar{q}$ as $\bar{q} \leftarrow \frac{1}{n} \sum_{i=1}^{n} q_i$. Finally, at the last step, the mean shape $\bar{f}$ is computed by SRNF inversion, i.e. finding a shape $\bar{f}$ such that $Q(\bar{f}) = \bar{q}$. This is computationally very efficient since C_q is a Euclidean space. Algorithm 7.1 summarizes the entire procedure.

Figure 7.9 shows eight human body shapes (three males and five females). We use the approach presented in this section to compute their Karcher mean (the highlighted shape within the box) and simultaneously register them to the computed Karcher mean. Observe that despite the large elastic deformation, the approach is able to compute a plausible mean, which is an indication of the quality of the correspondences and registrations.

Algorithm 7.1 Sample Karcher mean by SRNF inversion.

Input: A set of surfaces $\{f_1, \ldots, f_n\} \in C_f$ and their SNRFs $\{q_1, \ldots, q_n\} \in C_q$.

Output: Karcher mean surface $\bar{f}$.

1: Let $\bar{q} = Q(\bar{f})$ with $\bar{f}$ set to f_1 as the initial estimate of the Karcher mean. Set $j = 0$.

2: For $i = 1, \ldots, n$, set $O_i = I_{3\times 3}$, the 3×3 identity matrix, and $\gamma_i = \gamma_{Id}$, the identity diffeomoprhism.

3: For each $i = 1, \ldots, n$, register q_i to $\bar{q}$ resulting in $q_i^* = O_i^*(q_i, \gamma_i^*)$, where O_i^* and γ_i^* are the optimal rotation and reparameterization respectively.

4: Set $q_i \leftarrow q_i^*$.

5: For $i = 1, \ldots, n$, set $O_i \leftarrow O_i^* \times O_i$ and $\gamma_i \leftarrow \gamma_i \circ \gamma_i^*$.

6: Update the average $\bar{q} \leftarrow \frac{1}{n} \sum_{i=1}^{n} q_i$.

7: If the change in $\|\bar{q}\|$, compared to the previous iteration, is large go to Step 3.

8: Find $\bar{f}$ by SRNF inversion such that $Q(\bar{f}) = \bar{q}$.

9: Return $\bar{f}$, O_i and γ_i, $i = 1, \ldots, n$.

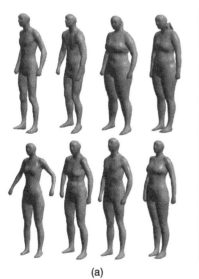

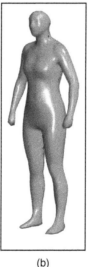

Figure 7.9 Example of coregistration: the input human body shapes are simultaneously coregistered to the Karcher mean computed using Algorithm 7.1. (a) Input 3D human body shapes. (b) Computed mean human body shape.

(a)　　　　　　(b)

7.6　Summary and Further Reading

We have presented in this chapter a set of methods that find correspondences between 3D objects that undergo elastic deformations, i.e. bending and stretching. We have focused on methods that formulate the problem of finding correspondences between a pair of 3D objects as the problem of optimally reparameterizing a source object so that its shape distance to the target object is minimized. We have seen that when the type of deformations is restricted to bending (also known as isometric motion), then the problem is reduced to the problem of finding a Möbius transform which aligns one surface onto the other. Since a Möbius transform is fully defined with three pairs of corresponding points, finding correspondences becomes the problem of finding such pairs of point triplets. This approach, which was originally introduced in [187], is relatively simple. It has, however, a few limitations. **First**, it requires the surfaces to be conformally parameterized. Complex surfaces, which contain many extruded parts (e.g. human bodies), cannot be conformally parameterized to a sphere without introducing large distortions. As such, the approach of Lipman and Funkhouser [187] finds only correspondences between a few points on the two surfaces. **Second**, the method also cannot handle surfaces with large stretch. Kim et al. [188] addressed this problem by first finding a set of consistent maps and then combining the maps using blending weights.

The second class of methods that we covered in this chapter do not put any restriction on the type of elastic deformations that the 3D objects undergo. The key to solving this general problem is an elastic metric that is reparameterization invariant. This metric quantifies both bending and stretching. Although the metric is complex, Jermyn et al. [18] showed that by carefully choosing the weights of the bending and stretching terms, the metric reduces to an $\mathbb{L}^2$ metric in the space of SRNF representations. This observation, along with the mechanism which maps elements in the space of SRNFs to 3D objects [193], reduces the problem of elastic shape analysis (e.g. correspondence, registration, and statistical summarization) into a much simpler problem, which can be solved using simple vector calculus in the Euclidean space of SRNFs.

The SRNF representation is a generalization of the square-root velocity function (SRVF) [195] proposed for the elastic analysis of the shape of planar curves, and used in many applications including biology and botany [196, 197]. It is also a special case of the family of square-root maps (SRMs). Another example of SRMs is the Q-maps introduced by Kurtek et al. [184, 189–192, 198, 199]. Unlike SRNF maps, Q-maps have no geometric or physical motivation. They have been solely designed to allow statistical analysis using the $\mathbb{L}^2$ metric.

Elastic 3D shape analysis has a wide range of applications in medicine [184, 191, 200], plant biology [197], and computer graphics [193, 198].

Finally, note that this chapter is not intended to provide an exhaustive survey of the rich literature of elastic 3D shape analysis. Readers can refer to the survey by Biasotti et al. [201] in the case of surfaces which undergo nearly isometric deformations, to the survey of Laga [202] in the case of 3D objects which undergo both bending and stretching, and to the book of Jermyn et al. [19] for an in-depth discussion of the Riemannian elastic metrics in 3D shape analysis.

8

Semantic Correspondences

8.1 Introduction

The approaches presented so far for shape correspondence are geometry driven, focusing solely on the geometrical and topological similarities between shapes. However, in many situations, corresponding shape parts may significantly differ in geometry or even in topology. Take the 3D models shown in Figure 8.1. Any human can easily put in correspondence and match parts of the chairs of Figure 8.1a despite the fact that they differ significantly in geometry and topology. For instance, some chairs have one leg while others have four. In these examples, correspondence is simply beyond pure geometric analysis; it requires understanding the semantics of the shapes in order to reveal relations between geometrically dissimilar yet functionally or semantically equivalent shape parts.

This fundamental problem of putting in correspondence 3D shapes that exhibit large geometrical and topological variations has been extensively studied in the literature, especially in the past five years [183, 203–205]. Beyond standard applications, such as blending and morphing, this problem is central to many modern 3D modeling tools. For instance, the modeling-by-example paradigm, introduced by Funkhouser et al. [206], enables the creation of new 3D models by pasting and editing 3D parts retrieved from collections of 3D models. Its various extensions [204, 207], which lead to smart 3D modeling techniques, create new 3D shape variations by swapping parts that are functionally equivalent but significantly different in geometry and topology.

In this chapter, we review some of the methods that incorporate prior knowledge in the process of finding semantic correspondences between 3D shapes. We will show how the problem can be formulated as the optimization of an energy function and compare the different variants of the formulation that have been proposed in the literature. We conclude the chapter with a discussion of the methods that are not covered in detail and outline some challenges to stimulate future research.

3D Shape Analysis: Fundamentals, Theory, and Applications, First Edition.
Hamid Laga, Yulan Guo, Hedi Tabia, Robert B. Fisher, and Mohammed Bennamoun.
© 2019 John Wiley & Sons, Inc. Published 2019 by John Wiley & Sons, Inc.

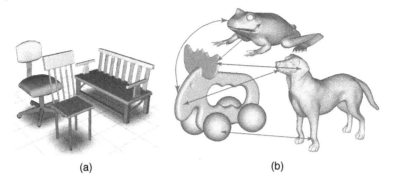

(a)   (b)

Figure 8.1 Partwise correspondences between 3D shapes in the presence of significant geometrical and topological variations. (a) Man-made 3D shapes that differ in geometry and topology. The partwise correspondences are color coded. (b) A set of 3D models with significant shape differences, yet they share many semantic correspondences.

8.2 Mathematical Formulation

Let M_s and M_t be two input 3D objects, hereinafter referred to as the source and target shapes, respectively. The goal of correspondence is to find for each point on M_s its corresponding point(s) on M_t, and for each point on M_t its corresponding point(s) on M_s. This is not a straightforward problem for several reasons. **First**, this correspondence is often not one-to-one due to differences in the structure and topology of shapes. Consider the example of Figure 8.1a where a desk chair has one central leg, a table chair has four legs, and the bench, which although has four legs, their geometry and structure are completely different from those of the table chair. Clearly, the one leg on the source chair has four corresponding legs on the target chair. **Second**, even when the source and target shapes have the same structure, matching parts can have completely different topologies (e.g. the backrests of the seats in Figure 8.1a), and geometries (e.g. the heads in the objects of Figure 8.1b). Thus, descriptors, which are only based on geometry and topology, are not sufficient to capture the similarities and differences between such shapes.

One approach, which has been extensively used in the literature, is to consider this correspondence problem as a semantic labeling process. More formally, we seek to assign labels to each point on M_s and M_t such that points that are semantically similar are assigned the same label l chosen from a predefined set of possible (semantic) labels. At the end of the labeling process, points, faces, patches, or parts from both objects that have the same label are then considered to be in correspondence. This labeling provides also a segmentation of the input objects. If performed jointly, this segmentation will be consistent across 3D objects belonging to the same shape family.

In this section, we lay down the general formulation of this semantic labeling process, which has been often formulated as a problem of optimizing an energy function. The remaining sections will discuss the details such as the choice of the terms of the energy function and the mechanisms for optimizing the energy function.

We treat a 3D mesh M as a graph $G = \{V, E\}$. Each node in V represents a surface patch. An edge $e \in E$ connects two nodes in V if their corresponding patches are spatially adjacent. In practice, a patch can be as simple as a single face, a set of adjacent faces grouped together, or a surface region obtained using some segmentation or mesh partitioning algorithms [208]. Let $V = \{v_1, \ldots, v_n\}$ and $E = \{e_{ij} = (v_i, v_j) \in V \times V\}$. We represent each node v_i using a descriptor $\mathbf{x}_i$, which characterizes the shape of the surface patch represented by v_i. We also characterize the relation between two adjacent nodes v_i and v_j using a descriptor $\mathbf{x}_{ij}$. Let $C = \{c_1, \ldots, c_n\}$ denote a labeling of the nodes of G where c_i is the label assigned to node $v_i \in V$. Each $c \in C$ takes its value from a predefined set of m (semantic) labels $\mathcal{L} = \{l_1, \ldots, l_m\}$. We seek to find the optimal labeling C^* of the graph G, and subsequently the mesh M. This can be formulated as a problem of minimizing, with respect to C, an objective function of the form:

$$E(C) = \sum_{i=1}^{n} U_d(c_i; \mathbf{x}_i) + \sum_{i,j} U_s(c_i, c_j; \mathbf{x}_{ij}). \tag{8.1}$$

This objective function is the sum of two terms; a data term U_d and a smoothness term U_s. The data term measures how likely it is that a given node v_i has a specific label c_i given its shape properties $\mathbf{x}_i$. The smoothness term U_s is used to impose some constraints between the labels c_i and c_j of the nodes v_i and v_j, given the properties $\mathbf{x}_{ij}$. It is also used to incorporate some prior knowledge in order to guide the labeling. Table 8.1 summarizes the different choices of these terms and the different combinations that have been used in the literature.

Now, solving the optimization problem of Eq. (8.1), for a given shape M, will provide a segmentation and a labeling of M. Often, however, we will be dealing

Table 8.1 Potential functions used in the literature for semantic labeling.

References	Data term $U_d(c_i; x_i)$	Smoothness term $U_s(c_i, c_j; x_{ij})$
Kalogerakis et al. [209]	$-a_i \ \log(P(c_i = l \vert x_i)$	$L(c_i, c_j)\Big[-\mu \ \log(P(c_i \neq c_j \vert x_{ij}) + \eta)$ $-\lambda \ \log\left(1 - \min\left(\frac{\alpha_{ij}}{\pi}, 1\right) + \epsilon\right)\Big]$
van Kaick et al. [203]	$-a_i \ \log(P(c_i = l \vert x_i)$	$L(c_i, c_j)[\mu d_{ij} + \lambda \alpha_{ij}]$ with $x_{ij} = [d_{ij}, \alpha_{ij}]$.
Sidi et al. [210]	$-a_i \ \log(P(c_i = l \vert x_i)$	0 if $c_i = c_j$, and $-d_{ij} \ \log \frac{\alpha_{ij}}{\pi}$ if $c_i \neq c_j$. with $x_{ij} = [d_{ij}, \alpha_{ij}]$.

$L(l_i, l_j)$ is the label compatibility term, α_{ij} is the dihedral angle between the nodes v_i and v_j, d_{ij} is the distance between v_i and v_j, a_i is the area of the surface patch corresponding to the ith node, and ϵ, η, λ, and μ are some constants.

with multiple shapes and we would like them to be labeled and segmented in a consistent way. Thus, instead of labeling, say a pair of, 3D models separately from each other, one can do it simultaneously in a process called *joint labeling*. The goal is to extract parts from the two (or more) input shapes while simultaneously considering their correspondence.

Let M_s and M_t be the source and the target shapes. We represent M_s with a graph $G_S = \{V_S, E_S\}$ where the nodes V_S are surface patches and the edges in E_S connect pairs of adjacent patches. A similar graph $G_T = \{V_T, E_T\}$ is also defined for the target shape M_t. To perform joint semantic labeling, we define a new graph $G = \{V, E\}$ where $V = V_S \cup V_T$ and the connectivity E of the graph is given by two types of arcs, i.e. $E = E_{intra} \cup E_{inter}$, where

- The intramesh arcs E_{intra} are simply given by $E_{intra} = E_S \cup E_T$.
- The intermesh arcs E_{inter} connect nodes in G_S to nodes in G_T based on the similarity of their shape descriptors.

Let also $C = C_S \cup C_T$ where C_S is a labeling of G_S, and C_T is a labeling of G_T. With this setup, the joint semantic labeling process can be written as the problem of optimizing an energy function of the form:

$$E(C) = \sum_{v_i \in V} U_d(c_i; \mathbf{x}_i) + \sum_{e_{ij} \in E_{intra}} U_s(c_i, c_j; \mathbf{x}_{ij}) + \sum_{e_{ij} \in E_{inter}} U_{inter}(v_i, v_j, c_i, c_j). \quad (8.2)$$

The first two terms are the same as the data term and the smoothness term of Eq. (8.1). The third term is the intermesh term. It quantifies how likely it is that two nodes across the two meshes have a specific pair of labels, according to a pairwise cost.

Implementing the labeling models of Eqs. (8.1) and (8.2) requires:

- Constructing the graph G and characterizing, using some unary and binary descriptors, the geometry of each of (i) its nodes, which represent the surface patches, and (ii) its edges, which represent the spatial relations between adjacent patches (Section 8.3).
- Defining the terms of the energy functions and setting their parameters (Section 8.4).
- Solving the optimization problem using some numerical algorithms (Section 8.5).

The subsequent sections describe these aspects in detail. In what follows, we consider that the input shapes have been normalized for translation and scale using any of the normalization methods described in Chapter 2.

8.3   Graph Representation

There are several ways of constructing the intramesh graph G for a given input shape M. The simplest way is to treat each face of M as a node. Two nodes will be connected with an edge if their corresponding faces on M are adjacent. When dealing with large meshes, this method produces large graphs and thus the subsequent steps of the analysis will be computationally very expensive. Another solution is to follow the same concept as superpixels for image analysis [211]. For instance, one can decompose a 3D shape into patches. Each patch will be represented with a node in G. Two nodes will be connected with an edge if their corresponding patches are spatially adjacent on M. Figure 8.2 shows an example of a 3D object decomposed into disjoint patches.

When jointly labeling a pair of 3D models M_s and M_t, one can incorporate additional intermesh edges to link nodes in the graph of M_s to nodes in the graph of M_t. This will require pairing, or putting in correspondence, patches across M_s and M_t.

8.3.1   Characterizing the Local Geometry and the Spatial Relations

The next step is to compute a set of unary descriptors $\mathbf{x}_i$, which characterize the local geometry of the graph nodes v_i, and a set of pairwise descriptors $\mathbf{x}_{ij}$ that characterize the spatial relation between adjacent nodes v_i and v_j.

Figure 8.2 An example of a surface represented as a graph of nodes and edges. First, the surface is decomposed into patches. Each patch is represented with a node. Adjacent patches are connected with an edge. The geometry of each node v_i is represented with a unary descriptor $\mathbf{x}_i$. The geometric relation between two adjacent nodes v_i and v_j is represented with a binary descriptor $\mathbf{x}_{ij}$.

8.3.1.1 Unary Descriptors

The descriptors $\mathbf{x}_i$ at each node $v_i \in V$ characterize the local geometry of the patch represented by v_i. They provide cues for patch labeling. In principle, one can use any of the local descriptors described in Chapter 5. Examples include:

- **The shape diameter function (SDF)** [212], which gives an estimate of the thickness of the shape at the center of the patch.
- **The area of the patch** normalized by the total shape area.
- **The overall geometry of the patch**, which is defined as:

$$\mu_l = \frac{\lambda_1 - \lambda_2}{\lambda_1 + \lambda_2 + \lambda_3}, \quad \mu_p = \frac{2(\lambda_2 - \lambda_3)}{\lambda_1 + \lambda_2 + \lambda_3}, \quad \mu_s = \frac{3\lambda_3}{\lambda_1 + \lambda_2 + \lambda_3},$$

where $\lambda_1 \geq \lambda_2 \geq \lambda_3 \geq 0$ are the three eigenvalues obtained when applying principal component analysis to all the points on the surface of the patch.

Other descriptors that can be incorporated include the principal curvatures within the patch, the average of the geodesic distances [213], the shape contexts [135], and the spin images [95].

8.3.1.2 Binary Descriptors

In addition to the unary descriptors, one can define for each pair of adjacent patches v_i and v_j, a vector of pairwise features $\mathbf{x}_{ij}$, that provides cues to whether adjacent nodes should have the same label [209]. This is used to model the fact that two adjacent patches on the surface of a 3D object are likely to have similar labels. Examples of such pairwise features include:

- **The dihedral angles** α_{ij}, i.e. the angle between the normal vectors of adjacent patches. The normal vectors can be computed at the center of each patch or as the average of the normal vectors of the patch faces.
- **The distance** d_{ij} between the centers of the patches v_i and v_j. This distance can be Euclidean or geodesic.

These pairwise descriptors are computed for each pair of adjacent nodes v_i and v_j that are connected with an edge e_{ij}.

8.3.2 Cross Mesh Pairing of Patches

The intermesh edges E_{inter}, which connect a few nodes in V_S to nodes in V_T, can be obtained in different ways. For instance, they can be manually provided, e.g. in the form of constraints or prior knowledge. They can be also determined by pairing nodes that have similar geometric descriptors. This is, however, prone to errors since a 3D shape may have several parts that are semantically different but geometrically very similar (e.g. the legs and the tail of a quadruple animal). Since the purpose of these intermesh edges is to guide the joint labeling to ensure consistency, only a few reliable intermesh edges is usually sufficient. These can be efficiently computed in a supervised manner if training data are available.

Let $G_S = (V_S, E_S)$ and $G_T = (V_T, E_T)$ be the graphs of two training shapes M_s and M_t. Let also $\mathcal{T}$ be a set of matching pairs of nodes (v_i^s, v_j^t), where $v_i^s \in V_S$ and $v_j^t \in V_T$. We assign to each pair of nodes (v_i^s, v_j^t) in $\mathcal{T}$ a label 1 if the pairing is correct, i.e. the node v_j^t is a true correspondence to the node v_i^s, and 0 otherwise. The set of nodes in $\mathcal{T}$ and their associated labels can then be used as training samples for learning a binary classifier, which takes a pair of nodes and assigns them the label 1 if the pair is a correct correspondence, and the label 0 otherwise. This is a standard binary classification problem, which, in principle, can be solved using any type of supervised learning algorithm, e.g. Support Vector Machines [214] or AdaBoost [215, 216]. van Kaick et al. [203], for example, used GentleBoost [215, 216] to learn this classifier. GentleBoost, just like AdaBoost, is an ensemble classifier, which combines the output of weak classifiers to build a strong classifier. Its advantage is the fact that one can use a large set of descriptor types to characterize the geometry of each node. The algorithm learns from the training examples the descriptors that provide the best performance. It also assigns a confidence, or a weight, to each of the selected descriptors.

The main difficulty with a training-based approach is in building the list of pairs $\mathcal{T}$ and their associated labels. Since these are used only during the training stage, one could do it manually. Alternatively, van Kaick et al. [203] suggested an automatic approach, which operates as follows:

- **First**, for a given pair of shapes represented with their graphs G_S and G_T, collect all their nodes and compute m types of geometric descriptors on each of their nodes. These can be any of the unary descriptors described in Section 8.3.1.
- **Next**, for each node in V_S, select the first k pairwise assignments with the highest similarity and add them to $\mathcal{T}$. The similarity is given by the inverse of the Euclidean distance between the descriptors of the two nodes.
- **Finally**, build a matrix A of m rows and $|V_S| \times |V_T|$ columns where each entry $A(i, j)$ stores the dissimilarity between the jth pair of nodes in $V_S \times V_T$ but computed using the ith descriptor type.

The matrix A will be used as input to the GentleBoost-based training procedure, which will learn a strong classifier as a linear combination of weak classifiers, each one is based on one descriptor type. The outcome of the training is a classifier that can label a candidate assignment with true or false. At run time, given a pair of query shapes, select candidate pairs of nodes with the same procedure used during the training stage. Then, using the learned classifier, classify each of these pairs into either a valid or invalid intermesh edge. This procedure produces a small, yet reliable, set of intermesh edges.

8.4   Energy Functions for Semantic Labeling

The energy to be minimized by the labeling process, i.e. Eqs. (8.1) and (8.2), is composed of two types of terms: the unary (or data) term and the pairwise (or smoothness) term. The data term takes into consideration how likely it is that a given node has a specific label. The smoothness term considers the intramesh (Eq. (8.1)) and intermesh (Eq. (8.2)) connectivity between nodes. It quantifies, according to a pairwise cost, how likely it is that two neighboring nodes (i.e. two nodes which are connected with an edge) have a specific pair of labels. Below, we discuss these two terms in more detail. We also refer the reader to Table 8.1.

8.4.1   The Data Term

The data term is defined as the negative log likelihood that a node v_i is part of a segment labeled $c_i = l$:

$$U_d(c_i = l; \mathbf{x}_i) = -a_i \, \log(P(c_i = l | \mathbf{x}_i)), \tag{8.3}$$

where a_i is the area of the surface patch corresponding to the ith node. This likelihood can be also interpreted as a classifier, which takes as input the feature vector $\mathbf{x}_i$ of a node v_i, and returns $P(c_i | \mathbf{x}_i)$, the probability distribution of labels for that node. This classifier, which provides a probabilistic labeling of the nodes on the query shapes, can be learned either in a supervised or in a nonsupervised manner, see Section 8.5. It provides a good initial labeling since it captures very well the part interiors but not the part boundaries. This initial labeling can be also interpreted as an initial segmentation where the set of nodes with the same label form a segment.

8.4.2   Smoothness Terms

To refine the result of the probabilistic labeling of Eq. (8.3), one can add other terms to the energy function to impose additional constraints and to incorporate some a-priori knowledge about the desired labeling. The main role of these terms is to improve the quality of the boundaries between the segments and to prevent incompatible segments from being adjacent. Examples of such constraints include:

8.4.2.1   Smoothness Constraints
In general, adjacent nodes in a 3D model are likely to belong to the same segment. Thus, they should be assigned the same label. The smoothness can then have the following form:

$$U_s(c_i, c_j; \mathbf{x}_{ij}) = 0 \text{ if } c_i = c_j, \quad \text{and } C_{ij} \text{ otherwise.} \tag{8.4}$$

Here, C_{ij} is a positive constant, which can be interpreted as the cost of having two adjacent nodes labeled differently. For instance, two adjacent nodes are likely to have the same label if their dihedral angle is close to zero, i.e. the patches are coplanar. They are likely to have different labels if they are far apart (large d_{ij}) or their corresponding dihedral angle is large. Using these observations, Sidi et al. [210] define C_{ij} as:

$$C_{ij} = -d_{ij} \, \log \frac{\alpha_{ij}}{\pi}. \tag{8.5}$$

Here, $\mathbf{x}_{ij} = [d_{ij}, \alpha_{ij}]$ where d_{ij} is the distance between the centers of the patches v_i and v_j, and $\alpha_{ij} \in [0, \pi)$, also known as the dihedral angle, is the angle between the normals to the two patches at their centers.

8.4.2.2 Geometric Compatibility

In addition to the smoothness constraints, one can incorporate a geometry-dependent term G, which measures the likelihood of there being a difference in the labels between two adjacent nodes as a function of their geometry. For instance, one can penalize boundaries in flat regions and favor boundaries between the nodes that have a high dihedral angle (i.e. concave regions). One can also use it to penalize the boundary length, which will help in preventing jaggy boundaries and in removing small, isolated segments. van Kaick et al. [203] define this term as:

$$G(\mathbf{x}_{ij}) = \mu d_{ij} + \lambda \alpha_{ij}, \tag{8.6}$$

where λ and μ are some real weights that control the importance of each of the two geometric properties (i.e. d_{ij}, which is the distance between the centers of the patches v_i and v_j, and the dihedral angle α_{ij}).

Kalogerakis et al. [209] used a more complex term, which is defined as follows;

$$G(\mathbf{x}_{ij}) = -\mu \, \log(P(c_i \neq c_j | \mathbf{x}_{ij}) + \eta) - \lambda \, \log \left(1 - \min \left(\frac{\alpha_{ij}}{\pi}, 1 \right) + \epsilon \right). \tag{8.7}$$

The first term, $P(c_i \neq c_j | \mathbf{x}_{ij})$, measures the probability of two adjacent nodes having distinct labels as a function of their pairwise geometric features $\mathbf{x}_{ij}$. These consist of the dihedral angles between the nodes v_i and v_j, and the differences of the various local features computed at the nodes, e.g. the differences of the surface curvatures and the surface third-derivatives, the shape diameter differences, and the differences of distances from the medial surface points. The log term of Eq. (8.7) penalizes boundaries between the faces with a high exterior dihedral angle α_{ij}, following Shapira et al. [217]. ϵ and η are small constants, which are set to 0.001. They are added to avoid computing $\log 0$ and to avoid numerical instabilities [217, 218].

Note that there are similarities between Eqs. (8.6) and (8.7). For instance, the second term of both equations is defined in terms of the dihedral angle α_{ij}. For the first term, although in both equations it is defined using the distance d_{ij},

Eq. (8.7) uses, in addition to the dihedral angle, more pairwise features and is defined as the log probability of two adjacent nodes having different labels given their pairwise geometric features. It is more general since any type of probability can be used. However, these probabilities should be learned, in a supervised manner, from training data. This will be discussed in Section 8.5.

8.4.2.3 Label Compatibility

Finally, one can impose penalties to some combinations of labeling. For instance, when labeling human body shapes, one can never find two adjacent nodes that are labeled head and torso, respectively. This label-compatibility constraint can be represented as an $m \times m$ symmetric matrix L of penalties for each possible pair of labels (m here is the number of possible semantic labels). Note that, $L(l_i, l_i) = 0$ for all i since there is no penalty when there is no discontinuity. The entries of the matrix L can be either manually specified or learned in a supervised manner from training data, see Section 8.5.

8.4.3 The Intermesh Term

Finally, the intermesh term U_{inter} is given by

$$U_{\text{inter}}(v_i, v_j, c_i, c_j) = vL(c_i, c_j)\sigma_{ij}, \tag{8.8}$$

where v_i is a node on the graph of the source mesh, v_j is a node on the graph of the target mesh, and L_{ij} is the label compatibility defined in Section 8.4.2. σ_{ij} is the confidence that the assignment between nodes v_i and v_j is correct. It is computed using the procedure described in Section 8.3.2. The higher the confidence value attached to the assignment is, the more the cost is increased if the labels are different. The parameter v regulates the influence of the intermesh term to the total energy.

8.5 Semantic Labeling

Before we discuss the algorithms that can be used to optimize the energy function of Eqs. (8.1) and (8.2), we will first discuss how to set the different terms. In particular, both the data and smoothness terms of the labeling energy require the probability that a node v_i is part of a segment labeled $c_i = l$ (Eq. (8.3)), and the probability that two adjacent nodes have distinct labels (Eq. (8.7)). These two distributions can be learned if some training data is available.

Let us assume that for each semantic label l we have n_l training nodes coming from different models. By computing the descriptors of each node, we obtain a collection of descriptors, $D_l = \{x_k^l, k = 1, \ldots n_l\}$, for each label l. These training nodes can be obtained either manually, by labeling some training samples, or automatically, in a nonsupervised manner, by using some unsupervised clustering techniques. We will first look at the latter case, i.e. how to build the training data (i.e. a set of nodes and their corresponding labels) using unsupervised clustering. We will then detail how the probability distributions of Eqs. (8.3) and (8.7) are learned independently of how the training data are obtained.

8.5.1 Unsupervised Clustering

One approach to automatically generate training samples for learning the conditional probability of Eq. (8.3) is by unsupervised clustering in the descriptor space. Let us start with a collection of 3D shapes, each shape is represented with their graphs of nodes. **The first** step is to pre-segment, in a nonsupervised manner, each shape by clustering its nodes based on their distance in the descriptor space. **Second**, all the segments of all the shapes are put together into a single bag. They are then clustered into groups of segments based on their geometric similarities. Each cluster obtained with this procedure corresponds to one semantic component (e.g. legs, handles, etc.) for which we assign a semantic label. **Finally**, since each cluster contains multiple instances of the semantic class with rich geometric variability, one can learn the statistics of that class. In particular, one can learn $P(c = l|\mathbf{x})$. This process is summarized in Figure 8.3. Below, we describe in detail each of these steps.

Step 1 – Presegmentation of individual shapes. After computing a shape descriptor $\mathbf{x}_i$ for each node v_i of a given 3D model, one can obtain an initial segmentation of the input 3D shape by clustering in the descriptor space all the nodes of the 3D shape using unsupervised algorithms such as k-means or mean-shift [219]. Both algorithms require the specification of one parameter; k-means requires setting in advance the number of clusters. Mean-shift, on the other hand, operates by finding the modes (i.e. the local maxima of the density or the cluster centers) of points in the descriptor space. The advantage of using mean-shift lies in that the number of clusters does not have

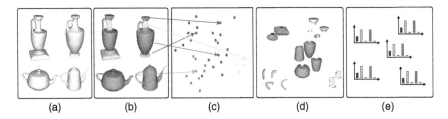

Figure 8.3 Unsupervised learning of the statistics of the semantic labels. (a) Training set. (b) Presegmentation of individual shapes. (c) Embedding in the descriptor space. (d) Clustering. (e) Statistical model per semantic label.

to be known in advance. Instead, the algorithm requires an estimation of the bandwidth or the radius of support around the neighborhood of a point. This can be estimated from the data. For instance, Shamir et al. [220] and Sidi et al. [210] fix the bandwidth to a percentage of the range of each descriptor. This procedure produces a pre-segmentation of each individual shape into a set of clusters. The last step is to break spatially disconnected clusters into separate segments.

Step 2 – Cosegmentation. Next, to produce the final semantic clusters, which will be used to learn the probability distribution of each semantic label, we cluster, in an unsupervised manner, all the segments produced in Step 1. These segments are first embedded into a common space via diffusion map. The embedding translates the similarity between segments into spatial proximity so that a clustering method based on the Euclidean distance will be able to find meaningful clusters in this space.

Let $S = \{s_i, i = 1, \dots, N\}$ be the set of segments obtained for all shapes and let $D(s_i, s_j)$ be the dissimilarity between two segments s_i and s_j. First, we construct an affinity matrix W such that each of its entries W_{ij} is given as:

$$W_{ij} = \exp(-D(s_i, s_j)^2/2\sigma). \tag{8.9}$$

The parameter σ can be compute either manually or automatically from the data by taking the standard deviation of all possible pairwise distances $D(s_i, s_j)$.

Now, let D be a diagonal matrix such that $D_{ii} = \sum_j W_{ij}$ and $D_{ij} = 0$ for $i \neq j$. We define the matrix $M = D^{-1}W$, which can be seen as a stochastic matrix with M_{ij} being the probability of a transition from the segment s_i to the segment s_j in one time step. The transition probability can be interpreted as the strength of the connection between the two segments.

Next, we compute the eigendecomposition of the matrix M and obtain the eigenvalues $\lambda_0 = 1 > \lambda_1 \geq \lambda_2 \geq \dots \geq \lambda_{N-1} \geq 0$ and their corresponding eigenvectors $\psi_0, \dots, \psi_{N-1}$. The diffusion map at time t is given by:

$$\Psi(s, t) = [\lambda_1^t \psi_1(s), \dots, \lambda_{N-1}^t \psi_{N-1}(s)]. \tag{8.10}$$

Here, $\psi_i(s)$ refers to the s-th entry of the eigenvector ψ_i, and λ_i^t refers to λ_i power t (Sidi et al. [210] used $t = 3$). The point $\Psi(s, t)$ defines the coordinates of the segment s in the embedding space. In practice, one does not need to use all the eigenvectors. The first leading ones (e.g. the first three) are often sufficient. Note that the eigenvector ψ_0 is constant and is thus discarded.

Finally, a cosegmentation is obtained by clustering the segments in the embedded space. Here, basically, any clustering method, e.g. k-means or fuzzy clustering, can be used. However, the number of clusters, which corresponds to the number of semantic parts (e.g. base, body, handle, or neck of a vase) that constitute the shapes, is usually manually specified by the user.

The output of this procedure is a set of node clusters where each cluster corresponds to one semantic category (e.g. base, body, handle, or neck of a vase) to which we assign a label l.

8.5.2 Learning the Labeling Likelihood

The next step is to learn a statistical model of each semantic cluster. For each semantic cluster labeled l, we collect into a set D_l the descriptor values for all of the nodes (surface patches) which form the cluster. One can then learn a conditional probability function $P(c = l|\mathbf{x})$ using the descriptors in D_l as positive examples and the remaining descriptors (i.e. the descriptors of the nodes whose labels are different from l) as negative ones. This conditional probability defines the likelihood of assigning a certain label for an observed vector of descriptors.

To learn this conditional probability distribution, van Kaick et al. [203] first use *GentleBoost* algorithm [215, 216] to train one classifier per semantic label. Each classifier returns the confidence value $\mathcal{K}_l(\mathbf{x})$, i.e. the confidence of $\mathbf{x}$ belonging to the class whose label is l. These unnormalized confidence values returned by each classifier are then transformed into probabilities with the softmax activation function as follows:

$$P(c = l|\mathbf{x}) = \frac{\mathcal{K}_l(\mathbf{x})}{\sum_{m \in C} \mathcal{K}_m(\mathbf{x})}. \tag{8.11}$$

For a given 3D shape, which is unseen during the training phase, this statistical model is used to estimate the probability that each of its nodes v belongs to each of the classes in C.

8.5.2.1 GentleBoost Classifier

Here, we briefly summarize the GentleBoost classifier and refer the reader to [216] for more details. GentleBoost takes an input feature $\mathbf{x} = (x_1, \ldots, x_d)$ and outputs a probability $P(c = l|\mathbf{x})$ for each possible label l is C. It is composed of decision stumps. A decision stump of parameters Φ is a very simple classifier that returns a score $h(\mathbf{x}, l; \Phi)$ of each possible class label l, given a feature vector $\mathbf{x} = (x_1, \ldots, x_d)$, based only on thresholding its fth entry x_f. It proceeds as follows:

- First, a decision stump stores a set of classes $C_S \subset C$.
- If $l \in C_S$, then the stump compares x_f against a threshold τ, and returns a constant a if $x_f > \tau$, and another constant b otherwise.
- If $l \notin C_S$, then a constant r_l is returned. There is one r_l for each $l \notin C_S$.

This can be written mathematically as follows;

$$h(\mathbf{x}, l; \Phi) = \begin{cases} a & \text{if } x_f > \tau \text{ and } l \in C_S. \\ b & \text{if } x_f \leq \tau \text{ and } l \in C_S. \\ r_l & \text{if } l \notin C_S. \end{cases} \tag{8.12}$$

The parameters Φ of a single decision stump are a, b, τ, f (the entry of $\mathbf{x}$ which is thresholded), the set C_S, and r_l for each $l \notin C_S$. The probability of a given class l is computed by summing the decision stumps:

$$H(\mathbf{x}, l) = \sum_j h_j(\mathbf{x}, l; \Phi_j), \tag{8.13}$$

where h_j is the jth decision stump and Φ_j are its parameters. The unnormalized confidence value returned by the classifier $\mathcal{K}_l$ for a given feature vector $\mathbf{x}$ is then given by:

$$\mathcal{K}_l(\mathbf{x}) = \exp(H(\mathbf{x}, l)). \tag{8.14}$$

Finally, the probability of $\mathbf{x}$ belonging to the class of label l is computed using the softmax transformation of Eq. (8.11).

8.5.2.2 Training GentleBoost Classifiers

The goal is to learn, from training data, the appropriate number of decision stumps in Eq. (8.13) and the parameters Φ_j of each decision stump h_j. The input to the training algorithm is a collection of m pairs $(\mathbf{x}_i, c_i)$, where $\mathbf{x}_i$ is a feature vector and c_i is its corresponding label value. Furthermore, each training sample $(\mathbf{x}_i, c_i)$ is assigned a per-class weight $w_{i,c}$. These depend on whether we are learning the unary term of Eq. (8.3) or the elements of the binary term of Eq. (8.7);

- For the unary term of Eq. (8.3), the training pairs are the per-node feature vectors and their labels, i.e. $(\mathbf{x}_i, c_i)$, for all the graph nodes in the training set. The weight $w_{i,c}$ is set to be the area of the ith node.
- For the pairwise term of Eq. (8.7), the training pairs are the pairwise feature vectors (see Section 8.3.1) and their binary labels. That is $(\mathbf{x}_{ij}, c_i \neq c_j)$. The weight $w_{i,c}$ is used to reweight the boundary edges, since the training data contain many more nonboundary edges. Let N_B and N_{NB} be, respectively, the number of boundary edges and the number of nonboundary edges. Then, $w_{i,c} = l_e N_B$ for boundary edges and $w_{i,c} = l_e N_{NB}$ for nonboundary edges, where l_e is the corresponding edge length.

GentleBoost learns the appropriate number of decisions stumps and the parameters of each decision stump which minimize the weighted multiclass exponential loss over the training set:

$$J = \sum_{i=1}^m \sum_{l \in C} w_{i,c} \exp(-I(c_i, l)H(\mathbf{x}_i, l)), \tag{8.15}$$

where $H(\cdot, \cdot)$ is given by Eq. (8.13), and $I(c_i, l)$ is an indicator function. It takes the value of 1 when $c = c'$ and the value -1 otherwise. The training is performed iteratively as follows:

1) Set $\tilde{w}_{i,c} \leftarrow w_{i,c}$.
2) At each iteration, add one decision stump (Eq. (8.12)) to the classifier.
3) Compute the parameters $\Phi_j = (a, b, r_l, f, \tau, C_S)$ of the newly added decision stump h_j by optimizing the following weighted least-squares objective:

$$J_{wse}(\Phi_j) = \sum_{l \in C} \sum_{i=1}^{m} \tilde{w}_{i,l}(I(c_i, l) - h_j(\mathbf{x}_i, l; \Phi_j))^2. \tag{8.16}$$

4) Update the weights $\tilde{w}_{i,c}$, $i = 1, \ldots, m$ as follows;

$$\tilde{w}_{i,c} \leftarrow \tilde{w}_{i,c} \exp(-I(c_i, l)h(\mathbf{x}_i, l; \Phi_j)). \tag{8.17}$$

5) Repeat from Step 2. The algorithms stops after a few hundred iterations.

Following Torralba et al. [216], the optimal a, b, r_l are computed in closed-form, and f, τ, C_S are computed by brute-force. When the number of labels is greater than 6, the greedy heuristic search is used for C_S.

Belonging to the family of boosting algorithms, GentleBoost has many appealing properties: it performs automatic feature selection and can handle large numbers of input features for multiclass classification. It has a fast learning algorithm. It is designed to share features among classes, which greatly reduces the generalization error for multiclass recognition when classes overlap in the feature space.

8.5.3 Learning the Remaining Parameters

The remaining parameters, μ, λ, η (Eqs. (8.6) and (8.7)), L (the label compatibility, see the second row of Table 8.1), and v (Eq. (8.8)), can be learned by performing a grid search over a specified range. Given a training dataset, and for each possible value of these parameters, one can evaluate the classification accuracy and select those which give the best score. One possible measure of the classification accuracy is the Segment-Weighted Error of Kalogerakis et al. [209], which weighs each segment equally:

$$E_S = \sum_i \frac{a_i}{A_{c_i}}(I(c_i, c_i^*) + 1)/2, \tag{8.18}$$

where a_i is the area of the ith node, c_i is the ground-truth label for the ith node, c_i^* is the output of the classifier for the ith node, A_{c_i} is the total area of all nodes (patches) within the segment S that has ground-truth label c_i, and $I(c_i, c_i^*) = 1$ if $c_i = c_i^*$, and zero otherwise.

To improve the computation efficiency, these parameters can be optimized in two steps. **First**, the error is minimized over a coarse grid in the parameter space by brute-force search. **Second**, starting from the minimal point in the grid, the optimization continues using the MATLAB implementation of Preconditioned Conjugate Gradient with numerically-estimated gradients [209].

8.5.4 Optimization Using Graph Cuts

Now that we have defined all the ingredients, the last step is to solve the optimizations of Eqs. (8.1) and (8.2). One can use for this purpose multilabel graph-cuts to assign labels to the nodes in an optimal manner. In general, graph-cuts algorithms require the pairwise term to be a metric. This is not necessarily the case in our setup. Thus, one can use the $\alpha - \beta$ swap algorithm [221]. By minimizing the given energy, we obtain the most likely label for each graph node (and thus surface patch) while also avoiding the creation of small disconnected segments. Algorithm 8.1 summarizes the graph cut's $\alpha - \beta$ swap algorithm. We refer the reader to the paper of Boykov et al. [221] for a detailed discussion of the algorithm.

Algorithm 8.1 Graph cut's $\alpha - \beta$ swap algorithm, see [221] for more details.

Input: The graph G, the set of semantic labels $\mathcal{L}$, and an initial labeling C^0, which can be arbitrary.

Output: Optimal labeling C.

1: Set $C \leftarrow C^0$.
2: Set *success* = 0;.
3: For each pairs of labels $\{\alpha, \beta\} \subset \mathcal{L}$

- Find C^* that minimizes $E(C')$ among C' within $\alpha - \beta$ swap of C.
- If $E(C^*) < E(C)$ then set $C \leftarrow C^*$, and *success* = 1.

4: If *success* = 1 then goto 2.
5: Return C.

8.6 Examples

In this section, we show a few semantic labeling and correspondence examples obtained using the techniques discussed in this chapter. In particular, we will discuss the importance of each of the energy terms of Eqs. (8.1) and (8.2).

First, Figure 8.4 shows that $P(c|\mathbf{x})$, and subsequently the data term of Eqs. (8.1) and (8.2), captures the main semantic regions of a 3D shape but fails to accurately capture the boundaries of the semantic regions. By adding the intramesh term, the boundaries become smoother and well localized (Figure 8.5).

Figure 8.6 shows samples from the training datasets, which were used as prior knowledge for learning the labeling. Note that, although at runtime the shapes are processed individually, the approach is able to label multiple shapes in a consistent manner. Thus, the labeling provides simultaneously a semantic segmentation and also partwise correspondences between the 3D models. Figure 8.6 also shows the effect of the training data on the labeling. In fact, different segmentation and labeling results are obtained when using different training datasets.

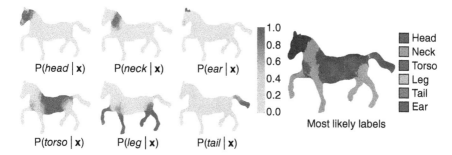

Figure 8.4 Effects of the unary term $P(c|x)$ on the quality of the segmentation. Source: The image has been adapted from the powerpoint presentation of Kalogerakis et al. [209] with the permission of the author.

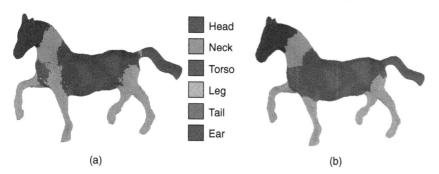

Figure 8.5 Semantic labeling by using (a) only the unary term of Eq. (8.1) and (b) the unary and binary terms of Eq. (8.1). The result in (b) is obtained using the approach of Kalogerakis et al. [209], i.e. using the unary term and binary term defined in the first row of Table 8.1. Source: The image has been adapted from the powerpoint presentation of Kalogerakis et al. [209] with the permission of the author.

Finally, Figure 8.7 demonstrates the importance of the intermesh term, and subsequently the joint labeling process. In Figure 8.7a, the correspondence is computed based purely on the prior knowledge, i.e. by only using the unary and intramesh terms of Eq. (8.1). Observe that part of the support of the candle on the right is mistakenly labeled as a handle and, therefore, does not have a corresponding part on the shape on the left. However, when the intramesh term (i.e. Eq. (8.2)), is taken into consideration, we obtain a more accurate correspondence, as shown in Figure 8.7b, since the descriptors of the thin structures on both shapes are similar. We observe an analogous situation in the chairs, where the seat of the chair is mistakenly labeled as a backrest, due to the geometric distortion present in the model. In Figure 8.7b, the joint labeling is able to obtain the correct part correspondence, guided by the similarity of the seats. This demonstrates that both knowledge and content are essential for finding the correct correspondence in these cases.

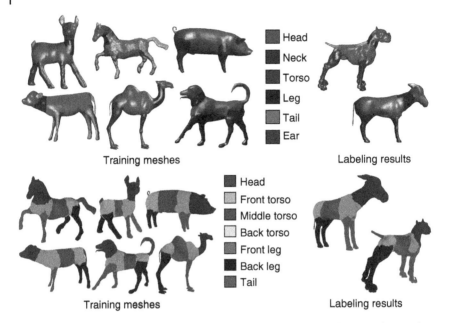

Figure 8.6 Supervised labeling using the approach of Kalogerakis et al. [209]. Observe that different training sets lead to different labeling results. The bottom row image is from Reference [209] ©2010 ACM, Inc. Reprinted by permission. http://doi.acm.org/10.1145/1778765.1778839.

8.7 Summary and Further Reading

We have presented in this chapter some mathematical tools for computing correspondences between 3D shapes that differ in geometry and topology but share some semantic similarities. The set of solutions, which we covered in this chapter, treat this problem as a joint semantic labeling process, i.e. simultaneously and consistently labeling the regions of two or more surfaces. This is formulated as a problem of minimizing an energy function composed of multiple terms; a data term, which encodes the a-priori knowledge about the semantic labels, and one or more terms, which constrain the solution. Most of existing techniques differ in the type of constraints that are encoded in these terms and the way the energy function is optimized.

The techniques presented in this chapter are based on the work of (i) Kalogerakis et al. [209], which presented a data-driven approach to simultaneous segmentation and labeling of parts in 3D meshes, (ii) van Kaick et al. [203], which used prior knowledge to learn semantic correspondences, and (iii) Sidi et al. [210], which presented a unsupervised method for jointly segmenting and putting in correspondence multiple objects belonging to the same family

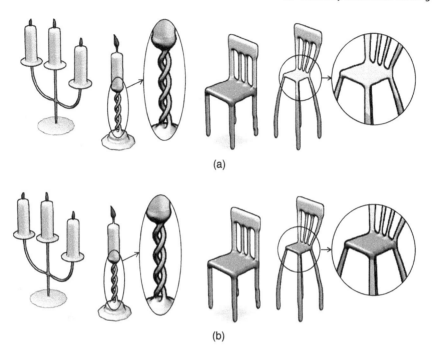

(a)

(b)

Figure 8.7 Effect of the intermesh term on the quality of the semantic labeling. Results are obtained using the approach of van Kaick et al. [203], i.e. using the energy functions of the second row of Table 8.1. (a) Semantic labeling without the intermesh term (i.e. using the energy of Eq. (8.1)). (b) Semantic labeling using the intermesh term (i.e. using the energy of Eq. (8.2)). Source: The image has been adapted from [203].

of 3D shapes. These are just examples of the state-of-the-art. Other examples include the work of Huang et al. [222], which used linear programming for joint segmentation and labeling, and Laga et al. [204, 223], which observed that structure, i.e. the way shape components are interconnected to each other, can tell a lot about the semantic of a shape. They designed a metric that captures structural similarities. The metric is also used for partwise semantic correspondence and for functionality recognition in man-made 3D shapes. Feragen et al. [224] considered 3D shapes that have a tree-like structure and developed a framework, which enables computing correspondences and geodesics between such shapes. The framework has been applied to the analysis of airways trees [224] and botanical trees [225].

Finally, semantic analysis of 3D shapes has been extensively used for efficient modeling and editing of 3D shapes [226–228] and for exploring 3D shape collections [229]. For further reading, we refer the reader to the surveys of Mitra et al. [205] on structure-aware shape processing and to the survey of van Kaick et al. [183] on 3D shape correspondence.

Part IV

Applications

9

Examples of 3D Semantic Applications

9.1 Introduction

The goal of this chapter is to illustrate the range of applications involving 3D data that have been annotated with some sort of meaning (i.e. semantics or label). There are many different semantics associated with 3D data, and some applications consider more than one category of labels, However, many of the current applications lie in the four following categories:

- **Shape or status**: What is the shape of the 3D object, and whether this is a normal shape or not (see Section 9.2).
- **Class or identity**: What sort of a 3D object is being observed, or even which specific 3D object (e.g. person) (See Section 9.3).
- **Behavior**: What are the 3D objects doing (see Section 9.4).
- **Position**: Where are the 3D objects (see Section 9.5).

This chapter briefly introduces examples of all of these, and the following chapters go more deeply into four of the applications: 3D face recognition (Chapter 10), object recognition in 3D scenes (Chapter 11), 3D shape retrieval (Chapter 12), and cross domain image retrieval (Chapter 13).

9.2 Semantics: Shape or Status

In this section, we consider applications where the 3D data are analyzed to extract usable shapes or properties from the 3D data. A key advantage of having 3D data is the ability to measure real 3D scene properties, such as the sizes of objects or the distance between locations. 2D images allow measurement of image properties (such as the distance between two pixels), which may allow estimation of an associated 3D measurement. On the other hand, 3D data allow direct measurement.

3D Shape Analysis: Fundamentals, Theory, and Applications, First Edition.
Hamid Laga, Yulan Guo, Hedi Tabia, Robert B. Fisher, and Mohammed Bennamoun.
© 2019 John Wiley & Sons, Inc. Published 2019 by John Wiley & Sons, Inc.

Example semantic 3D measurement applications include health monitoring (e.g. of commercial farm animals), estimating typical human body shape statistics, industrial part shape analysis (e.g. for quality control or robot grasping), or navigation planning (e.g. for free path and obstacle detection). The information extracted from the 3D data includes estimates of the size or parameters of standard models of the observed shapes, the position and size of different features on a part, and the flatness (or otherwise) of the ground surface ahead of an autonomous vehicle. More details of these examples are given below.

Figure 9.1 shows an example of a depth image of a cow, as scanned from above. The goal of the scanning is to assess the health of the cow [230] as the cow walks toward a milking station. The depth images are analyzed to: (i) detect when a cow is fully in the field of view, (ii) extract the cow from the depth image, and (iii) assess the angularity of the rear of the cow using a 3D morphological method (rolling ball algorithm [231]).

The CAESAR (Civilian American and European Surface Anthropometry Resource) project collected data from 2400 North American and 2000 European people aged 18–65. The data were collected using several full body 3D scanners, producing complete body point clouds. Three scans of the person in standing and seated poses were acquired. 3D landmarks were also extracted. From this dataset, researchers and companies are able to acquire better population information about the range of body sizes and shapes, leading to improved clothing, equipment, office, and work ergonomics. Figure 9.2a shows an example of the data and poses in the dataset.

Depth or 3D data greatly enhance the ability of a robot to grasp or manipulate a part. From the data, one can identify what parts are closer to the observer, identify graspable contact points on the surface of the object, estimate the mass distribution of the object, and identify subsets of the points that potentially satisfy graspability criteria, such as force closure. Figure 9.2b shows a depth image of a set of parts (in gray), with potential grasp locations and gripper finger positions shown in white.

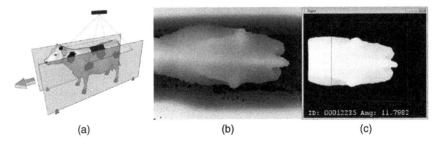

(a) (b) (c)

Figure 9.1 (a) The scanning configuration that produced the raw (b) and segmented (c) depth images. The images are used for cow health monitoring. Source: This figure was previously published in [230].

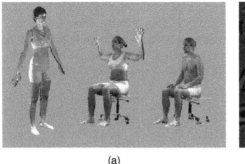
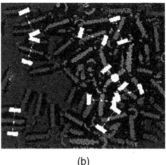

(a)   (b)

Figure 9.2 (a) This image shows three views of scanned people. The scans were used to collect statistics about human body shape and size. Source: This figure was previously published in [232]. (b) This image shows a depth image of a collection of parts from an industrial circuit breaker. Potentially graspable objects and gripper positions are marked in white. Source: This figure was previously published in [233].

A common example of practical 3D data shape analysis is Ground Plane Obstacle Detection, part of the technological underpinning needed for autonomous vehicles. The core idea is to fit a near-horizontal, near planar surface to the 3D points in front of the observer, which defines a ground plane. 3D points that lie significantly above the extracted surface are considered obstacles. Noise removal is often needed because of the presence of outliers. An example of ground plane and obstacle detection is shown in Figure 9.3a, and further details of the process can be found in [234].

Three-dimensional shape analysis and quantification is commonly used as a component in inspection and process control applications. The analysis can be

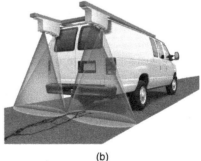

(a)   (b)

Figure 9.3 (a) This image shows the labeling of pixels based on a binocular stereo process [234], where white pixels are traversable areas, gray are obstacle boundaries, and black are obstacles. (b) An approach to laser-based 3D scanning of road surfaces from a moving vehicle. Source: Figure used with permission by PAVEMETRICS, http://www.pavemetrics .com.

more "semantic," rather than simply measuring some quantity, such as surface roughness or distance. For example, one application is in the detection of road surface cracks and other types of defects, as needed for wide-scale road quality surveys. An example of such a system is the vehicle-mounted system by Pavemetrics illustrated in Figure 9.3b, which detects and quantifies a variety of road defects, including cracks and potholes.

9.3   Semantics: Class or Identity

This section gives some example applications where the 3D data (usually along with color or motion data) contribute to the recognition of the general nature of the observed scene (e.g. it is a tree instead of grass), the recognition of the type of object that is being observed, the specific instance of the object, which is most often applied to recognizing specific people based, in part, on the shape of their face, or even specific properties of that face (e.g. its expression).

There has been an enormous amount of research into object recognition, particularly from 2D color image data. One well-known series of developments was stimulated by the PASCAL Visual Object Classes project (http://host .robots.ox.ac.uk/pascal/VOC/), which hosted a number of object recognition and localization challenges. Results on the 2012 challenge for recognition of 20 classes now reach 86.5% of objects correctly recognized. A successor set of developments is based on the Large Scale Visual Recognition Challenge dataset (http://www.image-net.org/) and associated challenges. A typical challenge (from 2017) is to detect objects from 200 categories in images, and current best results are over 0.73 mean average precision.

Here, the 3D data are analyzed to extract specific properties, which are then generally used in a 3D database matching process. The advantage of having 3D data is the 3D measurements, e.g. of distances, relative orientation, and local and global shapes, which cannot be directly measured from a 2D image. These measurements allow direct comparison of the properties of the observed objects to those of objects recorded in a database. More details of these examples are given below.

One of the most common applications of 3D shape to classification is for face-based person recognition or identity verification. (For a more detailed exploration of this application, see Chapter 10.) One application is for the identification of desired persons (e.g. in a law enforcement context), but this application is generally better satisfied when using color or video images, because 3D data capture generally requires the person to be at a fixed position in front of a 3D scanner. On the other hand, person reidentification or verification is the process of confirming that a person is who they have claimed themselves to be. In this class of application, the person usually presents themselves in front of the scanner, e.g. at building or passport security control.

In this case, one needs only to verify that the presenting person looks similar to the person recorded in the database. While much 2D research has been done in this area, issues that make the process more difficult are lighting and variable head pose. Consequently, researchers are interested in using the 3D shape of the face for verification that is invariant to lighting and small variations in head pose. General methods for 3D shape retrieval are presented in Chapter 12.

Person verification could be based on 3D video as well as a 3D static image, or 2D video. Zhang and Fisher [235] investigated using both static and dynamic measurements and shape properties to verify the identity of a person speaking a short passphrase (audio password). For example, they used the distance between the eyes, commonly used for recognizing people from static 3D images, but also information about the width and height of the mouth opening during the speaking of the passphrase. An example of face motion can be seen in Figure 9.4. A deeper exploration of 3D face recognition can be found in Chapter 10.

Another interesting face-based application considers recognition of a person's facial expressions, as a cue to their emotional state. As with personal identity verification, emotion recognition algorithms are also affected by illumination and head pose. The VT-KFER dataset of 7 expressions by 32 subjects allows the exploration of the benefits of 3D shape as well as color data. An example of the expressions in both intensity and depth can be seen in Figure 9.5. Expression analysis based on 3D data is a widely explored area because the pose and lighting invariant 3D information takes account of the person's 3D head shape rather than simply the person's appearance. Typical results can be found in [236] and [237], which also includes an extensive literature survey.

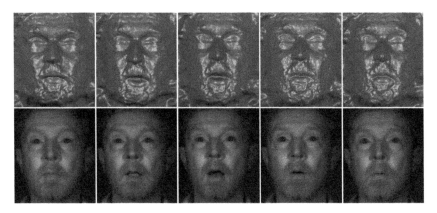

Figure 9.4 Five frames of a person speaking a passphrase with the infrared intensity image below and a cosine shaded depth image above. More information can be found in [235].

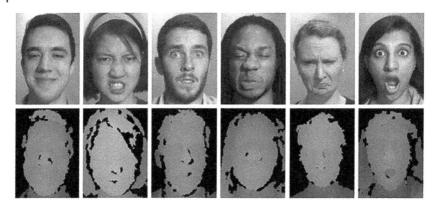

Figure 9.5 Six common facial expressions captured in both color and depth. More information can be found in [238].

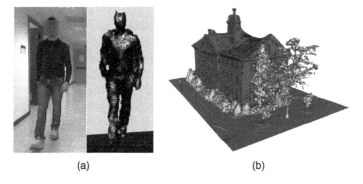

(a) (b)

Figure 9.6 (a) An intensity and processed depth image of a person walking, which is used for person reidentification. More information can be found in [239]. (b) A volumetric (3D) shape from stereo image data reconstructed from multiple color images, along with a semantic labeling (medium gray: building, dark gray: ground, bright gray: vegetation). More information can be found in [240].

Another approach to person reidentification is to use the whole 3D body shape, or details of the motion, as the person's face may not be seen with sufficient resolution [239]. This approach helps to deal with variations in clothing and body position, as well as lighting. An example intensity and processed depth image can be seen in Figure 9.6a.

An unusual research project [241] has investigated how one might use 3D information to detect when a person is using a mask to attempt to mimic another person, e.g. to gain access improperly. An example image showing two real images of the person, plus a third image of a person wearing a mask

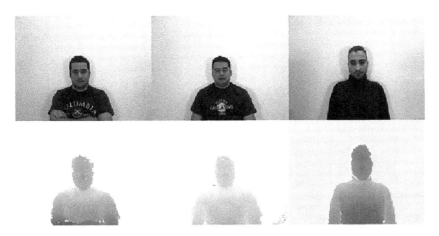

Figure 9.7 The left and middle intensity and depth images are of the same person, whereas the right image is of a second person wearing a mask imitating the first person. More information can be found in [241].

is shown in Figure 9.7. Although 2D masks can be easily detected by using 3D measurements, there is a risk that 3D masks with a shape similar to the target person could spoof a face recognition or verification system. However, research by Erdogmus and Marcel [241] shows that it may be possible to detect 3D spoofs as well.

Research into semantic segmentation (grouping related image data according to the class of structure as well as image properties) has found that using 3D shape information improves both shape and label recovery. The types of image regions give support for 3D relationships, and 3D information and relationships help reduce the ambiguity of image pixel values. Figure 9.6b shows an example of joint 3D shape from stereo and pixel class labeling [240].

An active research theme is the recognition of 3D objects (i.e. finding the semantic label of the object), with many approaches using both 2D color data and 3D point data (perhaps augmented with color as well). Many approaches to this problem have been attempted, using a variety of object and data descriptors. Mian et al. [93] developed a successful approach to model-based recognition of complex 3D shapes, even when observed in scenes with other objects which potentially also occluded the target object. Their approach was based on identifying, describing, and matching distinctive locations in the 3D point clouds from both the stored models and the scene data. Examples of the objects used and their 3D shapes can be seen in Figure 9.8. There has been substantial research into the recognition of 3D objects when using 3D data – Chapter 11 presents several current methods.

Figure 9.8 Examples of the color images and associated 3D models of objects that are recognized using specialized keypoints extracted from both the models and a scene containing these shapes. More information can be found in [93].

9.4 Semantics: Behavior

Behavior analysis typically starts from video data, and with the availability of 3D video sensors such as the Kinect©, additional time-varying information can be extracted from the 3D data to improve accuracy or provide statistics for e.g. health or performance monitoring. Examples of these applications are given below.

The most commonly explored area is that of activity recognition, where algorithms are developed to classify short segments of activity into one of a set of elementary action classes. The analysis might be based on the full 3D data, or from a "skeleton" extracted from the 3D point cloud or depth image. Typical examples of this classification approach come from researchers using the Cornell Activity Datasets [242] (see Figure 9.9) with 12 different activities represented in the dataset: rinsing mouth, brushing teeth, wearing contact lens, talking on the phone, drinking water, opening pill container, cooking (chopping), cooking (stirring), talking on couch, relaxing on couch, writing on whiteboard, and working on computer. There are many other action datasets based on 3D data, at least 20 of which can be found in the CVonline image and video database collection: http://homepages.inf.ed.ac.uk/rbf/CVonline/Imagedbase .htm.

While there has also been considerable research on identifying actions from color videos, there are several benefits to having the depth data for classification of activities, where it is easier to: (i) segment humans from background, (ii) fit

Figure 9.9 Examples of six activities and one frame from the color and depth videos, with the associated 3D skeleton overlaid. More information can be found at [242]. Source: http://pr.cs.cornell.edu/humanactivities/data.php.

a parameterized human skeleton to the depth data, and (iii) classify the action based on the skeleton's pattern of motion. Some examples of the actions with the associated depth images and skeletons can be seen in Figure 9.9.

As well as whole body activities, researchers have looked at identifying hand gestures [243] and also different deaf language signing [244]. Hand gestures may have skeletons fitted to the depth data of the hands, which allows classification methods similar to those used in whole-body activity classification. But often depth image resolution is insufficient for precise finger and joint angle recovery, in which case more holistic shape and motion analysis is used. Examples of the 10 gestures from the Sheffield KInect Gesture (SKIG) Dataset [243] can be seen in Figure 9.10. Another example of research on hand sign recognition (from Singapore's A*STAR agency) can be seen in Figure 9.11, which shows an example depth image and a corresponding view of the data glove that was used to verify the estimated hand poses [243].

Depth images captured over time create a depth video that can be used for analyzing human behavior, including their health and posture. Evidence for and

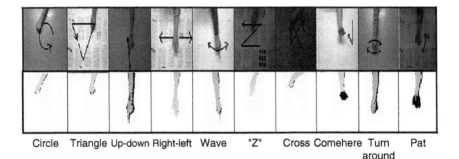

Circle Triangle Up-down Right-left Wave "Z" Cross Comehere Turn around Pat

Figure 9.10 Examples of the 10 hand gestures from the Sheffield KInect Gesture (SKIG) Dataset, showing both an example of the gesture and a frame of the corresponding depth image. More information can be found in [243].

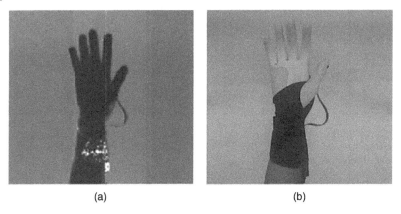

Figure 9.11 Examples of a depth image of a (a) hand pose and (b) of the corresponding data glove used to verify the estimated finger and hand positions. These were used in experiments investigating the recognition of some characters from Chinese number counting and American Sign Language. More information can be found in [244].

Figure 9.12 Examples of intensity and corresponding depth images for people carrying different weights. One can see how their posture changes. Source: See https://www.mmk.ei .tum.de/verschiedenes/tum-gaid-database/ and [245] for more information.

progression of neurodegenerative conditions such as Parkinson's disease can be estimated using analysis of the characteristic gait patterns. Similarly, one can look for atypical postures and gaits, such as when a person is carrying a heavy load – e.g. weapons or explosives. Figure 9.12 shows examples of how posture changes when carrying a load.

3D behavior analysis can involve multiple interacting people, for example, using either the depth data or the estimated stick-figure skeleton from the

Figure 9.13 Several frames of a punch interaction between two people showing the intensity image, the depth image and the stick figures extracted from the depth image [246]. Other approaches to skeleton or stick figure extraction use RGB images, point clouds, or various combinations of these data.

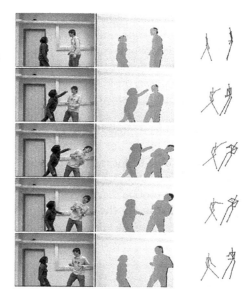

Kinect sensor and associated software. Like with identifying an individual person's actions, one can also attempt to recognize the nature of an interaction, such as the punching action seen in Figure 9.13.

9.5 Semantics: Position

This final section looks at applications where the information extracted from the 3D data is related to position. Examples include estimating an observer's position, such as on a mobile robot, an external object's position (e.g. for tracking the object), or the configuration of the tracked object (e.g. a person's body position when playing an interactive game). The most common observer pose estimation applications are based on the SLAM (simultaneous localization and mapping) family of algorithms (*c.f.* [247]), which match the currently acquired 3D point cloud to a previously acquired point cloud map. In this section, we present some more complex variants of the object position estimation problem.

The applications in the previous sections concentrated on classifying the activity, whereas here, the analysis concentrates on the positions taken by the people (and possible joint parameters) when doing the activity. A typical approach is to match known shapes, positions, or models to the currently observed 3D data. The relative orientation and translation (including any shape parameters) between the model and data then defines the position. We give some examples of these applications below.

Software associated with the Microsoft Kinect sensor is well known for its skeletal pose estimation and tracking, which underpins many popular video games. The skeleton joint angles and joint velocities can be used for representing and analyzing a number of different activities, such as fighting, driving, and bowling. For example, the G3D dataset [248] contains data for 20 different gaming actions. Figure 9.14 shows an example of a single person making a punching movement.

Given the skeletal pose and joint motion that can be estimated from a depth video sequence, one can do a variety of nongame static and dynamic analysis. This could be for medical purposes (as mentioned above), ergonomic purposes or sports analysis. A popular application is reachability analysis (i.e. can an object or tool be reached by a sitting person), for example as shown in Figure 9.15a illustrating the 3D hand positions in the space around a seated person [249].

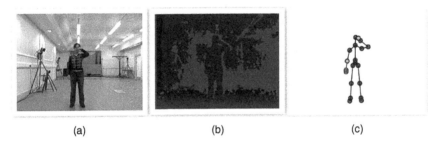

(a)   (b)   (c)

Figure 9.14 An example of a (a) fight punching movement, (b) its depth image and (c) associated skeleton, as captured by a Kinect sensor. See [248] for more information.

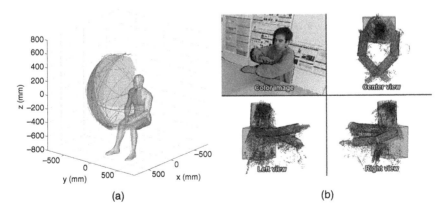

(a)   (b)

Figure 9.15 (a) Example of 3D tracking of a seated person reaching to different locations [249]. (b) 3D pose fitted to a set of 3D points captured by a Kinect sensor and used in an action analysis project. See [250] for more information.

While the Kinect sensor skeletons are useful, alternative approaches could be based on fitting shape models to the 3D point or range data. Figure 9.15b shows a simple articulated cylinder model that is fitted to the body, head, upper and lower arms, which is then used to identify the arm actions [250]. In this example, the set of 3D points, that the models are fitted to, is found by fusing points from four synchronized and calibrated Kinect sensors.

9.6 Summary and Further Reading

This chapter has presented an overview of some applications of 3D data processing that have more of a semantic flavor, as contrasted with applications that focus on property and distance measurement. In particular, the chapter looked at shape classification (Section 9.2), inference of object class or individual identity (Section 9.3), person behavior (Section 9.4), and finally estimating the object's pose or internal joint angle measurements (Section 9.5).

In the chapters that follow, there will be a more detailed exploration of different approaches to some specific applications of 3D semantic data analysis. In particular, Chapter 11 will look at 3D object recognition, and Chapter 10 will look at 3D face recognition. Another useful application is locating similar or identical 3D shapes in a database, for example, based on partial 3D data (Chapter 12) or non-3D data such as sketches or photos (Chapter 13).

10

3D Face Recognition

10.1 Introduction

The automatic recognition of human faces has many potential applications in various fields including security and human–computer interaction. An accurate and robust face recognition system needs to discriminate between the faces of different people under variable conditions. The main challenge is that faces, from a general perspective, look similar and their differences can be very subtle. They all have the same structure and are composed of similar components (e.g. nose, eyes, and mouth). On the other hand, the appearance of the same face can considerably change under variable extrinsic factors, e.g. the camera position and the intensity and direction of light, and intrinsic factors such as the head position and orientation, facial expressions, age, skin color, and gender. On that basis, face recognition can be considered to be more challenging than the general object recognition problem discussed in Chapter 11.

Pioneer researchers initially focused on 2D face recognition, i.e. how to recognize faces from data captured using monocular cameras. They reported promising recognition results, particularly in controlled environments. With the recent popularity of cost-effective 3D acquisition systems, face recognition systems are starting to benefit from the availability, advantages, and widespread use of 3D data. In this chapter, we review some of the recent advances in 3D face recognition. We will first present, in Section 10.2, the various 3D facial datasets and benchmarks that are currently available to researchers and then discuss the challenges and evaluation criteria. Section 10.3 will review the key 3D face recognition methods. Section 10.4 provides a summary and discussions around this chapter.

3D Shape Analysis: Fundamentals, Theory, and Applications, First Edition.
Hamid Laga, Yulan Guo, Hedi Tabia, Robert B. Fisher, and Mohammed Bennamoun.
© 2019 John Wiley & Sons, Inc. Published 2019 by John Wiley & Sons, Inc.

10.2   3D Face Recognition Tasks, Challenges and Datasets

There are two scenarios of 3D face recognition, namely, 3D face verification and 3D face identification.

10.2.1   3D Face Verification

3D face verification, also called 3D face authentication, refers to the one-to-one comparison of a probe 3D face with a specific 3D face from the 3D faces (gallery) stored on the system. If the comparison score is higher than a threshold then the probe is assigned the identity of the gallery face. Figure 10.1a presents a schematic diagram of an automatic 3D face verification system. The performance of such systems is generally measured in terms of one or many of the following evaluation criteria:

- The false acceptance rate (FAR) is the likelihood that a verification system will incorrectly accept an access attempt by an unauthorized user. It is defined as the ratio of the number of false acceptances divided by the number of identification attempts.
- The false rejection rate (FRR) is the likelihood that a verification system will incorrectly reject an access attempt by an authorized user. It is defined as the percentage of probes that have been falsely rejected.

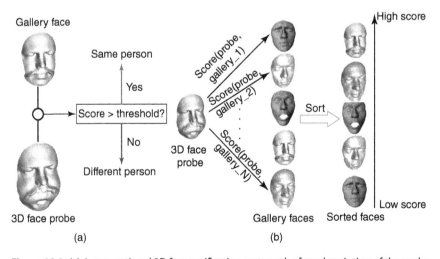

Figure 10.1 (a) A conventional 3D face verification system: the face description of the probe is compared to the gallery representation. If the matching score between the two faces is greater than a predefined threshold, then the individual is identified as the claimed entity; otherwise, it is considered as an imposter. (b) A conventional 3D face identification system: The 3D face description of the unknown subject is compared to all the descriptions of all 3D faces in the gallery. The unknown subject takes the identity of the highest matching score.

- The verification rates (VR) corresponds to the percentage of probes that have been correctly accepted.
- The receiver operating characteristic (ROC) curve plots the FAR versus the FRR. Another type of ROC curves plots the VR versus the FAR.
- The equal error rate (EER) is the error rate at the point on the ROC curve whereby FRR equals FAR. Both the EER and the VR at point 0.1 of the FAR (denoted by VR@0.1%FAR) give an important scalar measure of the verification performance using a single number.

The VR, FRR, and FAR change with the acceptance threshold, which is selected based on the application at hand.

10.2.2 3D Face Identification

3D face identification refers to the problem of matching the 3D face of an unknown person with all the 3D faces stored in a database. Figure 10.1b shows a schematic diagram of an automatic 3D face identification system. There are two main criteria that are used to evaluate the performance of such systems, namely:

- The cumulative match characteristic curve (CMC) provides the percentage of correctly recognized probes with respect to the rank number that is considered to be a correct match. Each probe is compared against all the gallery faces. The resulting scores are sorted and ranked. The CMC plots the probability of observing the correct identity within the top K ranks.
- The rank-1 recognition rate (R1RR) is the percentage of all probes for which the best match (rank one) in the gallery belongs to the same subject. R1RR is the most frequently used measure. It provides a single number for the evaluation of face identification performance.

10.2.3 3D Face Recognition Challenges

3D face recognition systems may come up with different types of challenges that can limit their performance. The most important ones are listed below.

10.2.3.1 Intrinsic Transformations
The performance of a 3D face recognition system can be greatly affected by intrinsic facial variations such as muscle movements and skin elasticity, which result in geometrical changes of the facial shape. Two of the major challenges are as follows:

- *Facial expressions*: Facial expressions can result in large deformations of the geometrical structure and texture of the human face. Different facial expressions may also result in different types of deformations. As such, the recognition accuracy can decrease due to the difficulty in differentiating between expression deformations from interpersonal disparities.

- *Age*: Face recognition under age variations (when the time period between the saved 3D face gallery and the captured 3D probe is long) is very challenging. This is because the geometrical structure and texture of human faces may change over time in an abrupt and unexpected fashion which is difficult to predict.

10.2.3.2 Acquisition Conditions

The most common acquisition conditions that can limit the effectiveness of 3D facial recognition systems include:

- Lighting: Unlike 2D face recognition, lighting or illumination conditions do not really affect the performance of 3D face recognition methods. This is because 3D methods rely on the geometry of the face, which is more robust under illumination than the texture information. Note, however, that lighting conditions may impact on the 3D acquisition process (e.g. when using stereo or structured light-based techniques), resulting in noisy 3D facial scans, which in turn complicate the 3D face recognition task.
- Pose: Usually, the probe and gallery images, acquired at the enrolment and during the recognition stage, have different poses. Although 3D face recognition methods are more robust to pose variations than their 2D counterparts, their accuracy can drop due to the missing data caused by self-occlusions.
- Occlusions: The occlusions such as those caused by caps, sunglasses, and scarves may also cause a broad range of discrepancies in 3D face recognition systems, especially when using global face recognition algorithms, which consider the face as a whole.
- Missing parts: Often 3D acquisition sensors capture 3D surfaces with missing parts (e.g. holes). An effective 3D face recognition system should be robust to these partiality transformations.

10.2.3.3 Data Acquisition

There are two main classes of techniques that are used to capture the 3D surface of a human face, namely: active and passive techniques. Among the active techniques, triangulation and structured light are the most popular, while stereo matching is the most used passive technique. Chapter 3 provides a detailed description of the various 3D surface acquisition approaches of objects (including faces). We also refer the reader to [251] for additional reading. Here, we only distinguish between data captured using cost effective sensors and 3D laser scanners.

- Cost-effective sensors: These 3D acquisition tools are generally based on structured light or stereo matching technologies. They are not able to achieve the same level of detail as 3D laser scanning devices. This limitation may highly affect the effectiveness of 3D face recognition systems.

- Laser scanners: The main benefit of 3D scanners is that they provide higher resolution and more detailed surfaces than cost-effective sensors. Therefore, facial details are potentially captured more intricately, making the resulting 3D face representation more detailed. The downside of having detailed 3D face representations is that they require a significant storage space. Also, laser scanners are intrusive and slow.

10.2.3.4 Computation Time

Computation time is an important factor when assessing the effectiveness of a 3D face recognition system, especially in identification scenarios when dealing with millions of 3D faces.

10.2.4 3D Face Datasets

In the last few years, several 3D datasets with annotated faces have been released to the face recognition community. These datasets have facilitated the evaluation of the proposed recognition methods. In this section, we present some of the most popular ones.

- **FRGCv2.0 [252]**. One of the most widely used 3D face recognition dataset is the second version of Face Recognition Grand Challenge (FRGCv2.0). It is also one of the largest publicly available 2D and 3D human face dataset. It contains 4007 nearly frontal facial scans belonging to 466 individuals. About Two thousand four hundred and ten scans out of the 4007 are with a neutral expression. The remaining are with nonneutral expressions such as surprise, happy, puffy cheeks, and anger.
- **GavabDB [253]**. The GavabDB dataset includes 549 scans of the facial surfaces of 61 different persons (45 male and 16 female). Each person has 9 scans: 2 frontal ones with neutral expression, 2 x-rotated views (30° up and 30° down, respectively) with neutral expression, 2 y-rotated views (90° to the left and 90° to right, respectively) with neutral expression, and 3 frontal with facial expressions (laugh, smile, and a random facial expression chosen by the user).
- **FRAV3D [254]**. The FRAV3D dataset includes 106 subjects, with approximately one woman for every three men. The data were acquired with a Minolta VIVID 700 scanner, which generates a 2D image and a 3D scan. The acquisition process followed a strict protocol, with controlled lighting conditions. The person sat down on an adjustable stool opposite to the scanner and in front of a blue wall. No glasses, hats, or scarves were allowed. In every acquisition session, a total of 16 scans per person were captured, with different poses and lighting conditions, to cover all possible variations, including turns in different directions and gestures and lighting changes. In every case, only one parameter was modified between two captures. This is one of the main advantages of this database compared to the others.

- **BJUT [255].** The BJUT-3D database contains 500 persons (250 females and 250 males). Every person has a 3D face data in a neutral expression and without accessories. The original high-resolution 3D face data were acquired with the CyberWare 3D scanner in a controlled environment. Each 3D face was preprocessed, and the redundant parts were cropped.
- **Bosphorus [256].** The Bosphorus dataset includes 4666 scans of the facial surfaces of 105 subjects. This dataset contains facial expressions that are composed of a judiciously selected subset of Action Units and the six basic emotions. Many actors/actresses have contributed to this dataset to obtain a realistic facial expression data. The dataset also contains a rich set of head pose variations as well as different types of face occlusions.
- **UMB [257].** The University of Milano Bicocca 3D face database is a collection of multimodal (3D + 2D color images) facial datasets. The UMB-DB was acquired with a particular focus on facial occlusions, i.e. scarves, hats, hands, eyeglasses, and other types of occlusions, which can occur in real-world scenarios. This dataset contains a total number of 1473 acquisitions (3D scans + 2D images) corresponding to 143 subjects (98 male, 45 female; a pair of male twins and a baby have been also included).

10.3 3D Face Recognition Methods

3D face recognition (identification and verification) has been well studied and several survey papers can be found in the literature [258–260]. Most of the state-of-the-art approaches follow the conventional recognition pipeline shown in Figure 10.2. Table 10.1 lists some of the most relevant 3D face verification approaches that are found in the literature. Table 10.2, on the other hand, lists some of the identification approaches. These approaches can be categorized according to the features they use, or based on the type of classification approach that they use. In this chapter, we group existing methods into two categories: *holistic* and *local* approaches. Holistic approaches describe the face as a whole and compute similarity measures directly based on that description. Local approaches extract, compute, and fuse local features to form global descriptors, which in turn are compared using similarity measures. Below, we detail each of these categories of methods.

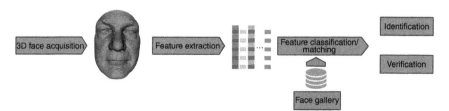

Figure 10.2 Conventional 3D face recognition pipeline.

Table 10.1 Examples of 3D face verification approaches.

Methods	Matching	Dataset Subjects—Faces	Performance VR or EER
Zhong et al. [114]	Learning visual codebook	466-4950	4.9% EER
Kelkboom et al. [261]	Template protection	FRGCv2.0	3.2% EER
Wang et al. [262]	Constraint deformation	FRGCv2.0	6.2% EER
Heseltine et al. [263]	Multi features	280-1770	7.2% ERR
Jahanbin et al. [264]	Euclidean distance	119-1196	2.79% ERR
Zhong et al. [265]	Quadtree clustering	FRGCv2.0	2.6% ERR
Kakadiaris et al. [266]	Wavelet	FRGCv2.0	97.3% at 0.001 FAR
Faltemier et al. [267]	Region ICP	FRGCv2.0	94.5% at 0.01 FAR
Lin et al. [268]	Similarity fusion	FRGCv2.0	90% at 0.1 FAR
Xu et al. [269]	Gabor-LDA	FRGCv2.0	97.5% at 0.1 FAR
Jin et al. [270]	LSDA	FRGCv2.0	82.8% at 0.1 FAR
Marras et al. [271]	Cosine-based distance	FRGCv2.0	93.4% at 0.1 FAR
Berretti et al. [272]	Graph	FRGCv2.0	97.7% at 0.1 FAR

Table 10.2 Some of the key 3D face identification approaches.

Methods	Matching	Dataset Subjects—faces	Performance R1RR (%)
Cook et al. [273]	Mahalanobis Cosine	FRGCv2.0	92.93
Cook et al. [274]	Mahalanobis Cosine	FRGCv2.0	94.63
Wang et al. [275]	Partial ICP	40-360	96.88
Mian et al. [276]	ICP	FRGCv2.0	98.03
Gokberk and Akarun [277]	Classifier fusion	106-	97.93
Zhang et al. [278]	Symmetry profile	164-213	90
Shin and Sohn [279]	SVM	300-2400	98.6
Mian et al. [280]	Modified ICP	FRGCv2.0	99.02
Kakadiaris et al. [281]	Haar-Pyramid metric	FRGCv2.0	97.3
Feng et al. [282]	Euclidean transformation	FRGCv2.0	95
Li and Zhang [283]	Weighting function	FRGCv2.0	95.56
Mpiperis et al. [284]	Geodesic Polar	100-2500	84.4
Mahoor and Abdel-Mottaleb [285]	Hausdorff	GavabDB	82
Mpiperis et al. [286]	Point signatures	20-800	91.36
Mian et al. [287]	PCA	FRGCv2.0	96.1
Amberg et al. [288]	Nonrigid ICP	61-427	99.7

(Continued)

Table 10.2 (Continued)

Methods	Matching	Dataset Subjects—faces	Performance R1RR (%)
Alyüz et al. [289]	Local ICP	81-3396	95.87
Faltemier et al. [290]	Region ICP	FRGCv2.0	97.2
Al-Osaimi et al. [291]	PCA	FRGCv2.0	93.78
Alyüz et al. [292]	Shape texture fusion	81-3396	98.01
Mpiperis et al. [293]	Deformable model	100-2500	86
Lu and Jain [294]	Deformable model	FRGCv2.0	—
Faltemier et al. [295]	Multi-instance	888-13450	98
Yang et al. [296]	Canonical correlation	200-	85.19
Theoharis et al. [297]	Deformable model	324-3259	99.4
Wang et al. [298]	Local shape difference	FRGCv2.0	98.22
Harguess et al. [299]	Average-half-face	104-1126	-
Gokberk et al. [300]	Fusion	FRGCv2.0	95.45
Fabry et al. [301]	Kernel correlation	61-427	90.6
Zhong et al. [302]	L1 distance	311-742	80.51
Yan-Feng et al. [303]	Feature fusion	50-150	86.7
Mousavi et al. [304]	PCA	61-427	97
ter Haar and Veltkamp [305]	Deformation model	876	97.5
Llonch et al. [306]	Sparse	277-	—
Al-Osaimi et al. [307]	PCA-ICP	FRGCv2.0	96.52
Paysan et al. [308]	Generative model	953-	—
Al-Osaimi et al. [309]	PCA-ICP	FRGCv2.0	96.52
Li et al. [310]	Sparse	120-600	94.68
Boehnen et al. [311]	Vector representation	410-3939	—
Huang et al. [312]	Asymmetric Scheme	200-1000	82.36
Dibeklioğlu et al. [313]	ICP	47-2491	—
Cadoni et al. [314]	Differential invariants	106-617	—
Daniyal et al. [315]	inter-landmark distances	100-2500	96.5
Yunqi et al. [316]	PCA	36-324	94.5
ter Haar and Veltkamp [317]	Curve matching	61-427	92.5
Queirolo et al. [318]	Surface Interpenetration	FRGCv2.0	98.4
Wang et al. [319]	Signed Shape Difference Map	FRGCv2.0	98.39
Maes et al. [320]	RANSAC	105-4666	93.7
Drira et al. [321]	Riemannian framework	61-427	96.99
Llonch et al. [322]	LDA	121-308	98,70
Huang et al. [323]	Hybrid	FRGCv2.0	96.1

Table 10.2 (Continued)

Methods	Matching	Dataset Subjects—faces	Performance R1RR (%)
Huang et al. [324]	Hybrid	FRGCv2.0	96.83
Tang et al. [325]	Vote LBP	60-360	93.1
Smeets et al. [326]	Vote	100-900	94.48
Li et al. [327]	ICP	90-2700	91.1
Wang et al. [328]	Point Direction Measure	123-1845	95.6
Zhang et al. [329]	Weighted sum	38-380	99.34
Miao and Krim [330]	Iso-geodesic	50-200	96.82
Ming et al. [331]	Spectral Regression	FRGCv2.0	—
Passalis et al. [332]	Weighted L_1 metric	118- 236	86.4
Spreeuwers [333]	PCA-LDA	FRGCv2.0	99
Zhang and Wang [334]	Fusion	FRGCv2.0	96.2
Li et al. [335]	Cosine distance	105-466	94.10
Huang et al. [336]	Sparse	50-250	92
Ocegueda et al. [337]	LDA	FRGCv2.0	99.0
Sharif et al. [338]	Hybrid	61-427	96.67
Smeets et al. [339]	Angle	130-730	98.6
Tang et al. [340]	Fisher LDA	350-2100	95.04
Li et al. [341]	Sparse	FRGCv2.0	94.61
Ming et Ruan [342]	Euclidean distance	FRGCv2.0	95.79
Huang et al. [343]	Hybrid matching	FRGCv2.0	97.6
Li and Da [344]	Weighted sum rule	FRGCv2.0	97.8
Taghizadegan et al. [345]	Euclidean distance	123-4674	98
Ballihi et al. [346]	Reimannian matching	FRGCv2.0	98
Zhang et al. [347]	SVM classifier	FRGCv2.0	93.1
Inan and Halici [348]	Cosine distance	FRGCv2.0	97.5
Belghini et al. [349]	Neural network	10-160	89
Drira et al. [350]	Reimannian matching	FRGCv2.0	97
Lei et al. [351]	SVM	FRGCv2.0	95.6
Smeets et al. [352]	Angle	FRGCv2.0	89.6
Tang et al. [353]	Fusion based	FRGCv2.0	94.89
Alyuz et al. [354]	Adaptively-selected-model	105-381	83.99
Lei et al. [88]	SVM	FRGCv2.0	94.1
Zhang et al. [355]	Sparse	466-2748	98.29
Tabia et al. [138]	optimal match	GavabDB	94.91
Elaiwat et al. [356]	Cosine dist	FRGCv2.0	97.1
Emambakhsh and Evans [357]	Mahalanobis Cosine dist	FRGCv2.0	97.9

10.3.1 Holistic Approaches

Holistic-based 3D face recognition approaches, also called global approaches, treat the face as a whole and work directly on its entire 3D surface. Researchers have been interested in holistic approaches since the early nineties. Examples of 3D holistic features include eigenfaces (principal component analysis, PCA), Fisherfaces (linear discriminant analysis, LDA), iterative closest point (ICP), Extended Gaussian Images (EGI), Canonical Forms, spherical harmonic (SH) features, and the tensor representation. Here, we review a few of them to give the reader some insight.

In what follows, we assume that a 3D face is represented as a triangulated mesh M with a set of vertices $V = \{v_i, i = 1, \dots, n\}$ and a set of triangular faces $T = \{t_i, i = 1, \dots, m\}$.

10.3.1.1 Eigenfaces and Fisherfaces

Eigenfaces and Fisherfaces were originally introduced for 2D face recognition, and have been successfully extended to 3D face recognition. Let $\{M_1, \dots, M_N\}$ be a set of N facial surfaces that belong to N_c different subjects. Here, $N_c \leq N$. Each facial surface M_k has n vertices $v_i^k, i = 1, \dots, n$. Let us assume that all the 3D faces that are being analyzed have the same number of vertices and the same triangulation. We also assume that all the 3D faces are in full correspondence, i.e. for every pair of facial surfaces M_k and M_l, the ith vertex of M_k is in correspondence with the ith vertex of M_k.

10.3.1.1.1 Eigenfaces This approach, which is based on PCA, takes each facial surface M and represents it with a long column vector $\mathbf{x} = (x_1, y_1, z_1, \dots, x_n, y_n, z_n)^\top$ of size $3n$. (Here $(x_i, y_i, z_i)^\top$ are the coordinates of the ith vertex of M.) Thus, M can been seen as a point $\mathbf{x} \in \mathbb{R}^{3n}$. Now, given a set of N points $\{\mathbf{x}_1, \dots, \mathbf{x}_N\}$, their average facial shape μ and covariance matrix C can be computed as:

$$\mu = \frac{1}{N} \sum_{i=1}^{N} \mathbf{x}_i \quad \text{and} \quad C = \frac{1}{N-1} \sum_{i=1}^{N} (\mathbf{x}_i - \mu)(\mathbf{x}_i - \mu)^\top. \tag{10.1}$$

The shape variability of the facial surfaces can be efficiently modeled using PCA. Let $\lambda_i, i = 1, \dots, d$, be the d-leading (largest) eigenvalues of the covariance matrix C and $\Lambda_i, i = 1, \dots, d$, be their corresponding eigenvectors, called also *modes of variation*. Then, any facial surface $\mathbf{x}$ can be approximated by $\hat{\mathbf{x}}$, which is the linear combination of the mean and the eigenvectors, i.e.

$$\hat{\mathbf{x}} = \mu + \sum_{i=1}^{d} \alpha_i \Lambda_i, \alpha_i \in \mathbb{R}. \tag{10.2}$$

This is equivalent to representing the facial surface $\mathbf{x}$ using a point of coordinates $(\alpha_1, \dots, \alpha_d)^\top$ in the subspace of $\mathbb{R}^{3n}$ that is spanned by the d eigenvectors

$\Lambda_i, i = 1, \ldots, d$, hereinafter referred to as *the eigenface space*. It is also equivalent to projecting $\mathbf{x}$ onto this subspace. This projection can be written in the matrix form as follows:

$$\hat{\mathbf{x}} = W_{pca}^{\top}(\mathbf{x} - \mu), \tag{10.3}$$

where $W_{pca} = [\Lambda_1 | \ldots | \Lambda_d]$ is a $3n \times d$ matrix whose ith column is the ith eigenvector Λ_i.

During the recognition phase, we take the probe facial surface f, represented with a vector $\mathbf{x} \in \mathbb{R}^{3n}$, compute its projection $\hat{\mathbf{x}}$ onto the eigenface space using Eq. (10.3), and then compare its representation in the eigenface space to the gallery faces $\hat{\mathbf{x}}_i$ using various distance metrics, e.g. $\mathbb{L}^2$.

10.3.1.1.2 Fisherfaces 3D Fisherfaces are computed by performing 3D LDA on the 3D facial data, as follows:

- **First,** each facial surface M_i, represented with its vector $\mathbf{x}_i$, is mapped onto the PCA subspace using Eq. (10.3). The dimension of the PCA subspace is often fixed to $d = N - N_c$, where N is the total number of facial surfaces in the training set, and N_c is the number of the different subjects (also referred to as identities or classes) in the training set.
- **Second,** the intraclass and interclass covariance matrices, hereinafter denoted by C_{itra} and C_{iter}, respectively, of the training set are then computed as follows:

$$C_{itra} = \sum_{i=1}^{N_c} C_i \quad \text{and} \quad C_{iter} = \sum_{i=1}^{N_c} N_i(\overline{\mathbf{x}_i} - \overline{\mathbf{x}})(\overline{\mathbf{x}_i} - \overline{\mathbf{x}})^T, \tag{10.4}$$

where C_i is the covariance matrix of the ith class, $\overline{\mathbf{x}}$ is the mean of all the vectors $\{\mathbf{x}_i, i = 1, \ldots, N\}$, $\overline{\mathbf{x}_i}$ is the mean of the vectors of the ith class, and N_i is the number of facial surfaces in the ith class. Note that, in the literature, C_{itra} is also known as the *within-class scatter matrix* and denoted by S_W, and C_{iter} is known as the *between-class scatter matrix* and is denoted by S_B.
- **Third,** the optimal mapping W_{lda}, which maps the data onto the Fisher subspace, is defined as the matrix which maximizes the ratio of the determinant of the inter-class scatter matrix of the projected points to the determinant of the intraclass scatter matrix of the projected points. That is,

$$W_{lda} = \underset{W}{\arg\max} \left\{ \frac{W^{\top} C_{iter} W}{W^{\top} C_{itra} W} \right\}. \tag{10.5}$$

- **Finally,** the projection $\hat{\mathbf{x}}$ of a facial surface M, represented with its vector $\mathbf{x}$, onto the Fisherface subspace is computed as follows;

$$\hat{\mathbf{x}} = W_{lda}^{\top} W_{pca}^{\top} \mathbf{x}. \tag{10.6}$$

Similar to eigenfaces, recognition is performed by first projecting all the probe faces as well as the query face onto the Fisher subspace, and then comparing the projected points using various distance measures, e.g. the $\mathbb{L}^2$ metric. Heseltine et al. [358] showed that the 3D Fisherface algorithm provides significantly better results than the 3D eigenface algorithm.

10.3.1.2 Iterative Closest Point

The ICP algorithm, which is described in detail in Chapter 6, has been widely used for face recognition [276, 290, 327]. The ICP algorithm considers the original 3D facial data as features and does not extract any specific high-level features from the 3D facial data. It starts with a search of the closest points between two point clouds. These closest point pairs (i.e. correspondences) are used to compute a rigid transformation between the two point clouds. The estimated transformation is then applied to one of the point clouds to bring it as close as possible to the other point cloud. This procedure is iterated until the mean square error (MSE) of the correspondences is minimized. The MSE is generally used as a similarity measure between 3D faces.

10.3.1.3 Hausdorff Distance

The Hausdorff distance, also called Pompeiu-Hausdorff distance, has been extensively used to measure the similarity between two 3D faces, considered as two subsets of 3D points. Let $P = \{p_i\}_{i=1}^{N_p}$ and $Q = \{q_j\}_{j=1}^{N_q}$ be two nonempty subsets of a metric space. Their Hausdorff distance is defined as:

$$d_H(P, Q) = \max \left\{ \sup_{p \in P} \inf_{q \in Q} d(p, q), \sup_{q \in Q} \inf_{p \in P} d(p, q) \right\}. \tag{10.7}$$

Here, sup represents the supremum, inf the infimum, and d is the Euclidean distance between the points p and q. Note that the Hausdorff distance as defined by Eq. (10.7) is not invariant to the similarity-preserving transformations (i.e. translation, scale, and rotation) of the point sets P and Q. Thus, in practice, the points sets P and Q are first normalized for translation, scale, and rotation in a preprocessing step, using the approaches described in Section 2.2.

Russ et al. [359] used this Hausdorff-based approach for 3D face recognition. Achermann and Bunke [360] used the Hausdorff distance to classify range images of human faces and reported a 100% recognition rate on a dataset composed of 240 range images. To do so, they quantified the facial 3D points using voxels and used Eq. (10.7) to compute the facial similarities. Lee and Shim [361] proposed a curvature-based human face recognition using depth weighted Hausdorff distance.

10.3.1.4 Canonical Form

The main purpose of the canonical form representation proposed by Bronstein et al. [362, 363] is to make 3D face recognition algorithms robust under expression variations. The canonical form representation of a 3D facial surface is computed based on the assumption that the geodesic distance between any two points on the surface remains invariant under expression variations. Considering this, Bronstein et al. [362, 363] proposed to embed the pairwise geodesic distances of a set of points $\{p_1, p_2, \ldots, p_N\}$ sampled from a 3D facial surface S into a low-dimensional Euclidean space. This results is a canonical representation, which can be processed using standard methods for rigid shape matching (see Chapter 6).

Formally, let us consider two facial surfaces S_1 and S_2 of the same subject having different facial expressions. Bronstein et al. [362, 363] assume that the pairwise geodesic distances between points in each surface are equivalent. So the idea is to find an embedding space onto which the two surfaces correspond to the same representation (i.e. a canonical form). This is done by applying a dimensionality-reduction technique called multidimensional scaling (MDS). Once the two surfaces are embedded in a common canonical space, one can use various rigid surface analysis techniques to compare and classify them.

10.3.2 Local Feature-Based Matching

Local feature-based matching methods use local surface proprieties, such as points, curves, or regions, of the 3D faces. A recognition measure is then defined based on the aggregation of the different local proprieties. Compared to the holistic approaches, the main advantage of local feature-based approaches is their robustness to pose variations, facial expressions, and occlusions. In this subsection, we group the methods of this class into three categories: keypoint-based, curve-based, and patch-based features [251].

10.3.2.1 Keypoint-Based Methods

Keypoints are 3D points of interest extracted according to some geometric proprieties of the 3D faces. The keypoints should be detected with high repeatability in different range images of the same 3D face in the presence of noise and pose variations. Chapter 5 discusses various methods for keypoint detection on 3D shapes in general. Here, we focus on those that are specific to 3D faces, which we classify into two categories: landmark-based techniques, which extract facial points according to anatomical studies of the face, and SIFT-like methods, which are derived from the popular 2D SIFT features used in image analysis.

Note that these methods usually require a large number of keypoints. This makes them very expensive in terms of computational time, especially when using assignment algorithms in the matching phase.

10.3.2.1.1 Landmark-Based Methods Landmark-based methods describe specific properties that are computed on a set of precisely detected points. These proprieties include the original coordinates of the points, the length of an edge between pairs of points, the area of a region composed of several points, the angle between two edges, and the differential properties of the surface at that point. These proprieties, which are used for face matching, highly depend on the location of the detected landmarks. Thus, the precise locations of these landmarks are critical for the effectiveness of the subsequent recognition and identification methods. Examples of landmarks include the positions of the nose root, alar and tip, the eye interior, exterior, top, bottom and cavities, and the top and bottom lip. The feature vectors computed from the landmark proprieties are generally compact and thus the landmark locations have to be precise to ensure high discrimination power.

Landmark detection and location are generally based on the curvatures of the 3D facial surface. Besl and Jain [364], for example, used the mean and the Gaussian curvatures (Chapter 2). Lu et al. [365], on the other hand, used the shape index to detect the nose tip and the points corresponding to the inside and the outside of the eye regions.

Many other hand-crafted detectors have been used for landmark localization. Examples include spin images [366], the curvature properties of the facial profile curves [278], and the radial symmetry of the 3D facial surface [367]. Machine learning-based algorithms, e.g. AdaBoost, have also been used to detect keypoints from 3D faces [368].

Once the facial landmarks have been detected, a feature vector computed from their configuration can be constructed and used for 3D face matching. Gupta et al. [369] proposed the Euclidean and geodesic distances between the detected landmarks as a feature. Gordon [370] proposed a feature that encodes the left and right eye widths, the eye separation, the total width of the eyes, the nose height, width, and its depth, and the head width. Moreno et al. [371] added the areas, distances, angles, and curvatures averaged within the regions around the detected landmarks. Xu et al. [372] used Gaussian-Hermite moments to encode the areas around a set of landmarks (e.g. the mouth, nose, and the left and right eyes).

10.3.2.1.2 SIFT-Like Keypoints Mian et al. [287] proposed a 3D keypoint detector inspired by the Scale Invariant Feature Transform detector [107]. For each sample point p on the 3D facial surface, a region of influence $\mathbf{P} = \{p_j = (x_j, y_j, z_j)^\top, j = 1, \ldots, n_i\}$ is extracted using a cropping sphere centered at p and of radius r. Let $\mathbf{V}$ be the matrix whose columns are the eigenvectors, sorted in descending order of their associated eigenvalues, of the covariance matrix $\mathbf{C} = \frac{1}{n_i - 1} \sum_{j=1}^{n_i} (p_j - p_\mu)(p_j - p_\mu)^\top$. Here, p_μ is the mean of $\{p_j, j = 1, \ldots, n_i\}$. These eigenvectors form a local coordinate frame. The coordinates of the points $\mathbf{P}$ in this local coordinate frame are computed as follows:

$$\mathbf{P}_{new} = \mathbf{V}^\top [(p_1 - p_\mu), \ldots, (p_j - p_\mu), \ldots, (p_{n_i} - p_\mu)]. \tag{10.8}$$

This is a 3-by-n_i matrix. Let $\mathbf{P}_{new}^{(1)}$ and $\mathbf{P}_{new}^{(2)}$ be the first two rows of $\mathbf{P}_{new}$. Mian et al. [287] compared the quantity $\delta = (\max(\mathbf{P}_{new}^{(1)}) - \min(\mathbf{P}_{new}^{(1)})) - $

$(\max(\mathbf{P}^{(2)}_{new}) - \min(\mathbf{P}^{(2)}_{new}))$ and a fixed threshold τ. A point p is defined as a keypoint if $\delta > \tau$. The first, respectively second, term of δ measures the range of the data along the first, respectively second, principal direction. They are in fact related to the first and second eigenvalues of the covariance matrix $\mathbf{C}$.

Once all keypoints are extracted from probe and gallery faces, Mian et al. [287] proposed to compute a high-level feature per keypoint. A graph-based matching algorithm is finally used for face comparison.

Smeets et al. [352] proposed another keypoint detector called the Mesh Scale-Invariant Feature Transform (MeshSIFT). The algorithm detects curvature extrema between different smoothed versions (i.e. scale spaces) of the 3D facial surface. Formally, MeshShift starts by constructing scale spaces corresponding to smoothed versions of the facial surface S. A scale space is obtained by subsequent convolutions of the surface S with a binomial filter, which updates each vertex v_i as follows:

$$v_i^s = v_i + \frac{1}{|\mathcal{N}_i|} \sum_{j \in \mathcal{N}_i} v_j, \tag{10.9}$$

where $\mathcal{N}_i$ is the list of vertices that are neighbors to v. At each scale s, the mean curvature H is then computed for every vertex. Finally, the keypoints are detected as the extrema of the mean curvatures between subsequent scales. This means that the vertex with higher (or lower) mean curvature than its neighbors along the upper and lower scale is selected as a keypoint. Note that Smeets et al. [352] tested other alternatives to the mean curvature but concluded that the latter provides better repeatability than the Gaussian curvature, the shape index, or the curvedness.

In order to perform 3D face recognition, Smeets et al. [352] described the neighborhood of each keypoint using a feature vector. The feature vectors of two 3D facial surfaces are matched by comparing the angles between each probe feature and all the gallery features.

10.3.2.2 Curve-Based Features

Many researchers have focused their attention on the description of the 3D face using curves. A curve can be a closed contour that is extracted according to a function defined on the 3D facial surface [373]. It can also be an open curve, also called a radial curve, corresponding to the intersection of a cutting plane with the 3D surface. Such curves contain information located at different subregions of the 3D face. The curve-based representation is generally less compact than landmark-based ones. In order to be effective, a large number of curves are needed to cover most of the area of the face. On the other hand, some of curve-based representations may highly rely on nose tip location [350, 374].

Samir et al. [373] proposed to approximate the 3D facial surface using a set of indexed level curves. These curves have been defined as the level contours of the depth (i.e. z-coordinate) function defined on the 3D surface. The comparison between two 3D faces is then performed based on the distance between curves at the same level. Curve distances are computed using a differential geometric approach proposed by Klassen et al. [375]. An extension of Samir et al. work has been proposed in [374]. The extended version cope with 3D curves extracted according to the level set of the geodesic distance computed on the 3D face and starting from the nose tip. Drira et al. [350] extended the work of Samir et al. [374] by representing 3D faces using a set of radial curves emanating from the nose tip. Radial curves from two different 3D faces are matched using the elastic metric proposed by Srivastava et al. [195]. The main advantage of the curve-based techniques that use the elastic metric of Srivastava et al. [195] is the fact that they capture the elasticity of the face along the curves. These methods, however, assume that the correspondence between the curves of the probe face and the curves of the gallery face is given. Thus, they do not capture the full elasticity of the 3D human faces.

Beretti et al. [272] proposed a 3D face recognition approach that captures characterizing features of the face by measuring the spatial displacement between a set of stripes. The stripes are computed using iso-geodesics starting from the nose tip. Each stripe is described by some geometric features and used as nodes in a graph. 3D face recognition is then considered as a graph-matching process.

10.3.2.3 Patch-Based Features

These algorithms generate local features using the geometric information of several local patches of the 3D facial surface. A number of patch-based features have been proposed, with a few representative examples listed below. For additional 3D local feature description algorithms, the reader is referred to the review papers of Guo et al. [85] and Soltanpour et al. [376].

1) **LBP-based features.** The local binary pattern (LBP) descriptor has been introduced by Ojala et al. [377] for texture analysis. Later, it has been used for 2D face recognition [378] and 3D face recognition [319, 343, 353, 379, 380]. Figure 10.3 illustrates the conventional LBP operator. It is computed at each

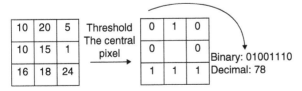

Figure 10.3 The conventional LBP operator computed on a pixel. The value (gray-level) of each pixel in the first ring neighborhood is compared to the value of the central pixel. If the value of the pixel is greater than the value of the central one, then a corresponding binary bit is assigned to one; otherwise, it is assigned to zero. The decimal value is used in the matching process.

pixel of the 2D image of the face. A histogram that counts the number of occurrences of each decimal value in the whole image is then computed and used to match faces. Werghi et al. [381] extended the LBP operator to 3D mesh surfaces by constructing rings of facets around a central one. The method relies on ideal mesh tessellation with six-valence vertices.

2) **Covariance matrices.** Tabia and Laga [139] proposed a 3D face recognition method based on covariance descriptors. Unlike feature-based vectors, covariance-based descriptors enable the fusion and the encoding of different types of features and modalities into a compact representation.

Let $\mathcal{P} = \{P_i, i = 1, \dots, m\}$ be the set of patches extracted from a 3D face. Each patch P_i defines a region around a feature point $p_i = (x_i, y_i, z_i)^t$. For each point p_j in P_i, compute a feature vector $\mathbf{x}_j$, of dimension d, which encodes the local geometric and spatial properties of the point. Different local properties have been tested including the 3D coordinates of the point p_j, the minimum and maximum curvatures, and the distance of p_j from the origin. The covariance matrix X_i computed for a 3D face patch is then given by:

$$X_i = \frac{1}{n} \sum_{j=1}^{n} (\mathbf{x}_j - \mu)(\mathbf{x}_j - \mu)^\top, \tag{10.10}$$

where μ is the mean of the feature vectors $\{\mathbf{x}_j\}_{j=1,\dots,n}$ computed on the patch P_i, and n is the number of points in P_i. The diagonal entries of X_i represent the variance of each feature and the nondiagonal entries represent their respective covariances.

The matching between covariance matrices is performed using geodesic distances on the manifold of symmetric positive definite (SPD) matrices. The Hungarian algorithm for matching unordered sets of covariance descriptors is used for 3D face recognition.

10.3.2.4 Other Features

Many other 3D patch-based features have been proposed for 3D face recognition. Examples includes point signature features, which have been initially proposed for 3D object recognition [85, 128] and then extended to the area of 3D face recognition [382–384], the geodesic polar representation which has been proposed for expression-invariant 3D face recognition [284], and the TriSI feature which copes with meshes with varying resolutions, occlusion, as well as clutter [385, 386].

10.4 Summary

3D face recognition is one of the most popular applications of 3D shape analysis. It is still a very active research topic in both computer vision and pattern recognition communities [387–389]. Several researchers have focused on facial surface description and comparison. We have seen that feature extraction and selection play a critically important role within any face recognition system. In this chapter, we have presented the main challenges and the main evaluation criteria for 3D face recognition. We have also discussed several face recognition methods. Although significant advances have been made in the past, efficient 3D face recognition is still a very challenging task. Most of the existing methods only consider 3D face surfaces of high quality and fail to recognize 3D faces captured with low-resolution scanners.

11

Object Recognition in 3D Scenes

11.1 Introduction

Object recognition has a large number of applications in various areas such as robotics, autonomous vehicles, augmented reality, surveillance systems, and automatic assembly systems. The task of 3D object recognition is to determine, in the presence of clutter and occlusion, the identity and pose (i.e. position and orientation) of an object of interest in a 3D scene [129]. Compared to 2D images, 3D data usually provide more accurate geometrical and pose information about 3D objects. Thus, 3D object recognition in cluttered scenes is always an important research topic in computer vision. Recently, the rapid development of low-cost 3D sensors (such as Microsoft Kinect, Asus Xtion, and Intel Realsense) has further boosted the number of applications of 3D object recognition systems.

Existing 3D object recognition methods can be classified into surface matching-based and machine learning-based methods. In this chapter, we first cover the surface registration-based 3D object recognition methods (Section 11.2). We then introduce some recent advances in machine learning-based 3D object recognition methods (Section 11.3). We finally summarize this chapter with some discussions in Section 11.4.

11.2 Surface Registration-Based Object Recognition Methods

Most traditional 3D object recognition methods follow a hypothesize-and-test surface registration scheme, which usually contains three modules: (i) a feature matching module, (ii) a hypothesis generation module, and (iii) a hypothesis verification module, see Figure 11.1. Specifically, given a scene and several models of interest, local features are first extracted and then matched against all the features of the model, resulting in a set of feature correspondences. Next, these feature correspondences are used to generate candidate models

3D Shape Analysis: Fundamentals, Theory, and Applications, First Edition.
Hamid Laga, Yulan Guo, Hedi Tabia, Robert B. Fisher, and Mohammed Bennamoun.
© 2019 John Wiley & Sons, Inc. Published 2019 by John Wiley & Sons, Inc.

Figure 11.1 A general scheme of hypothesize-and-test based methods. This scheme consists of three modules, namely: a feature matching module, a hypothesis generation module, and a hypothesis verification module.

and transformation hypotheses between the scene and each candidate model. Each candidate model and transformation hypothesis is then verified using a surface alignment approach. Finally, the identity and pose of each object in the scene is determined.

11.2.1 Feature Matching

Feature matching is the first step of surface registration-based 3D object recognition methods. It aims to generate a set of feature correspondences between a scene and a model. Similar to image feature matching [390], three types of feature matching strategies can be used, namely: threshold-based matching, nearest neighbor (NN)-based matching, and nearest neighbor distance ratio (NNDR)-based matching.

- *Threshold-based matching*: given a feature in the scene, all the features in the model with distances to the scene feature that are smaller than a threshold are retrieved. These retrieved model features and the scene feature form several pairs of feature correspondences.
- *NN-based matching*: for a feature in the scene, if a model feature has the smallest distance to the scene feature and the distance is smaller than a threshold, the model feature and the scene feature are then considered to be a feature correspondence.
- *NNDR-based matching*: for a feature in the scene, if a model feature is the nearest one to the scene feature, and the ratio of the nearest to the second nearest one is smaller than a threshold, the model feature and the scene feature are considered to be a feature correspondence.

For both NN and NNDR-based feature matching, the nearest neighbor in the model feature library has to be generated for a given scene feature. A naive approach is to calculate the distance between the scene feature and all the features of the model. This method is very accurate but also time-consuming. Alternatively, several feature indexing approaches, such as feature hashing, k-d tree, and locality sensitive tree [103, 129, 391], can be used to accelerate the feature matching process.

11.2.2 Hypothesis Generation

Once the feature matching process is performed, the feature correspondences $C = \{C_1, C_2, \ldots, C_{n_c}\}$ can be used to vote for candidate models that are present in the scene, and obtain, in addition, the transformation between the scene and the candidate model. The hypothesis generation process consists of two major tasks: (i) the generation of candidate models and (ii) the generation of transformation hypotheses.

To generate a transformation hypothesis between a scene and a model, three typical cases should be considered according to the availability of information on local reference frames (LRFs), positions, and surface normals.

- *Local reference frame*: if the LRF of each keypoint in the feature correspondences C is used, a transformation hypothesis can be obtained from each pair of feature correspondences. Specifically, given a LRF F_s of a keypoint p_s on the scene, and the LRF F_m of its corresponding keypoint p_m on the model, the rotation hypothesis between the scene and the model can be calculated as $O = F_s^{\mathsf{T}} F_m$, and the translation can be calculated as $t = p_s - O p_m$. Consequently, n_c transformation hypotheses can be generated from the feature correspondences C.
- *Position and surface normal*: if both the position and the surface normal of each keypoint in the feature correspondences C are used, a transformation hypothesis can be obtained from every two pairs of feature correspondences. Specifically, a candidate rotation matrix can be estimated by aligning the surface normals of two pairs of corresponding points. Then, a candidate translation can be estimated by aligning the two pairs of corresponding points (more details in Section 6.2.1.4). Therefore, $C_{n_c}^2$ transformation hypotheses can be generated from the feature correspondences C.
- *Position*: if only the position of each keypoint in the feature correspondences C is used, a transformation hypothesis can be obtained from every three pairs of feature correspondences by minimizing the least squared error of the coordinates of these corresponding points (more details in Section 6.2.1.3). Therefore, $C_{n_c}^3$ transformation hypotheses can be generated from the feature correspondences C.

Since there are false matches in the feature correspondences C, several false transformation hypotheses can also be generated. How to derive an accurate transformation hypothesis from C is an important issue. Different types of hypothesis generation methods have been proposed in the literature, including geometric consistency-based, pose clustering-based, constrained interpretation tree-based, RANdom SAmple Consensus (RANSAC)-based, game theory-based, and generalized Hough transform-based methods.

11.2.2.1 Geometric Consistency-Based Hypothesis Generation

In general, transformations which are generated from correspondences that are not geometrically consistent usually have larger errors. Therefore, geometric consistency can be used to group feature correspondences for the calculation of plausible transformations. A transformation, which is generated from a set of geometrically consistent feature correspondences, is more robust than the one which is generated from just a few raw feature correspondences.

A typical method is proposed in [95, 392]. Given a set of feature correspondences $C = \{C_1, C_2, \ldots, C_{n_c}\}$, a seed correspondence C_i is selected from the list C to initialize a group $G_i = \{C_i\}$. Then, if a correspondence C_j is the one which is the most geometrically consistent with G_i and the consistency measure is larger than a threshold, the correspondence C_j is added to G_i. The process for adding correspondences to G_i is continued until no feature correspondence can be added anymore. Once the grouping process for one correspondence C_i is completed, the process is then performed on another correspondence in C. Consequently, n_c groups of geometrically consistent correspondences can be generated from C. Note that, a feature correspondence can appear in multiple groups. Finally, each group G_i is used to calculate a transformation hypothesis T_i.

11.2.2.2 Pose Clustering-Based Hypothesis Generation

If a scene is correctly matched with a model, there should be a large number of transformations that are close to the ground-truth transformation between the scene and the model. Therefore, if clustering is performed on these transformations, there should be a transformation cluster that is very similar to the ground-truth transformation. Based on this assumption, Mian et al. [93] introduced a pose clustering approach. Given a set of feature correspondences $C = \{C_1, C_2, \ldots, C_{n_c}\}$, a transformation can be generated from every k pairs of feature correspondences. Here, k is the minimum number of feature correspondences which is required to estimate a transformation. It can be 1, 2, or 3 depending on the information used for transformation estimation. For these n_c feature correspondences, $C_{n_c}^k$ transformations can be obtained. For each transformation $\{O, t\}$, the rotation matrix O is first converted to three Euler angles and then combined with the translation vector t to obtain a six-dimensional vector. Consequently, $C_{n_c}^k$ transformation vectors can be generated in total. Then, these transformations are clustering using, for example, the k-means clustering algorithm. The centers of the transformation clusters are considered as transformation hypotheses. Also, the number of transformations falling into a cluster provides a measure for the confidence of the transformation hypothesis.

11.2.2.3 Constrained Interpretation Tree-Based Hypothesis Generation

The constrained interpretation tree-based hypothesis generation method, introduced in [105], performs as follows. For each model, an interpretation tree is generated. Specifically, there is no correspondence at the root of the tree. Then, each successive level of the tree is built by sequentially selecting a model feature and adding its corresponding scene features, with highly similar descriptors, to the tree as nodes. Therefore, each node in the tree represents a hypothesis, which is formed by the feature correspondences at that node and all its parent nodes. Searching all correspondences that are represented by the entire interpretation tree, for a complex scene, is extremely time-consuming. Therefore, several strategies are proposed to constrain and prune the tree. Below, we discuss a few of them:

- *Scale hierarchy*: for a model feature $\mathbf{x}_{m1}^{\sigma_1}$ with a scale σ_1, the successive levels of the tree are generated using the model features $\{\mathbf{x}_{m2}^{\sigma_2}\}$ with the scale $\sigma_2 \leq \sigma_1$. Consequently, a hierarchical structure of the interpretation tree is built.
- *Valid correspondences*: a correspondence between a model feature $\mathbf{x}_m^{\sigma_a}$ and a scene feature $\mathbf{x}_s^{\sigma_b}$ is considered valid if they have the same scale (i.e. $\sigma_a = \sigma_b$), and the similarity between these two features is larger than a threshold τ_s.
- *Geometric constraint*: for each node in the tree, a set of correspondences at that node and all its parent nodes are represented. Therefore, a transformation T can be calculated for that node to align the scene feature points and the model feature points which belong to these correspondences. Given a feature correspondence $(\mathbf{x}_m^{\sigma_a}, \mathbf{x}_s^{\sigma_b})$ and the associated feature points p_m and p_s, the correspondence can be added to the interpretation tree only if $\| Tp_m - p_s \| \leq \epsilon$, where ϵ is a threshold.
- *Pruning*: if the number of nodes at any level of the tree is larger than a threshold n_{max}, the tree is pruned, i.e. only the n_{pruned} nodes that correspond to the most confident hypotheses are kept. Here, the confidence of a hypothesis is defined by the cardinality of its correspondence set and the average transformation error produced by its corresponding transformation. Specifically, all nodes at the level to be pruned are sorted in a descending order according to the cardinalities of their correspondence sets. If several nodes have the same number of correspondences, they are ranked in an ascending order according to the average transformation error. Once these nodes are ranked, only the top n_{pruned} nodes are preserved, while the rest are pruned.

Finally, n_{pruned} transformation hypotheses between the scene and the model can be generated from the remaining nodes in the interpretation tree.

11.2.2.4 RANdom SAmple Consensus-Based Hypothesis Generation

RANSAC [164] is an iterative method for hypothesis generation. First, a minimal set of point correspondences are randomly selected to calculate a transformation between the scene and the model. The number of transformed model points, with their corresponding points on the scene having distances smaller than a threshold (i.e. inliers), is used to measure the score of the estimated transformation. This process is repeated for a predetermined number of iterations and the transformation with the highest score is considered as the transformation hypothesis. Given the probability $prob_1$ of recognizing the object in a single iteration and the number of iterations n_{itr}, the probability $prob_o$ of recognizing an object in the scene within n_{itr} trials can be calculated as:

$$prob_o = 1 - (1 - prob_1)^{n_{itr}}. \tag{11.1}$$

If the inlier rate of feature correspondences can be estimated, then the probability $prob_1$ is known. Specifically, given the inlier ratio γ and the number n_c of feature correspondences, then $prob_1 = C_{\gamma n_c}^k / C_{n_c}^k$, where C represents the combination operation, k is the number of feature correspondence pairs for the estimation of a candidate transformation between two surfaces.

Given the desired object recognition probability $prob_o$, the minimum number of iterations can be calculated as:

$$n_{itr} = \frac{\ln(1 - prob_o)}{\ln(1 - prob_1)}. \tag{11.2}$$

The RANSAC method is simple, general, and robust to outliers [393]. However, its direct application to 3D object recognition is computationally very expensive.

11.2.2.5 Game Theory-Based Hypothesis Generation

Game theory can be used to select a set of feature correspondences satisfying a global rigidity constraint. The survived feature correspondences after a noncooperative game competition are used to generate several transformation hypotheses [394, 395].

11.2.2.5.1 Preliminary on Game Theory
Game theory is used to formalize a system characterized by the actions of different individuals with competing objectives. Specifically, each player aims to use a set of strategies to maximize their payoff, where the payoff also depends on the strategies that are used by the other players. Assume that a set of available strategies $O = \{s_1, s_2, \ldots, s_{n_s}\}$ and a matrix $\Pi = \{\pi_{ij}\}$ that represents the payoff that an individual receives when that individual plays strategy s_i against the strategy s_j played by another individual. The probability distribution $\mathbf{p} = (\mathbf{p}_1, \mathbf{p}_2, \ldots, \mathbf{p}_{n_s})^\top$ over the strategy space O gives the mixed strategy, where $\mathbf{p}$ lies in the n_s-dimensional standard simplex Δ, that is:

$$\Delta = \left\{ \mathbf{p} \in \mathbb{R}^{n_s} : \mathbf{p}_i \geq 0 \quad \text{and} \quad \sum_{i=1}^{n_s} \mathbf{p}_i = 1 \right\}. \tag{11.3}$$

If an individual plays a strategy s_i against a mixed strategy $\mathbf{p}$ played by another player, the expected payoff for the strategy s_i is $(\Pi \mathbf{p})_i = \sum_j \pi_{ij} \mathbf{p}_j$ [394]. Note that, the mixed strategy $\mathbf{p}$ provides a probability distribution over the strategy space. Therefore, if a mixed strategy $\mathbf{q}$ is played against a mixed strategy $\mathbf{p}$, the expected payoff is $\mathbf{q}^\top \Pi \mathbf{p}$. The best replies against a mixed strategy $\mathbf{p}$ is the mixed strategy $\mathbf{q}$ which achieves the maximum value of $\mathbf{q}^\top \Pi \mathbf{p}$. If a strategy $\mathbf{p}$ is the best reply to itself, then the strategy $\mathbf{p}$ is considered to be a Nash equilibrium, i.e. $\forall \mathbf{q} \in \Delta, \mathbf{p}^\top \Pi \mathbf{p} \geq \mathbf{q}^\top \Pi \mathbf{p}$. In game theory, if no player can benefit from the game by changing its own strategies while the strategies of the other players remain unchanged, then the current set of strategies and the payoffs reaches a Nash equilibrium. That is, the payoffs of all available strategies are the same.

11.2.2.5.2 Matching Game for Transformation Hypothesis Generation Given a scene and a candidate model, keypoints are detected and their associated feature descriptors (e.g. SHOT [394]) are extracted both from the scene and the model (see Chapter 5). Then, a matching game can be built based on four entities: the set of scene keypoints P, the set of model keypoints Q, the set of feature correspondences $C \subseteq P \times Q$, and a pairwise compatibility function between feature correspondences $\Pi : C \times C \rightarrow \mathbb{R}_+$. The task of this gameplay is to find a set of compatible elements from C. In this matching game, the available strategies O are defined by the set of feature correspondences C, the payoffs between strategies are determined by the pairwise compatibility function Π, while the probability distribution $\mathbf{p} = (\mathbf{p}_1, \mathbf{p}_2, \ldots, \mathbf{p}_{n_s})^\top$ represents the population number that strategy s_i is played at a given time. To start the game, the probability distribution (i.e. population) is first initialized with its barycenter. The population is then evolved using the replicator dynamics equation [394]:

$$\mathbf{p}_i(t+1) = \mathbf{p}_i(t)\frac{(\Pi \mathbf{p}(t))_i}{(\mathbf{p}(t))^\top \Pi \mathbf{p}(t)}. \tag{11.4}$$

This dynamics will finally converge to a Nash equilibrium. However, there are two issues that remain to be resolved: (i) how to generate feature correspondences C from scene keypoints P and the model keypoints Q and (ii) how to define the pairwise compatibility function Π.

- *Feature correspondence generation*: to obtain the strategies C, each scene keypoint is associated with its k-nearest model keypoints in the descriptor space. If the distance between the descriptor of a scene keypoint and its closest descriptor of the model keypoint is too far, the scene keypoint and its associated model keypoints are excluded from C. Note that, different approaches for feature correspondence generation (as listed in Section 11.2.1) may be used to generate the strategies C.
- *Pairwise compatibility calculation*: If rigid transformation between scene keypoints and model keypoints is enforced, then strategies conforming to the rigidity constraint are more likely to lay on the same surface of both the scene and the model. Given two pairs of strategies $s_1 = (p_1, q_1)$ and $s_2 = (p_2, q_2)$, the compatibility can be calculated as [394]:

$$\delta((p_1, q_1),(p_2, q_2)) = \frac{\min(\|p_1 - p_2\|, \|q_1 - q_2\|)}{\max(\|p_1 - p_2\|, \|q_1 - q_2\|)}, \tag{11.5}$$

where p_1 and p_2 are scene keypoints, q_1 and q_2 are their corresponding model points. The compatibility is one if the distance between two keypoints in the scene is exactly the same as the distance between their corresponding keypoints on the model. In order to further remove many-to-many matches, the compatibility is set to 0 if the source or destination point is shared by multiple correspondences. Consequently, the final pairwise compatibility for the game is defined as:

$$\Pi = \begin{cases} \delta((p_1, q_1), (p_2, q_2)) & \text{if } p_1 \neq p_2 \text{ and } q_1 \neq q_2, \\ 0 & \text{otherwise.} \end{cases} \tag{11.6}$$

Once the feature correspondences C and the payoff matrix Π are generated, the matching game can be started from the barycenter of the simplex Δ (as defined in Eq. (11.3)). When a Nash equilibrium is reached, the strategies (feature correspondences) supported by a large percentage of the population are considered as correct matches and used to generate transformation hypotheses.

11.2.2.6 Generalized Hough Transform-Based Hypothesis Generation

This technique performs voting in a parametric Hough space using feature correspondences, where each point in the Hough space corresponds to a transformation between the scene and the model [85]. The peaks with values higher than a threshold in the Hough accumulator are used to generate transformation hypotheses. A typical method is proposed in [396], and is briefly described below.

During offline training, a reference point q_r is calculated for the model (usually the centroid of the model), and a LRF is generated for each keypoint in the model and scene. For each model keypoint q_i, a vector is calculated as

$$\boldsymbol{v}_{iG}^M = q_r - q_i. \tag{11.7}$$

To achieve invariance to rotation and translation, each vector $\boldsymbol{v}_{iG}^M$ is transformed to the coordinates defined by the LRF of the keypoint p_i:

$$\boldsymbol{v}_{iL}^M = O_{GL}^M \cdot \boldsymbol{v}_{iG}^M, \tag{11.8}$$

where O_{GL}^M is the rotation matrix defined by the LRF of q_i. That is, $R_{GL}^M = [\boldsymbol{L}_{ix}^M, \boldsymbol{L}_{iy}^M, \boldsymbol{L}_{iz}^M]^T$, where $\boldsymbol{L}_{ix}^M$, $\boldsymbol{L}_{iy}^M$, and $\boldsymbol{L}_{iz}^M$ are the X, Y, and Z axes of the LRF of point q_i. Once the above process is completed, each model keypoint q_i is associated with a vector $\boldsymbol{v}_{iL}^M$.

During the online stage, a scene keypoint p_i corresponding to the model keypoint q_i is obtained from the feature matching, $\boldsymbol{v}_{iL}^S$ is equal to $\boldsymbol{v}_{iL}^M$ due to the rotation invariance of the LRF. $\boldsymbol{v}_{iL}^S$ is then transformed to the global coordinates of the scene:

$$\boldsymbol{v}_{iG}^S = O_{LG}^S \boldsymbol{v}_{iL}^S + p_i, \tag{11.9}$$

where O_{LG}^S is the rotation matrix defined by the LRF of p_i. That is, $R_{LG}^S = [L_{ix}^S, L_{iy}^S, L_{iz}^S]$, where L_{ix}^S, L_{iy}^S, and L_{iz}^S are the X, Y, and Z axes of the LRF of point p_i.

For each scene keypoint p_i, a vote in the 3D space represented by v_{iG}^S can be cast. The presence of a particular model can be supported by thresholding the peaks in the Hough space. For a candidate model, the subset of feature correspondences voting for the presence of the model are used to generate the transformation hypotheses.

11.2.3   Hypothesis Verification

Once hypotheses are generated from feature matching, true hypotheses have to be distinguished from the false ones to correctly recognize an object in the scene. The existing hypothesis verification methods can broadly be classified into two categories: individual verification and global verification [86].

11.2.3.1   Individual Verification

The individual verification method considers a candidate model and a transformation hypothesis at a time. Hypothesis verification can be performed using different approaches including feature matching, surface registration, and surface consistency (such as active space violations).

Given each candidate model and a transformation hypothesis, the scene is first coarsely aligned with the model using the transformation hypothesis. The alignment between the model and the scene is then refined using a fine registration algorithm, such as iterative closest point (ICP) [158] described in Chapter 6. If part of the scene can be accurately registered with the candidate model, the candidate model is considered as being present in the scene. The pose of the object can then be determined by the transformation hypothesis after refinement. Subsequently, the scene points that are close to the transformed model can be recognized and segmented from the scene. If the scene cannot be aligned with the candidate model, the transformation hypothesis is rejected. Once a pair of candidate model and transformation hypothesis is verified, the next pair is checked individually by turn. The verification process is continued until all possible pairs of candidate models and transformation hypotheses are verified.

The major difference between the existing methods lies in the method for verification. Several approaches can be individually used or combined to verify the candidate model and the transformation hypothesis. For example:

- *Similarity between features*: since feature matching is used to generate the transformation hypotheses between the scene and the model, it is natural to use a feature matching similarity to verify the alignment between the scene and the model [391].

- *Surface registration residual*: once the scene and the candidate model are registered, the surface registration error (e.g. the average distance between model and scene surfaces) can be used to verify the hypothesis [129]. That is, a correct hypothesis usually produces a small residual error while a false hypothesis produces a high residual error.
- *Overlap ratio*: after surface registration between the scene and the candidate model, their overlap ratio can be calculated. If the overlap ratio is larger than a threshold, then one can assume that the candidate model is present in the scene. Specifically, in [95], the ratio between the number of corresponding points to the number of points on the candidate model is used as the overlap ratio. In [105], the overlap area between the scene and the candidate model is first calculated, and the overlap ratio is calculated by dividing the overlap area to the total visible surface area of the model.
- *Active space violations*: the hypotheses can further be verified by checking if the active space of the sensor has been violated by the transformed model [391, 397]. There are two types of active space violations: free space violation (FSV) and occupied space violation (OSV). Normally, the space along the line of sight from the depth sensor to the surface of the scene should be clear (as shown in Figure 11.2a). If a region of the model is transformed into the space between the depth sensor and the surface of the scene, i.e. the transformed model $\mathcal{M}$ blocks the visibility of the scene S from the depth sensor $\mathcal{V}_c$ (as shown in Figure 11.2b), a FSV occurs. If a region of the scene S that is visible to the depth sensor $\mathcal{V}_c$ is not observed (as shown in Figure 11.2c), that is, part of the model is transformed into a visible region of the scene where no surface is detected by the depth sensor, an OSV occurs. FSVs can only be caused by the misalignment between the scene and the model, while OSVs

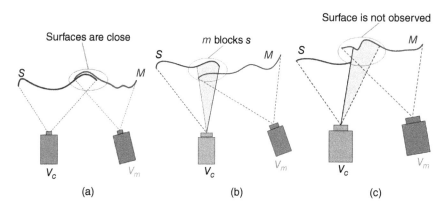

Figure 11.2 An illustration of active space violations. (a) Consistent surfaces: the space along the line of sight from the depth sensor $\mathcal{V}_c$ to the surface of the scene is clear. (b) Free space violation (FSV): the transformed model $\mathcal{M}$ blocks the visibility of the scene S from the depth sensor $\mathcal{V}_c$. (c) Occupied space violation (OSV): a region of the scene S that is visible to the depth sensor $\mathcal{V}_c$ is not observed.

can be caused by the misalignment between the scene and the model, or the presence of sensor errors (e.g. missing data on black surfaces). Therefore, if significant active space violations are detected, the hypothesis should be rejected.

An individual verification method considers a single hypothesis at a time by thresholding an alignment score. Determining an appropriate threshold is very challenging. That is, a high threshold can not only avoid most false positives but also produce a low recognition rate. In contrast, a low threshold can improve the recognition rate but it also produces a high false positive rate. Besides, the interaction between hypotheses is also disregarded by an individual verification method, while the interaction is potentially able to improve the performance of object recognition, especially in challenging scenarios (e.g. highly occluded objects).

11.2.3.2 Global Verification

A global hypothesis verification method considers all hypotheses simultaneously to select a subset of hypotheses which best explains the scene data. It can be formalized as a process of minimizing a cost function defined over the set of hypotheses and the scene. The cost function can be defined using both geometric and appearance cues.

The first global verification method is presented in [398]. Given a library $\mathbf{M}$ with n_m models (i.e. $\mathbf{M} = \{\mathcal{M}_1, \dots, \mathcal{M}_{n_m}\}$) and a scene S, the transformation T between a model and its instance in the scene is defined by a 6-DOF rigid transformation. Note that, multiple instances or no instance in $\mathbf{M}$ can be present in the scene S. Assume that n_h recognition hypotheses have been generated as $\mathcal{H} = \{h_1, \dots, h_{n_h}\}$ (Section 11.2.2), each hypothesis h_i consists of a candidate model $\mathcal{M}_{h_i}$ and a transformation hypothesis T_{h_i}. The task of a global hypothesis verification is to select a subset of hypotheses from $\mathcal{H}$ to maximize the recognition rate, while minimizing the false positive rate, using an optimization approach.

The selection of the subset of hypotheses can be represented by a set of Boolean variables $\mathcal{X} = \{x_1, \dots, x_{n_h}\}$, where $x_i = 1$ means that the hypothesis h_i is selected, $x_i = 0$ means that the hypothesis h_i is dismissed. A cost function can then be defined over the solution space $\mathcal{X}$:

$$\mathfrak{F}(\mathcal{X}) = f_s(\mathcal{X}) + \lambda f_m(\mathcal{X}) + \beta \|\mathcal{X}\|_0, \tag{11.10}$$

where f_s and f_m are the cost terms for the scene and model cues, respectively. λ controls the weight for the model cues, and $\beta \|\mathcal{X}\|_0$ is a regularization term for encouraging a sparse solution. Note that, the function $\mathfrak{F}(\mathcal{X}):\mathbb{B}^{n_h} \to \mathbb{N}$ is a pseudo-Boolean function, where $\mathbb{B} = \{0, 1\}$ is a Boolean domain. The cost term f_s for the scene cues is defined as:

$$f_s(\mathcal{X}) = \sum_{p \in S} (-\Omega_{\mathcal{X}}(p) + \Lambda_{\mathcal{X}}(p) + \Upsilon_{\mathcal{X}}(p)). \tag{11.11}$$

The cost term f_m for the model cues is defined as:

$$f_m(\mathcal{X}) = \sum_{i=1}^{n_h} |\Phi_{h_i}| x_i, \tag{11.12}$$

where $|\cdot|$ denotes the cardinality operator for a set. Note that, the terms $\Omega_{\mathcal{X}}$, Φ_{h_i}, $\Lambda_{\mathcal{X}}$, and $\Upsilon_{\mathcal{X}}$ in Eqs. (11.11) and (11.12) represent four different basic cues and are explained below. Specifically, the cue defined by $\Omega_{\mathcal{X}}$ is used to maximize the number of model instances that are recognized in the scene, while the other cues are used to minimize the number of false positive recognition results. The global verification method operates as follows;

1) **The Term Φ_{h_i}.** A model hypothesis $\mathcal{M}_{h_i}$ is transformed according to the transformation hypothesis T_{h_i} and its occluded points that are not visible in the scene are removed, resulting in a visible model $\mathcal{M}_{h_i}^v$. To check if a point on $\mathcal{M}_{h_i}$ is visible or not in the scene, it is first back-projected onto the rendered range image of the scene. If its projection falls onto a valid depth pixel and its depth is smaller than the depth of the corresponding pixel in the rendered range image, the point is considered to be visible. Otherwise, the point is considered to be occluded or self-occluded.

   Then, each point on $\mathcal{M}_{h_i}^v$ has to be checked to determine if it has a correspondence in the scene. If a correspondence can be found in the scene, the point supports the presence of the model instance in the scene. Consequently, the model point is considered as a *model inlier*, and the corresponding scene point is considered as an *explained* point. Otherwise, the model point does not support the presence of the object in the scene, the model point is considered as a *model outlier*, and the corresponding scene point is considered as an *unexplained* point.

   Given a model point $q \in \mathcal{M}_{h_i}^v$ and a small neighborhood on the scene $\mathcal{N}(q, S)$, a feature vector $\mathbf{x}_q$ is extracted using geometric and color cues to represent the fitting degree between q and $\mathcal{N}$. The Mahalanobis distance $D_m(\mathbf{x}_q)$ of the feature vector $\mathbf{x}_q$ to the feature distribution (μ, Σ) learned from inliers can be calculated as:

$$D_m(\mathbf{x}_q) = \sqrt{(\mathbf{x}_q - \mu)^\mathsf{T} \Sigma^{-1} (\mathbf{x}_q - \mu)}. \tag{11.13}$$

   If $D_m(\mathbf{x}_q) \leq \rho_e$, the model point q is considered to be an inlier. Otherwise, it is considered to be an outlier. Here, μ and Σ are the mean and covariance of the feature vector learned from inliers, respectively, and ρ_e is a predefined threshold. Then, Φ_{h_i} is used to represent the set of model outliers for hypothesis h_i. The set cardinality $|\Phi_{h_i}|$ is further used to define the model cost term in Eq. (11.12). An illustration of the scene and model hypotheses is shown in Figure 11.3a where the scene point cloud is shown in dark gray and the active model hypotheses are superimposed onto the scene. In Figure 11.3b,

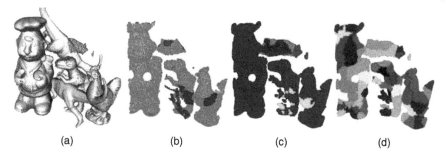

(a) (b) (c) (d)

Figure 11.3 Illustration of the different terms for global verification. (a) A scene and model hypotheses, where the scene point cloud is shown in dark gray and the active model hypotheses are superimposed onto the scene. (b) The model inliers (shown in light gray) and the model outliers (shown in dark gray). (c) The scene points with a single hypothesis, scene points with multiple hypotheses, and unexplained scene points are shown in different gray levels. (d) A segmented scene, where each gray level represents a segment label. Source: Reprinted with permission from Reference [398]. ©2015, IEEE.

the model inliers are shown in light gray, and the model outliers are shown in dark gray.

The term Φ_{h_i} is used to minimize the number of outliers associated with the active hypotheses defined by $\mathcal{X}$.

2) **The Term $\Omega_{\mathcal{X}}$.** A weight $\delta(q)$ for the model point q can be defined as:

$$\delta(q) = \begin{cases} 1 - \dfrac{D_m(\mathbf{x}_q)}{\rho_e} & \text{if } D_m(\mathbf{x}_q) \le \rho_e, \\ 0 & \text{otherwise.} \end{cases} \tag{11.14}$$

If a scene point $p \in S$ belongs to the neighborhood $\mathcal{N}(q, S)$ of one or more model inliers in $\mathcal{M}_{h_i}^v$, the scene point p is then considered as a point *explained* by h_i. Otherwise, it is considered as an *unexplained* point. That is,

$$\eta_{h_i}(p) = \begin{cases} 1 & \text{if } \exists q \in \mathcal{M}_{h_i}^v : p \in \mathcal{N}(q, S) \wedge \delta(q) > 0, \\ 0 & \text{otherwise.} \end{cases} \tag{11.15}$$

Note that, a scene point p could be explained by more than one model inliers in $\mathcal{M}_{h_i}^v$. Let $Q \in \mathcal{M}_{h_i}^v$ represent the set of model inlier points explaining the point p. A function to measure how accurately p is explained by h_i can be calculated as:

$$\omega_{h_i}(p) = \max_{q \in Q} \delta(q). \tag{11.16}$$

Therefore, the term $\Omega_{\mathcal{X}}(p)$ in Eq. (11.11) is calculated as:

$$\Omega_{\mathcal{X}}(p) = \max_{i=1,\dots,n_h} (x_i \cdot \omega_{h_i}(p)). \tag{11.17}$$

The term Φ_{h_i} is used to maximize the number of scene points explained by hypotheses defined in the solution set $\mathcal{X}$. That is, it is used to maximize the number of model instances that are recognized in the scene.

3) **The Term $\Lambda_{\mathcal{X}}$.** Ideally, a scene point p should be explained by no more than one hypothesis. If a scene point is explained by two or more hypotheses, there might be incoherent hypotheses within the solution $\mathcal{X}$. Therefore, given a scene point p and a solution $\mathcal{X}$, $\Lambda_{\mathcal{X}}$ can be defined as:

$$\Lambda_{\mathcal{X}}(p) = \begin{cases} \sum_{i=1}^{n_h} x_i \cdot \omega_{h_i}(p) & \text{if } \sum_{i=1}^{n_h} x_i \cdot \eta_{h_i}(p) > 1, \\ 0 & \text{otherwise.} \end{cases} \qquad (11.18)$$

Here, $\Lambda_{\mathcal{X}}$ provides the soft-weighted number of conflicting hypotheses for a scene point p. An illustration is given in Figure 11.3c where the scene points with a single hypothesis, scene points with multiple hypotheses, and unexplained scene points are shown in different gray levels.

The term $\Lambda_{\mathcal{X}}$ is used to minimize the number of scene points that are explained by multiple hypotheses.

4) **The Term $\Upsilon_{\mathcal{X}}$.** To handle clutter, a hypothesis which explains several scene points, but not their neighboring points, on the same smooth surface patch should be penalized. Specifically, a scene is first segmented, with each scene point being given a segment label $l(p)$, as shown in Figure 11.3d. For a solution $\mathcal{X}$, a clutter term $\Upsilon_{\mathcal{X}}(p)$ can be calculated at each unexplained point p:

$$\Upsilon_{\mathcal{X}}(p) = \sum_{i=1}^{n_h} x_i \cdot \gamma_{h_i}(p), \quad \text{where } \gamma_{h_i}(p) = k \frac{|S_{h_i}^{l(p)}|}{|S^{l(p)}|}. \qquad (11.19)$$

Here, $|S_{h_i}^{l(p)}|$ is the number of scene points with label $l(p)$ explained by h_i, $|S^{l(p)}|$ is the total number of scene points with label $l(p)$, and k is a weighting parameter for the clutter term.

The term $\Upsilon_{\mathcal{X}}(p)$ can be used to minimize the number of unexplained scene points.

Once the cost function $\mathfrak{F}(\mathcal{X})$ is defined using the four terms, the solution $\tilde{\mathcal{X}}$ can be obtained by minimizing the cost function $\mathfrak{F}(\mathcal{X})$ over the entire solution space $\mathbb{B}^{n_h}$:

$$\tilde{\mathcal{X}} = \arg \min_{\mathcal{X} \in \mathbb{B}^{n_h}} (\mathfrak{F}(\mathcal{X})). \qquad (11.20)$$

The cardinality of the entire solution space $\mathbb{B}^{n_h}$ is 2^{n_h}. Thus, an exhaustive search over the solution space to find the global minimum is computationally prohibited. Instead, a solver for the pseudo-Boolean optimization problem should be used. In [398], three popular meta-heuristic techniques were investigated, including the local search, simulated annealing, and Tabu search. It was demonstrated that Tabu search is more appropriate for this task due to its tradeoff between efficiency and accuracy for object recognition.

11.3 Machine Learning-Based Object Recognition Methods

Several machine learning-based methods have been proposed for 3D object recognition. Examples include Hough forest-based methods and deep learning-based methods.

11.3.1 Hough Forest-Based 3D Object Detection

Wang et al. [399] use a Hough forest to detect ground objects on traffic roads from 3D point clouds. Note that the problem of object detection in this context is quite similar to object recognition. In [399], the object detection task is actually to "recognize" a specific class of objects (e.g. cars, traffic signs, and street lamps) in the scene. The method consists of three major modules: 3D local patch extraction, 3D local patch representation, and Hough forest training and testing.

11.3.1.1 3D Local Patch Extraction

A 3D local patch is extracted as a cluster containing a supervoxel and its neighborhood. Given a point cloud, it is first over-segmented using the Voxel Cloud Connectivity Segmentation (VCCS) algorithm [400]. Then, an adjacency graph G is constructed for all supervoxels. Specifically, the vertices V of the adjacency graph G are formed by the centers of supervoxels, and the edges E of the adjacency graph G are generated by connecting neighboring supervoxels (i.e. the vertices V of graph). For a given supervoxel centered at a vertex v, its k-order neighborhood $\mathcal{N}_k(v)$ is defined as:

$$\mathcal{N}_k(v) = \{v_i | d(v, v_i) \leq k\}, \tag{11.21}$$

where the distance $d(v, v_i)$ between vertices v and v_i is defined as the minimum number of edges connecting v and v_i. Consequently, a local patch around the vertex v can be obtained using the k-order neighborhood $\mathcal{N}_k(v)$ of v.

To constrain the Hough voting used for object detection, a LRF is defined for each local patch using an approach proposed in [129]. Note that the LRF can be unambiguously and uniquely defined for an asymmetrical local patch only. In order to avoid the ambiguity of a LRF, the ratio between the two largest eigenvalues λ_1 and λ_2 is used to measure the asymmetry of the local patch, i.e. $\varepsilon = \frac{\lambda_2}{\lambda_1}$. For a symmetrical surface patch, ε is around 1. For an asymmetrical surface patch, $\varepsilon \leq 1$. A small value of ε means that the surface is highly asymmetrical. Therefore, the LRF F for a local patch is finally defined as:

$$F = \begin{cases} \{\tilde{\Lambda}_1, \tilde{\Lambda}_2, \tilde{\Lambda}_3\}^\top & \text{if } \varepsilon \leq \tau_\varepsilon, \\ \mathbf{0} & \text{if } \varepsilon > \tau_\varepsilon. \end{cases} \tag{11.22}$$

Here, τ_ε is a threshold used to define an asymmetrical surface, it is set to 0.9 in [399].

11.3.1.2 3D Local Patch Representation

Each extracted 3D local patch is then represented by geometric and reflectance features including spectral features, eigenvalues of the covariance matrix, 3D invariant moments, and the Fast Point Feature Histograms (FPFH) feature [137].

- *Eigenvalues of the covariance matrix.* Given a local patch, the eigenvalues $\{\lambda_1, \lambda_2, \lambda_3\}$ of its scatter matrix ($\lambda_1 \geq \lambda_2 \geq \lambda_3$) represent the extent of the local patch along three different directions.
- *Spectral features.* Using the eigenvalues $\lambda_1, \lambda_2, \lambda_3$, the saliency σ_s of the scatter, the linearity σ_l, and the planarity σ_p are calculated as $\sigma_s = \lambda_3$, $\sigma_l = \lambda_1 - \lambda_2$, and $\sigma_p = \lambda_2 - \lambda_3$.
- *3D invariant moments.* These moments are invariant to rotation and translation [399].
- *FPFH.* The FPFH feature describes the histogram of the geometric attributes (i.e. surface normals) of the local patch.
- *Reflectance feature.* The median of the reflectance intensities in the patch is used in [399].
- *Height and area.* The height from the lowest point of the point cloud to the center of the local patch and the area of the local patch in the horizontal plane are also used in this method.

Consequently, each local patch $\mathcal{P}$ can be represented by $(\mathbf{x}, F, l, \boldsymbol{d})$, where $\mathbf{x}$ is a descriptor which consists of the aforementioned features, F is the LRF for the patch $\mathcal{P}$, l is the class label, and $\boldsymbol{d}$ is the offset vector from the object center to the patch center. For a negative sample, $\boldsymbol{d} = \mathbf{0}$.

11.3.1.3 Hough Forest Training and Testing

Object detection is performed with a Hough forest algorithm, which includes offline training and online detection.

11.3.1.3.1 Offline Training The offline training process for the Hough forest-based 3D object detection algorithm is shown in Figure 11.4. The point cloud is divided into individual supervoxels using the VCCS over-segmentation algorithm. A local patch $\mathcal{P}_i$ is then extracted for each supervoxel using its first-order neighborhood and further represented by four components $(\mathbf{x}_i, F_i, l_i, \boldsymbol{d}_i)$. Next, the parameters of the split function on each branch node are learned using the features of the local patches. With the split function, the training patches arriving at a branch node are split into two subsets. This splitting process is repeated until the depth of the nodes reaches a predetermined number or the number of samples in the leaf node is smaller

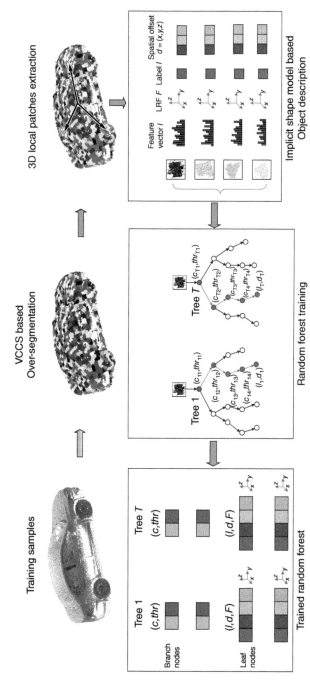

Figure 11.4 The training process for Hough forest-based 3D object detection. The point cloud is first divided into individual supervoxels. A local patch is then extracted for each supervoxel. The parameters of the split function on each branch node of the tree in the forest are learned using the features of the local patches.

than a given threshold. Once the splitting process is complete, each branch node stores the feature channel used for splitting and its associated splitting threshold. Meanwhile, each leaf node stores the offset vectors and the LRFs of the positive training patches in that node.

11.3.1.3.2 Online detection The online object detection process is shown in Figure 11.5. Given a scene point cloud, ground points are first removed to obtain individual segments of nonground objects such as vehicles, vegetation, and buildings. Then, local patches are extracted from the remaining point cloud. Each local patch is represented by the feature descriptor **x** and the LRF F, using the same method as the one used for forest training. The local patch is then passed down to a leaf node of each tree, using the splitting information stored in the branch nodes of the tree. The offset vectors **d** in that leaf node are used to vote for the object center. That is, a 3D Hough space can be formed by these votes and can further be used to determine the object center following a nonmaximum suppression approach. Note that, the 3D pose of an object in a scene can be different from the pose of the object used for training. The pose variation issue has to be addressed for robust 3D object detection (as explained next).

For 3D objects (such as vehicles and trees) on traffic roads, pose variation mainly lies in azimuth rotation. To achieve invariance to pose, the LRFs of two local patches are used to obtain the rotation transformation. Specifically, given the local patch P_s around a scene point p_s and its associated LRF F_s, the patch P_s is passed down through a trained tree to reach a node leaf. Suppose that the leaf node contains a patch P_m with a LRF F_m and an offset vector d_m. The rotation between patches P_s and P_m can be estimated as:

$$O = F_m^T F_s. \tag{11.23}$$

Then, the offset vector d_s of the scene patch P_s can be estimated as:

$$d_s = O d_m. \tag{11.24}$$

Consequently, the object center o_s in the scene can be calculated as:

$$o_s = p_s - d_s = p_s - F_m^T F_s d_m. \tag{11.25}$$

For each local patch on the scene, several votes can be cast to the object center o_s by its destination leaf nodes in all the trees of the forest. Since a number of the local patches can be extracted from the scene, the votes in the 3D Hough voting space can be used to determine the existence of an object of interest.

Note that the LRF cannot be uniquely defined if the local patch of the scene or the training model is symmetrical (i.e. $\varepsilon > \tau_\varepsilon$ in Eq. (11.22)). In this case, an alternative called circular voting is used. Specifically, the offset vector is rotated for all orientations θ along the azimuth direction. All points with a distance d_h to the center p_s of the local patch on the horizontal plane and a distance d_z to p_s along the vertical direction form potential positions of the object center.

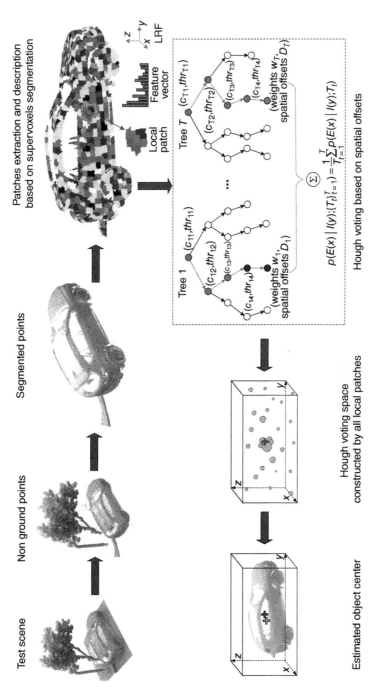

Figure 11.5 The online object detection process for Hough forest-based 3D object detection. Ground points are first removed from the scene to obtain individual segments of non-ground objects. Local patches are extracted from the remaining point cloud and then passed down to a leaf node of each tree using the splitting information. The offset vectors in the leaf nodes are finally used to vote for the object center.

Given a local patch of the scene and its center $p_s = \{p_{sx}, p_{sy}, p_{sz}\}$, the object center can be calculated as:

$$\begin{cases} o_{sx} = p_{sx} + d_h \cos(\theta), \\ o_{sy} = p_{sy} + d_h \sin(\theta), \\ o_{sz} = p_{sz} - d_z. \end{cases} \qquad (11.26)$$

In summary, if the local patches of both the scene and the training model are asymmetrical, then the LRF of the two local patches are used to obtain the object center in the scene. Otherwise, the circular voting strategy [399] is used to achieve invariance along rotation in the azimuth direction for symmetrical local patches.

Once a large number of possible object centers are generated, they are used to vote for the existence of a particular object. Experimental results on several point cloud datasets acquired by a RIEGL VMX-450 system show that this algorithm can effectively and efficiently detect cars, traffic signs, and street lamps.

11.3.2 Deep Learning-Based 3D Object Recognition

Although deep learning has been extensively used in object recognition from 2D images, performing 3D object recognition from point clouds with deep neural networks is still an open problem. Compared to 2D images, there are several challenges for the application of deep learning to 3D data. Below, we discuss a few of them.

1) **Data representation**. A 2D image has a regular structure (i.e. 2D lattice) while a 3D point cloud (or mesh) is represented by a set of irregular vertices. Therefore, a 3D point cloud (or mesh) cannot be directly fed into a neural network. Recently, several approaches have been proposed to handle the problem of irregular data representation.
   - The first approach extracts low-level hand-crafted features from the original 3D point cloud (or mesh), and then uses deep neural networks to learn high-level features from these low-level features (Section 11.3.2.1).
   - In the second approach, a point cloud (or mesh) is transformed into one or several 2D images, then a deep neural network is applied to these 2D images to learn discriminative features (Section 11.3.2.2).
   - In the third approach, a point cloud (or mesh) is transformed into a 3D voxel representation, which is then fed into a deep neural network (Section 11.3.2.3).
   - In the fourth approach, a specific neural network is designed to learn features from 3D point clouds (Section 11.3.2.4).
2) **Dataset**. Compared to large-scale 2D image datasets such as ImageNet, the existing 3D model datasets, which are publically available, are still very

small. Popular 3D model datasets include ModelNet [71] and ShapeNet [401] datasets. ModelNet dataset contains 1 27 915 models from 662 categories. It has two subsets, i.e. ModelNet10 and ModelNet40. ModelNet10 subset consists of 4899 models from 10 categories, while ModelNet40 subset consists of 12 311 models from 40 categories. The ShapeNet dataset contains about 3 million models, while its subset ShapeNetCore consists of 51 300 models from 55 categories. These datasets are small and therefore introduce challenges for the design and training of deep learning architectures. Note that both ModelNet and ShapeNet datasets are designed for 3D shape classification and retrieval. Only complete 3D models are included in these two datasets, without any occlusion and clutter. Consequently, these two datasets are not suitable for the applications of 3D object detection and recognition in cluttered scenes.

Recently, several large-scale 3D point cloud datasets have been released for various tasks. For example, the Semantic3D dataset [402] provides the largest labeled 3D point cloud dataset of natural scenes with over 3 billion points. This dataset includes different types of urban objects, such as churches, streets, squares, railroad tracks, villages, castles, and soccer fields. It is designed for 3D point cloud labeling and segmentation. The Sydney Urban Objects dataset [403] consists of 631 scans of different urban object classes, including signs, trees, vehicles, and pedestrians. This dataset was acquired by a Velodyne HDL-64E LIDAR and is designed for 3D object classification. With the release of these large-scale 3D point cloud datasets, it is now possible to develop advanced deep neural networks for 3D object detection and recognition.

3) **Computational cost.** Compared to 2D images, the dimensionality of 3D data is very high. For example, if a 3D shape is represented by a volumetric grid of size $30 \times 30 \times 30$, its computational cost for a network is already equivalent to a network for 2D images with a size of 165×165. Note that, a volumetric grid of size $30 \times 30 \times 30$ has already introduced a heavy computational burden to a deep neural network system [68]. However, its representation power is still very poor, with most of the shape details being lost. How to represent a 3D shape with sufficient details, while achieving an acceptable computational efficiency, is still an open problem.

Deep learning-based 3D object recognition algorithms are still in their infancy. Most existing 3D deep learning methods are developed for the task of 3D shape classification and retrieval (Chapter 12), where occlusion and clutter is almost not considered. These methods can be broadly divided into hand-crafted feature-based, 2D view-based, 3D voxel-based, and raw 3D data-based methods, according to their data representations. In this section, we only briefly introduce several attempts in the area to address different issues of 3D deep learning.

11.3.2.1 Hand-crafted Feature-Based Methods

These methods first represent a 3D shape with hand-crafted feature descriptors, and a deep neural network is then trained to extract high-level features from these hand-crafted feature descriptors. These methods are able to fully employ existing low-level feature descriptors, and well developed deep learning models.

For example, Bu et al. [404] first extract a scale-invariant heat kernel signature and an average geodesic distance as their low-level 3D shape descriptors for a 3D shape. These low-level features are then encoded into middle-level features using the geometric bag-of-words method. Finally, high-level shape feature are further learned from bag-of-words using deep belief networks (DBNs). The features learned from DBN are successfully used in applications of shape classification and retrieval. Xie et al. [405] propose a discriminative deep auto-encoder network to learn deformation invariant shape features. Specifically, a multiscale histogram of heat kernel signatures of a 3D shape is first extracted and used as an input to an auto-encoder. A deep discriminative auto-encoder is trained for each scale and the outputs from the hidden layers of these auto-encoders, with different scales, are further concatenated to generate the shape descriptor. The descriptors that are learned by the deep auto-encoder are finally used for 3D shape retrieval.

A major limitation of these methods is that the discrimination of the high-level features learned by deep neural networks highly relies on the hand-crafted low-level features. They cannot completely handle the problems faced by hand-crafted features. Since these methods cannot learn features directly from raw data, the advantages of deep learning methods are lost.

11.3.2.2 2D View-Based Methods

These methods represent each 3D shape with one or multiple 2D images, and then learn features from these 2D images using deep neural networks [68, 406]. These methods have several obvious advantages: **First**, they can fully employ standard neural network models that are pretrained on large-scale 2D images. This can overcome the limitation of insufficient data for 3D object recognition. **Second**, there are sufficient 2D images for the training of a new deep neural network.

For example, Su et al. [68] learn to recognize 3D shapes from a set of 2D images rendered from 3D shapes under different views. **First**, a standard CNN model (i.e. VGG-M) is trained to independently learn features from each rendered 2D view of the 3D shape. **Then**, another CNN model is trained to combine features learned from multiple views of a 3D shape, resulting in a compact overall feature descriptor. Experimental results show that the shape classification accuracy achieved by the features learned from a single view is better than existing hand-crafted 3D shape descriptors. Further, if features learned from multiple views are fused, the shape classification accuracy can further be improved. This means that a collection of 2D views provides highly discriminative information for 3D shape classification. Shi et al. [407] convert each 3D shape into a panoramic view using a cylindrical projection around the principle axis of the shape. Then, a CNN model is used to learn deep features from the panoramic views of the 3D shape. Experiments have been performed on tasks related to 3D shape retrieval and classification.

The above methods are mainly proposed for the retrieval and classification of complete 3D shapes without any occlusion and clutter. Recently, several view-based methods have also been proposed for object detection from incomplete 3D point clouds. For example, Chen et al. [408] propose Multi-View 3D (MV3D) networks for 3D object detection from point clouds for autonomous driving. Specifically, the network uses a multiview representation of a 3D point cloud and a 2D image as its input. The point cloud is first projected onto two views, i.e. the bird's eye view and the front view. For the bird's eye view, the point cloud is represented by height, intensity and density feature maps. For the front view, the point cloud is represented by height, distance, and intensity feature maps. A region proposal network (RPN) is then used to generate 3D object proposals from the bird's eye view map. Finally, a region-based fusion network is designed to combine features learned from multiple views to classify each 3D object proposal and to perform regression on each oriented 3D box. Experimental results on the KITTI benchmark show that this method achieves promising 3D object localization and detection performance.

Although promising results have been achieved, these methods still face several challenges. **First**, although part of the 3D geometric information is preserved, some specific detailed information is lost during the process of projection, which further limits the discrimination capability of the learned features. **Second**, these methods usually assume that the 3D scenes (or shapes) have already been aligned in the upright direction, which limits the application of these methods.

11.3.2.3 3D Voxel-Based Methods

These methods consider a 3D shape as a probability distribution in a 3D volumetric grid and represent each shape as a 3D tensor. Figure 11.6 shows the volumetric representation of a chair model under different resolutions. The volumetric representation preserves the 3D information of a shape and it is therefore possible to improve the discrimination capability of the learned features.

Wu et al. [71] represent a 3D shape as a distribution of binary variables in a 3D volumetric space. Specifically, each 3D shape is denoted by a binary tensor, where 1 means that the voxel is inside the 3D surface while 0 means that the voxel is outside the 3D surface. Then, a Convolutional Deep Belief Network (CDBN) is used to learn a feature representation for 3D shapes from this binary volumetric data. Actually, the DBN is a popular probabilistic model

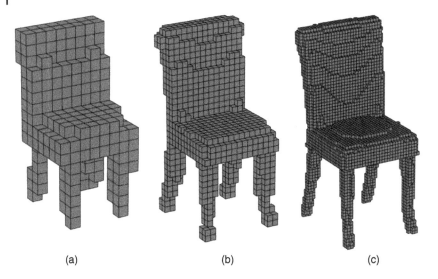

(a)   (b)   (c)

Figure 11.6 The volumetric representation of a chair model under different resolutions. (a) Chair with a resolution of 16 × 16 × 16. (b) Chair with a resolution of 32 × 32 × 32. (c) Chair with a resolution of 64 × 64 × 64.

for the modeling of a joint probabilistic distribution between voxels and their labels. Since the number of parameters in a fully connected DBN is too large, the convolution operation is used to reduce the model parameters in the CDBN. The proposed method can be used to recognize 3D objects from a single-view depth image. It can also be used to predict the next-best-view, in view planning for active vision tasks. Song and Xiao [409] introduce a 3D convolutional neural network (called Deep Sliding Shapes) to learn 3D object proposals and classifiers from an RGB-D image. **First**, a 3D RPN is proposed to learn the objectness of objects from geometric shapes at two different scales. **Then**, a joint Object Recognition Network (ORN) is proposed to learn color features from 2D images using a 2D CNN and to learn geometric features from depth maps using a 3D CNN. The 3D bounding boxes for objects are also regressed by the ORN net. Experimental results show that the proposed method can learn a discriminative representation for 3D shapes.

Since the computational and memory costs increase cubically with the volumetric resolution, most 3D volumetric networks are limited to low 3D resolutions. For example, in the order of 30 × 30 × 30 voxels. To address this problem, Riegler et al. [410] propose OctNet to enable deep 3D convolutional networks to work with high-resolution volumetric data. Specifically, the sparsity of 3D data is used to hierarchically partition the 3D space using a set of unbalanced octrees. Each octree divides the 3D space according to the data density, while each leaf node contains a pooled feature representation. Therefore, the majority of memory allocation and computation is given to the dense regions. Finally, the computational and memory requirements are reduced. The performance of OctNet has been demonstrated on several 3D vision tasks including 3D object classification, 3D orientation estimation and point cloud labeling. It is also shown that a higher resolution is particularly beneficial for the tasks of orientation estimation and point cloud labeling. Recently, Ma et al. [411] use input binarization and weight binarization to reduce the computational and memory cost of volumetric networks. It is shown that the binary volumetric convolutional networks outperform their float-valued networks by a large margin in terms of computational and memory cost, with a marginal decrease in object recognition performance.

11.3.2.4 3D Point Cloud-Based Methods

A few methods have also been designed to work directly on 3D point clouds. For example, Qi et al. [412] propose a PointNet architecture to perform object classification, part segmentation, and scene semantic parsing from 3D point clouds. Specifically, the point cloud is given as an input to PointNet and each point is processed independently using the same approach. To make the network invariant to the input permutation, several approaches can be considered. **First**, the input points are sorted into a canonical order. **Second**, the input is considered as a sequence and an RNN can be then trained using all permutations. **Third**, a symmetric function is used to aggregate the information learned from all points. In [412], the third approach is adopted and max pooling is used as the symmetric function. The global information that is aggregated from all points is then used to train a multilayer perceptron classifier for 3D object classification. In addition, segmentation and scene labeling can also be achieved by PointNet by concatenating both the global and the local features.

11.4   Summary and Further Reading

Early methods of 3D object recognition commonly follow a hypothesis-and-verification paradigm. That is, hypotheses of candidate objects and their poses are first generated through the matching between features of the scene and the models in the library. For each hypothesis, the scene and the candidate model can then be aligned using a plausible pose. The hypothesis is finally verified by checking the alignment quality and/or the geometric consistency of the alignment. Extensive investigations have been conducted to improve the performance of each part of this hypothesis-and-verification paradigm. For example, feature extraction, feature matching, hypothesis generation, surface registration, and hypothesis verification. The hypothesis-and-verification based methods can achieve a reasonably good performance on relatively small datasets for the task of 3D object instance recognition. However, for large datasets, it is very challenging or even impossible to perform feature matching and surface registration between the scene and all the objects in the library. Besides, these methods are hard to extend to the task of 3D object category recognition. That is because the variation of the shapes within a category can be very large, and it is impossible to store all the instances of a category in the library.

Recently, machine learning methods have increasingly attracted more attention for this application. Machine learning methods can be scaled to large datasets and are suitable for the tasks of both category recognition and instance recognition. Based on the success of deep learning on different computer vision tasks, deep learning has now been employed for 3D object recognition. Currently, most methods focus on the classification of segmented and clean 3D shapes. 3D object recognition in cluttered scenes still requires further investigation. Significant progress has been achieved by deep learning methods working on data with regular lattices, e.g. depth images, 2D projected views of 3D shapes, and voxels of 3D shapes. Deep learning methods working directly on unstructured 3D point clouds have also been investigated in last two years, examples include PointNet [412] and PointCNN [413]. However, more research still need be directed to deep learning methods working on 3D point clouds. Several major questions remain open in this direction. For example, how to improve the computational and storage efficiency when dealing with large-scale point clouds, how to fully employ the sparse characteristics of point clouds, and how to effectively handle the uneven distribution of point clouds (e.g. those acquired by a Velodyne LIDAR).

12

3D Shape Retrieval

12.1 Introduction

The need for efficient 3D search engines is mainly motivated by the large number of 3D models that are being created by novice and professional users, used and stored in public and in private domains, and shared at large scale. In a nutshell, a 3D search engine has the same structure as any image and text search engine. **In an offline step**, all the 3D models in a collection are indexed by computing, for each 3D model, a set of numerical descriptors which represent uniquely their shape. This is the *feature extraction stage.* (For details about 3D shape description methods, we refer the reader to Chapters 4 and 5.) The 3D models that are similar in shape should have similar descriptors, while the 3D models that differ in shape should have different descriptors. **At runtime**, the user specifies a query. The system then ranks the models in the collection according to the distances computed between their descriptors and the descriptor(s) of the query and returns the top elements in the ranked list. This is the *similarity computation stage.*

In general, the query can have different forms; the user can, for instance, type a set of keywords, draw (a set of) 2D sketch(es), use a few images, or even use another 3D model. We refer to the first three cases as *cross-domain* retrieval since the query lies in a domain that is different from the domain of the indexed 3D models. This will be covered in Chapter 13. In this chapter, we focus on the tools and techniques that have been developed for querying 3D model collections using another 3D model as a query. In general, these techniques aim to address two types of challenges. **The first type** of challenges is similar to what is currently found in text and image search engines and it includes:

- **The concept of similarity.** The user conception of similarity between objects is often different from the mathematical formulation of distances. It may even differ between users and applications.

3D Shape Analysis: Fundamentals, Theory, and Applications, First Edition.
Hamid Laga, Yulan Guo, Hedi Tabia, Robert B. Fisher, and Mohammed Bennamoun.
© 2019 John Wiley & Sons, Inc. Published 2019 by John Wiley & Sons, Inc.

- **Shape variability in the collection**. For instance, 3D shape repositories contain many classes of shapes. Some classes may have a large intra-class shape variability, while others may exhibit a low inter-class variability.
- **Scalability**. The size of the collection that is queried can be very large.

The second type of challenges is specific to the raw representation of 3D objects and the transformations that they may undergo. Compared to (2D) images, the 3D representation of objects brings an additional dimension, which results in more transformations and deformations. These can be:

- **Rigid**. The rigid transformations include translation, scale and rotation. An efficient shape retrieval method should be invariant to this type of transformations (see Chapter 2 for information about shape-preserving transformations).
- **Nonrigid**. 3D objects can also undergo nonrigid deformations, i.e. by bending and stretching their parts (see the examples of Figure 7.1). Some users may require search engines that are invariant to some types of nonrigid deformations, e.g. bending. For example, one may search for 3D human body models of the same person independently of their pose. In that case, the descriptors and similarity metrics for the retrieval should be invariant to this type of deformations.
- **Geometric and topological noise**. These types of noise are common not only in 3D models that are captured by 3D scanning devices but also in 3D models created using 3D modeling software. In the example of Figure 12.1, the shape of the hand may undergo topological deformations by connecting the thumb and index fingers. In that case, a hole appears and results in a change of the topology of the hand. An efficient retrieval method should not be affected by this type of deformations.

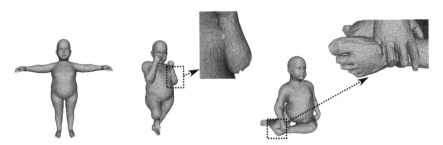

Figure 12.1 Some examples of topological noise. When the arm of the 3D human touches the body of the foot, the tessellation of the 3D model changes and results in a different topology. All the models, however, have the same shape (up to isometric deformations) and thus a good 3D retrieval system should be invariant to this type of topological noise. The 3D models are from the SHREC'2016 dataset [414].

- **Partial data.** Often, users may search for complete 3D objects using queries with missing parts. This is quite common in cultural heritage where the user searches for full 3D models with parts that are similar to artifacts found at historical sites.

To address these challenges, several techniques have been proposed in the literature. Most of them differ by the type of features, which are computed on the 3D models, the way these features are aggregated into descriptors, and the metrics and mechanisms used to compare the descriptors. A common pipeline of the state-of-the-art approaches is given in Figure 12.2. Early works use handcrafted global descriptors (see Chapter 4) and simple similarity measures. While they are easy to compute and utilize, global descriptors are limited to the coarse analysis and comparison of 3D shapes. Their usage is also limited to search in small databases.

Handcrafted local descriptors, which have been covered in Chapter 5, are particularly suitable for the registration and alignment of 3D models. They can also be used for 3D shape retrieval either by one-to-one comparison of the local descriptors after registration [415], or by aggregating them using histogram-like techniques such as bag of words (BoW) [154], or spatially-sensitive BoWs [123]. Aggregation methods have been discussed in Section 5.5. The performance of handcrafted global and local descriptors in 3D shape retrieval will be discussed in Section 12.4.1.

Recently, motivated by their success in many image analysis tasks, deep learning methods are gaining in popularity. In this category of methods, features and the similarity measure are jointly learned from training data. This allows to discard the domain expert knowledge needed when using handcrafted features. Another important advantage of deep learning-based methods is their scalability, i.e. their ability to handle very large collections of 3D models. These methods will be discussed in Section 12.4.2.

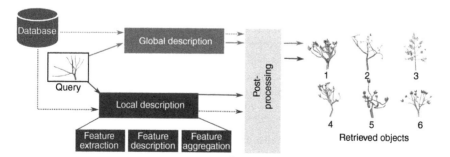

Figure 12.2 Global and local-based 3D shape retrieval pipelines. Dashed lines indicate the offline processes while solid lines indicate online processes.

Central to every 3D shape retrieval algorithm is a mechanism for computing distances, dissimilarities, or similarities between descriptors. Different similarity measures can lead to different performances even when using the same type of descriptor. Some of the similarity measures that have been used in the literature will be discussed in Section 12.3.

12.2 Benchmarks and Evaluation Criteria

Before describing the 3D shape retrieval techniques in detail, we will first describe the different datasets and benchmarks, which are currently available in the public domain. We will also present the metrics used to evaluate the quality and performance of a 3D retrieval algorithm. These will enable the reader to compare, in a quantitative manner, the different approaches.

12.2.1 3D Datasets and Benchmarks

There are several 3D model datasets and benchmarks that can be used to evaluate the performance of 3D shape retrieval algorithms. Some datasets have been designed to evaluate how retrieval algorithms perform on generic 3D objects. Others focus on more specific types of objects, e.g. those which undergo non-rigid deformations. Some benchmarks target partial retrieval tasks. In this case, queries are usually parts of the models in the database. Early benchmarks are of a small size (usually less than 10K models). Recently, there have been many attempts to test how 3D algorithms scale to large datasets. This has motivated the emergence of large-scale 3D benchmarks. Below, we discuss some of the most commonly used datasets.

- **The Princeton Shape Benchmark (PSB) [53]** is probably one of the most commonly used benchmarks. It provides a dataset of 1814 generic 3D models, stored in the Object File Format (.off), and a package of software tools for performance evaluation. The dataset comes with four classifications, each representing a different granularity of grouping. The base (or finest) classification contains 161 shape categories.
- **TOSCA [416]** is designed to evaluate how different 3D retrieval algorithms perform on 3D models which undergo isometric deformations (i.e. bending and articulated motion). This database consists of 148 models grouped into 12 classes. Each class contains one 3D shape but in different poses (1–20 poses per shape).
- **The McGill 3D Shape Benchmark [417]** emphasizes on including models with articulating parts. Thus, whereas some models have been adapted from the PSB, and others have been downloaded from different web repositories, a substantial number have been created by using CAD modeling tools. The 320 models in the benchmark come under two different representations: voxelized form and mesh form. In the mesh format, each object's surface

is provided in a triangulated form. It corresponds to the boundary voxels of the corresponding voxel representation. They are, therefore, "jagged" and may need to be smoothed for some applications. The models are organized into two sets. The **first** one contains objects with articulated parts. The **second** one contains objects with minor or no articulation motion. Each set has a number of basic shape categories with typically 20–30 models in each category.

- **The Toyohashi Shape Benchmark (TSB) [418]** is among the first large-scale datasets which have been made publicly available. This benchmark consists of 10 000 models manually classified into 352 different shape categories.
- **SHREC 2014 – Large Scale [80]** is one of the first attempts to benchmark large-scale 3D retrieval algorithms. The dataset contains 8978 models grouped into 171 different shape categories.
- **ShapeNetCore [419]** consists of 51 190 models grouped into 55 categories and 204 subcategories. The models are further split between training (70% or 35 765 models), testing (20% or 10 266 models), and validation (10% or 5 159 models). The database also comes in two versions. In the *normal* version, all the models are upright and front orientated. On the *perturbed* version, all the models are randomly rotated.
- **ShapeNetSem [419]** is a smaller, but more densely annotated, subset of ShapeNetCore, consisting of 12 000 models spread over a broader set of 270 categories. In addition, models in ShapeNetSem are annotated with real-world dimensions, and are provided with estimates of their material composition at the category level, and estimates of their total volume and weight.
- **ModelNet [73]** is a dataset of more than 127 000 CAD models grouped into 662 shape categories. It is probably the largest 3D dataset currently available and is extensively used to train and test deep-learning networks for 3D shape analysis. The authors of ModelNet also provide a subset, called **ModelNet40**, which contains 40 categories.

We refer the reader to Table 12.1 that summarizes the main properties of these datasets.

12.2.2 Performance Evaluation Metrics

There are several metrics for the evaluation of the performance of 3D shape descriptors and their associated similarity measures for various 3D shape analysis tasks. In this section, we will focus on the retrieval task, i.e. given a 3D model, hereinafter referred to as the query, the user seeks to find, from a database, the 3D models which are similar to the given query. The retrieved 3D models are then ranked and sorted in an ascending order with respect to their similarity to the query. The retrieved list of models is referred to as the ranked list. The role of the performance metrics is to quantitatively evaluate the quality of this ranked list. Below, we will look in detail to some of these metrics.

Table 12.1 Some of the 3D shape benchmarks that are currently available in the literature.

Name	No. of models	No. of categories	Type	Training	Testing
The Princeton Shape Benchmark [53]	1 814	161	Generic	907	907
TOSCA [416]	148	12	Nonrigid	—	—
McGill [417]	320	19	Generic + Nonrigid	—	—
The Toyohashi Shape Benchmark [418]	10 000	352	Generic	—	—
SHREC 2014 – large scale [80]	8 987	171	Generic	—	—
ShapeNetCore [75]	51 190	55	Generic	35 765	10 266
ShapeNetSem [75]	12 000	270	Generic	—	—
Princeton ModelNet [73]	127 915	662	Generic	—	—
Princeton ModelNet40 [73]	12 311	40	Generic	—	—

The last two columns refer, respectively, to the size of the subsets used for training and testing.

Let A be the set of objects in the repository that are relevant to a given query, and let B be the set of 3D models that are returned by the retrieval algorithm.

12.2.2.1 Precision

Precision is the ratio of the number of models in the retrieved list, B, which are relevant to the query, to the total number of retrieved objects in the ranked list:

$$\text{precision} = \frac{|A \cap B|}{|B|}, \tag{12.1}$$

where $|X|$ refers to the cardinality of the set X. Basically, precision measures how well a retrieval algorithm discards the junk that we do not want.

12.2.2.2 Recall

Recall is the ratio of the number, in the ranked list, of the retrieved objects that are relevant to the query to all objects in the repository that are relevant to the query:

$$\text{recall} = \frac{|A \cap B|}{|A|}. \tag{12.2}$$

Basically, recall evaluates how well a retrieval algorithm finds what we want.

12.2.2.3 Precision-Recall Curves

Precision or recall alone are not good indicators of the quality of a retrieval algorithm or a descriptor. For example, one can increase the recall to 100% by returning all the models in the database. This, however, will decrease the precision. When we perform a search, with any search engine, we are looking to find the most relevant objects (maximizing recall), while minimizing the junk that is retrieved (maximizing precision). The precision-recall curve measures precision as a function of recall and provides the trade-off between these two parameters. For a database of N models, the precision-recall curve for a query model Q is computed as follows:

- First, rank all the N models in the database in ascending order according to their dissimilarity to the query.
- For $m \in \{1, \ldots, N\}$, take the first m elements of the ranked list and then measure both the precision and recall. This gives a point on the precision-recall curve.
- Finally, interpolate these points to obtain the precision-recall curve.

The precision-recall curve is one of the most commonly used metrics to evaluate 3D model retrieval algorithms.

12.2.2.4 F- and E-Measures

These are two measures that combine precision and recall into a single number to evaluate the retrieval performance. The F-measure is the weighted harmonic mean of precision and recall. It is defined as:

$$F_\alpha = \frac{(1 + \alpha) \times \text{precision} \times \text{recall}}{\alpha \times \text{precision} + \text{recall}}, \tag{12.3}$$

where α is a weight. When $\alpha = 1$, then

$$F_1 \equiv F = 2 \times \frac{\text{precision} \times \text{recall}}{\text{precision} + \text{recall}}. \tag{12.4}$$

The E-Measure is defined as $E = 1 - F$. Note that the maximum value of the E-measure is 1.0 and the higher the E-measure is, the better is the retrieval algorithm. The main property of the E-measure is that it tells how good are the results which have been retrieved in the top of the ranked list. This is very important since, in general, the user of a search engine is more interested in the first page of the query results than in the later pages.

12.2.2.5 Area Under Curve (AUC) or Average Precision (AP)

By computing the precision and recall at every position in the ranked list of 3D models, one can plot the precision as a function of recall. The average precision

(AP) computes the average value of the precision over the interval from recall $=$ 0 to recall $= 1$:

$$AP = \int_0^1 \text{precision}(r)dr, \tag{12.5}$$

where r refers to recall. This is equivalent to the area under the precision-recall curve.

12.2.2.6 Mean Average Precision (mAP)
Mean average precision (mAP) for a set of queries is the mean of the AP scores for each query:

$$mAP = \frac{1}{N_q} \sum_{q=1}^{Q} AP(q), \tag{12.6}$$

where N_q is the number of queries and $AP(q)$ is the AP of the query q.

12.2.2.7 Cumulated Gain-Based Measure
This is a statistic that weights the correct results near the front of the ranked list more than the correct results that are down in the ranked list. It is based on the assumption that a user is less likely to consider elements near the end of the list. Specifically, a ranked list R is converted to a list G. An element $G_i \in G$ has a value of 1 if the element $R_i \in R$ is in the correct class and a value of 0 otherwise. The cumulative gain CG is then defined recursively as follows:

$$CG[i] = \begin{cases} G_1 \text{if } i = 1, \\ CG[i-1] + G_i \text{ for } i > 1. \end{cases} \tag{12.7}$$

The discounted cumulative gain (DCG), which progressively reduces the object weight as its rank increases but not too steeply, is defined as:

$$DCG[i] = \begin{cases} G_1 \text{if } i = 1, \\ DCG[i-1] + \frac{G_i}{\log_2(i)} \text{ for } i > 1. \end{cases} \tag{12.8}$$

We are usually interested in the *NDCG*, which scales the DCG values down by the average, over all the retrieval algorithms which have been tested, and shifts the average to zero:

$$NDCGA = \frac{DCGA}{AvgDCG} - 1, \tag{12.9}$$

where NDCGA is the normalized DCG for a given algorithm A, DCGA is the DCG value computed for the algorithm A, and AvgDCG is the average DCG value for all the algorithms being compared in the same experiment. Positive/negative normalized DCG scores represent the above/below average performance, and the higher numbers are better.

12.2.2.8 Nearest Neighbor (NN), First-Tier (FT), and Second-Tier (ST)

These performance metrics share a similar idea, that is, to check for the ratio of models in the query's class, which appear within the top k matches. In the case of the nearest neighbor (NN) measure, k is set to 1. It is set to the size of the query's class for the first-tier (FT) measure, and to twice the size of the query's class for second-tier (ST).

These measures are based on the assumption that each model in the dataset has only one ground truth label. In some datasets, e.g. the ShapeNetCore [419], a model can have two groundtruth labels (one for the category and another for the subcategory). In this case, one can define a modified version of the NDCG metric that uses four levels of relevance:

- Level 3 corresponds to the case where the category and subcategory of the retrieved shape perfectly match the category and subcategory of the query.
- Level 2 corresponds to the case where both the category and subcategory of the retrieved 3D shape are the same as the category of the query.
- Level 1 corresponds to the case where the category of the retrieved shape is the same as the category of the query, but their subcategories differ.
- Level 0 corresponds to the case where there is no match.

Meanwhile, since the number of shapes in different categories is not the same, the other evaluation metrics (MAP, F-measure and NDCG) will also have two versions. The first one, called *macro-averaged* version, is used to give an unweighted average over the entire dataset. The second one, called *micro-averaged* version, is used to adjust for model category sizes, which results in a representative performance metric averaged across categories. To ensure a straightforward comparison, the arithmetic mean of macro version and micro version is also used.

12.3 Similarity Measures

Retrieval algorithms are based on similarity search, i.e. they compute the similarity of the query's descriptor to the descriptors of the models in the database. There are several methods that have been used to compute such similarity. Here, we will look at three classes of methods. **The first** class of methods use standard distance measures (Section 12.3.1). **The second** class of methods use Hamming distances on hash codes (Section 12.3.2). **The final** class of methods use manifold ranking techniques (Section 12.3.3).

12.3.1 Dissimilarity Measures

Table 12.2 lists a selection of 12 measures used for 3D shape retrieval. When the measure is a distance, then the models in the database are ranked in ascending

Table 12.2 A selection of 12 distance measures used in 3D model retrieval algorithms.

Name	Type	Measure
City block ($\mathbb{L}^1$)	Distance	$\sum_{i=1}^{d} |x_i - y_i|$
Euclidean ($\mathbb{L}^2$)	Distance	$\sum_{i=1}^{d} (x_i - y_i)^2$
Canberra metric	Distance	$\dfrac{1}{d} \sum_{i=1}^{d} \dfrac{|x_i - y_i|}{|x_i| + |y_i|}$
Divergence coefficient	Distance	$\dfrac{1}{d} \sum_{i=1}^{d} \dfrac{(x_i - y_i)^2}{|x_i| + |y_i|}$
Correlation coefficient	Similarity	$\left(\sum_{i=1}^{d}(x_i - \mu_x)(y_i - \mu_y) \right) \left(\sum_{i=1}^{d}(x_i - \mu_x)^2 \sum_{i=1}^{d}(y_i - \mu_y)^2 \right)^{-\frac{1}{2}}$
Profile coefficient	Similarity	$\dfrac{\sum_{i=1}^{d} x_i y_i + dm^2 - m \sum_{i=1}^{d}(x_i + y_i)}{\left(\left(dm^2 + \sum_{i=1}^{d}(x_i^2 - 2mx_i) \right) \left(dm^2 + \sum_{i=1}^{d}(y_i^2 - 2my_i) \right) \right)^{-\frac{1}{2}}}$
Intraclass coefficient	Similarity	$\dfrac{1}{d\sigma^2} \sum_{i=1}^{d}(x_i - \mu)(y_i - \mu).$
Catell 1949	Similarity	$\dfrac{\sqrt{2d} - \|\mathbf{x} - \mathbf{y}\|}{\sqrt{2d} + \|\mathbf{x} - \mathbf{y}\|}$
Angular distance	Similarity	$\left(\sum_{i=1}^{d} x_i y_i \right) \left(\sum_{i=1}^{d} x_i^2 \sum_{i=1}^{d} y_i^2 \right)^{-1}$
Meehl index	Distance	$\sum_{i=1}^{d-1} (|x_i - x_{i+1}| - |y_i - y_{i+1}|)$
Kappa	Distance	$\sum_{i=1}^{d} \dfrac{y_i - x_i}{\ln(y_i) - \ln(x_i)}$
Inter-correlation coefficient	Similarity	$\left(\sum_{i=1}^{d}(x_i - \mu_y)(y_i - \mu_x) \right) \left(\sum_{i=1}^{d}(x_i - \mu_y)^2 \sum_{i=1}^{d}(y_i - \mu_x)^2 \right)^{-\frac{1}{2}}$

Here, $\mathbf{x} = (x_1, \ldots, x_d)$ and $\mathbf{y} = (y_1, \ldots, y_d)$ are two descriptors, $d \geq 1$ is the size, or dimension, of the descriptors, $\mu_x = \frac{1}{d} \sum_{i=1}^{d} x_i$, $\mu_y = \frac{1}{d} \sum_{i=1}^{d} y_i$, $\sigma_x^2 = \frac{1}{d-1} \sum_{i=1}^{d} (x_i - \mu_x)^2$, $\mu = \frac{1}{d} \sum_{i=1}^{d}(x_i + y_i)$, $\sigma^2 = \frac{1}{2d-1} \left\{ \sum_{i=1}^{d}(x_i - \mu)^2 + \sum_{i=1}^{d}(y_i - \mu)^2 \right\}$, $M = \max(x_i, i = 1, \ldots, d) - \min(x_i, i = 1, \ldots, d)$, $m = \frac{M}{2}$.

order of their distance to the query. When the measure is a similarity, then the models are ranked in a descending order of their similarity to the query.

12.3.2 Hashing and Hamming Distance

Hashing is the process of encoding high-dimensional descriptors as similarity-preserving compact binary codes (or strings). This can enable large efficiency gains in storage and computation speed when performing similarity search such as the comparison of the descriptor of a query to the descriptors of the elements in the data collection being indexed. With binary codes, similarity search can also be accomplished with much simpler data structures and algorithms, compared to alternative large-scale indexing methods [420].

Assume that we have a set of n data points $\{\mathbf{x}_i \in \mathbb{R}^d, i = 1, \ldots, n\}$ where each point is a descriptor of a 3D shape, and d is the dimensionality of the descriptors, which can be very large. We also assume that $\mathbf{x}_i$ is a column vector. Let $\mathbf{X}$ be a $n \times d$ matrix whose ith row is set to be $\mathbf{x}_i^\top$ for every i. We assume that the points are zero centered, i.e. $\sum_{i=1}^n \mathbf{x}_i = \mathbf{0}$ where $\mathbf{0}$ is the vector whose elements are all set to 0. The goal of hashing is to learn a binary code matrix $\mathbf{B} \in \{-1, 1\}^{n \times c}$, where c denotes the code length. This is a two step process:

The first step applies linear dimensionality reduction to the data. This is equivalent to projecting the data onto a hyperplane of dimensionality c, the size of the hash code, using a projection matrix $\mathbf{W}$ of size $d \times c$. The new data points are the rows of the $n \times c$ matrix $\mathbf{V}$ given by:

$$\mathbf{V} = \mathbf{X}\,\mathbf{W}. \tag{12.10}$$

Gong et al. [420] obtain the matrix $\mathbf{W}$ by first performing principal component analysis (PCA). In other words, for a code of c bits, the columns of $\mathbf{W}$ are formed by taking the top c eigenvectors of the data covariance matrix $\mathbf{X}^\top \mathbf{X}$.

The second step is a binary quantization step that produces the final hash code $\mathbf{B}$ by taking the sign of $\mathbf{V}$. That is:

$$\mathbf{B} = \text{sign}(\mathbf{V}), \tag{12.11}$$

where $\text{sign}(x) = 1$ if $x \geq 0$, and 0 otherwise. For a matrix or a vector, $\text{sign}(\cdot)$ denotes the result of the element-wise application of the above sign function. The ith row of $\mathbf{B}$ represents the final hash code of the descriptor $\mathbf{x}_i$.

A good code, however, is the one that minimizes the quantization loss, i.e. the distance between the original data points and the computed hash codes. Instead of quantizing directly $\mathbf{V}$, Gong et al. [420] starts by rotating the data points, using a $c \times c$ rotation matrix O, so that the quantization loss:

$$Q(\mathbf{B}, O) = \|\mathbf{B} - \mathbf{V}O\|_F^2 = \|\text{sign}(\mathbf{V}O) - \mathbf{V}O\|_F^2 \tag{12.12}$$

is minimized. (Here $\| \cdot \|_F$ refers to the Frobenius norm.) This can be solved using the iterative quantization (ITQ) procedure, as detailed in [420]. That is:

- **Fix O and update B.** Let $\tilde{\mathbf{V}} = \mathbf{V}O$. For a fixed rotation O, the optimal code **B** is given as $\mathbf{B} = \text{sign}(\tilde{\mathbf{V}})$.
- **Fix B and update O.** For a fixed code **B**, the objective function of Eq. (12.12) corresponds to the classic Orthogonal Procrustes problem in which one tries to find a rotation to align one point set to another. In our case, the two point sets are given by the projected data **V** and the target binary code matrix **B**. Minimizing Eq. (12.12) with respect to O can be done as follows: **first** compute the SVD of the $c \times c$ matrix $\mathbf{B}^{\mathsf{T}}\mathbf{V}$ as $\mathbf{U}\Omega\mathbf{S}^{\mathsf{T}}$ and **then** let $O = \mathbf{U}\mathbf{S}^{\mathsf{T}}$.

At runtime, the descriptor **x** of a given query is first projected onto the hyperplane, using the projection matrix **W**, and then quantized to form a hash code **b** by taking the sign of the projected point. **Finally**, the hash code **b** is compared to the hash codes of the 3D models in the database using the Hamming distance, which is very efficient to compute. In fact, the Hamming distance between two binary codes of equal length is the number of positions at which the corresponding digits are different.

Hashing has been extensively used for large-scale image retrieval [420–424] and has been recently used for large-scale 3D shape retrieval [425–427].

12.3.3 Manifold Ranking

Traditional 3D retrieval methods use distances to measure the dissimilarity of a query to the models in the database. In other words, the perceptual similarity is based on pair-wise distances. Unlike these methods, manifold ranking [428] measures the relevance between the query and each of the 3D models in the database (hereinafter referred to as data points) by exploring the relationship of all the data points in the feature (or descriptor) space. The general procedure is summarized in Algorithm 12.1. Below, we describe the details.

Consider a set of n 3D shapes and their corresponding global descriptors $\mathbf{x}_i, i = 1, \dots, n$. Let S denote the space of such descriptors. Manifold ranking starts by building a graph whose nodes are the 3D shapes (represented by their descriptors), and then connecting each node to its k-NNs in the descriptor space. Each edge e_{ij}, which connects the ith and jth node, is assigned a weight w_{ij} defined as

$$w_{ij} = \exp \frac{-d^2(\mathbf{x}_i, \mathbf{x}_j)}{2\sigma^2}, \tag{12.13}$$

where $d(\cdot, \cdot)$ is a measure of distance in the descriptor space, and σ is a user-specified parameter. We collect these weights into an $n \times n$ matrix W and set w_{ij} to 0 if there is no edge which connects the ith and jth nodes.

Now, let us define a ranking vector $\mathbf{r} = (r_1, \dots, r_n)^{\mathsf{T}} \in \mathbb{R}^n$, which assigns to each point $\mathbf{x}_i$ a ranking score r_i. Let $\mathbf{y} = [y_1, \dots, y_n]^{\mathsf{T}}$ be a vector such that $y_i = 1$

Algorithm 12.1 Manifold ranking algorithm.

Input:

- A set of n 3D shapes and their corresponding global descriptors $\mathbf{x}_i, i = 1, \ldots, n$.
- The index $l \in \{1, \ldots, n\}$ of the model which will be used as a query.
- A vector $\mathbf{y} = [y_1, \ldots, y_n]^\top$ such that $y_l = 1$ and $y_i = 0$ for $i \neq l$.

Output: A vector $\mathbf{r} = (r_1, \ldots, r_n)$ such that r_i indicates how relevant is the ith database model to the input query.

1: Build an adjacency graph G whose nodes are the data points $\mathbf{x}_i$. The graph connects each node to its k-NNs using a measure of distances $d(\cdot, \cdot)$.
2: Form the $n \times n$ affinity matrix W such that $w_{ii} = 0$, and w_{ij} is given by Eq. (12.13) if there is an edge between the ith and jth point, and 0 otherwise.
3: Symmetrically normalize W by setting $A = D^{-1/2}WD^{-1/2}$, where D is the diagonal matrix with its diagonal element D_{ii} equal to the sum of the ith row of W.
4: Iterate $\mathbf{r} \leftarrow \alpha A\mathbf{r} + (1 - \alpha A)\mathbf{y}$, where $\alpha \in [0, 1)$ until convergence. Let $\mathbf{r}^*$ denote the limit.
5: Rank each data point $\mathbf{x}_i$ according to its ranking score $\mathbf{r}_i^*$. The larger the ranking score is, the more relevant is that model to the query.

if $\mathbf{x}_i$ is the query and $y_i = 0$ otherwise. The cost associated with the ranking vector $\mathbf{r}$ is defined as:

$$\text{Cost}(\mathbf{r}) = \frac{1}{2}\left(\sum_{i,j=1}^n w_{ij} \left\| \frac{r_i}{\sqrt{D_{ii}}} - \frac{r_j}{\sqrt{D_{jj}}} \right\|^2 + \mu \sum_{i=1}^n \|r_i - y_i\|^2 \right), \qquad (12.14)$$

where $\mu > 0$ is a regularization parameter and D is an $n \times n$ diagonal matrix, whose diagonal elements $D_{ii} = \sum_{j=1}^n w_{ij}$. **The first term** of Eq. (12.14) is a smoothness constraint. It ensures that adjacent nodes will likely have similar ranking scores. **The second term** is a fitting constraint to ensure that the ranking result fits to the initial label assignment. If we have more prior knowledge about the relevance or confidence of each query, we can assign different initial scores to the queries.

Finally, the optimal ranking is obtained by minimizing Eq. (12.14), which has a closed-form solution. That is:

$$\mathbf{r}^* = \underset{\mathbf{r}}{\text{argmin}}\{\text{Cost}(\mathbf{r})\} = (I_n - \alpha A)^{-1}\mathbf{y}, \qquad (12.15)$$

where $\alpha = \frac{1}{1+\mu}$, I_n is the $n \times n$ identity matrix, and $A = D^{-1/2}WD^{-1/2}$.

Equation (12.15) requires the computation of the inverse of the square matrix $I_n - \alpha A$, which is of size $n \times n$. When dealing with large 3D datasets (i.e. when n is very large), computing such matrix inverse is computationally very expensive. In such cases, instead of using the closed-form solution, we can solve the optimization problem using an iterative procedure that starts by initializing the ranking r. Then, at each iteration t, the new ranking $r(t)$ is obtained as

$$\mathbf{r}(t) = \alpha A\mathbf{r}(t - 1) + (I_n - \alpha A)\mathbf{y}. \tag{12.16}$$

This procedure is then repeated until convergence, i.e. $\mathbf{r}^* = \lim_{t \to \infty} \mathbf{r}(t)$. The ranking score $\mathbf{r}_i^*$ of the ith 3D model in the database is proportional to the probability that it is relevant to the query, with large ranking scores indicating a high probability.

12.4 3D Shape Retrieval Algorithms

12.4.1 Using Handcrafted Features

Early works in 3D shape retrieval are based on handcrafted descriptors, which can be global, as described in Chapter 4, or local but aggregated into global descriptors using bag of features (BoF) techniques, as described in Section 5.5.

The use of global descriptors for 3D shape retrieval is relatively straightforward; in an **offline step**, a global descriptor is computed and stored for each model in the database. At **runtime**, the user specifies a query, and the system computes a global descriptor, which is then compared, using some distance measures, to the descriptors of the models in the database.

Table 12.3 summarizes and compares the performance of some global descriptors. These results have been already discussed in some other papers, e.g. [53, 61, 68], but we reproduce them here for completeness. In this experiment, we take the descriptors presented in Sections 4.2, 4.3, and 4.4 and compare their retrieval performance, using the performance metrics that are presented in Section 12.2, on the 907 models of the test set of the PSB [53]. We sort the descriptors in a descending order with respect to their performance on the DCG measure. Table 12.3 also shows the distance metrics that have been used to compare the descriptors.

The interesting observation is that, according to the five evaluation metrics used in Table 12.3, the light field descriptor (LFD), which is view-based, achieves the best performance. The LFD also offers a good trade-off between memory requirements, computation time, and performance compared to other descriptors such as the Gaussian Euclidean Distance Transform (GEDT) and SHD, which achieve a good performance but at the expense of significantly higher memory requirements. Note also that spherical-wavelet descriptors,

Table 12.3 Performance evaluation of handcrafted descriptors on the test set of the Princeton Shape Benchmark [53].

Descriptor	Size (bytes)	Distance	NN	FT	ST	E-measure	DCG
SWC_d [61]	512	$\mathbb{L}^2$	0.469	0.314	0.397	0.205	0.654
LFD [52]	4 700	min $\mathbb{L}^1$ difference	0.657	0.38	0.487	0.280	0.643
SWEL1 [61]	76	$\mathbb{L}^1$	0.373	0.276	0.359	0.186	0.626
REXT [53, 59]	17 416	$\mathbb{L}^2$	0.602	0.327	0.432	0.254	0.601
SWEL2 [61]	76	$\mathbb{L}^2$	0.303	0.249	0.315	0.161	0.594
GEDT [53, 60]	32 776	$\mathbb{L}^2$	0.603	0.313	0.407	0.237	0.584
SHD [53, 60, 76]	2 184	$\mathbb{L}^2$	0.556	0.309	0.411	0.241	0.584
EXT [53, 58]	512	$\mathbb{L}^2$	0.549	0.286	0.379	0.219	0.562
SECSHELLS [51, 53]	32 776	$\mathbb{L}^2$	0.546	0.267	0.350	0.209	0.545
D2 [48, 429]	136	$\mathbb{L}^2$	0.311	0.158	0.235	0.139	0.434

"Distance" refers to the distance measure used to compare the descriptors.

such as SWC_d, SWEL1, and SWEL2 (see Section 4.4.2.2), are very compact, achieve good retrieval performance in terms of DCG, but their performance in terms of Nearest Neighbor (NN), First Tier (FT), Second Tier (ST), and E-measure (E-) is relatively low compared to the LFD, REXT, and SHD.

Table 12.4 summarizes the performance of some 3D shape retrieval methods that use local descriptors aggregated using BoFs techniques. We can see from Tables 12.3 and 12.4 that local descriptors outperform the global descriptors on benchmarks that contain generic 3D shapes, e.g. the PSB [53] and on datasets that contain articulated shapes (e.g. McGill dataset [417]).

Table 12.4 Examples of local descriptor-based 3D shape retrieval methods and their performance on commonly used datasets.

Method	PSB [53]			WM-SHREC07 [430]			McGill [417]		
	NN	FT	ST	NN	FT	ST	NN	FT	ST
VLAT coding [431]	—	—	—	—	—	—	0.969	0.658	0.781
Cov.-BoC [139]	—	—	—	0.930	0.623	0.737	0.977	0.732	0.818
2D/3D hybrid [432]	0.742	0.473	0.606	0.955	0.642	0.773	0.925	0.557	0.698
CM-BoF + GSMD [141]	0.754	0.509	0.640	—	—	—	—	—	—
CM-BoF [141]	0.731	0.470	0.598	—	—	—	—	—	—
Improved BoC [427]	—	—	—	0.965	0.743	0.889	0.988	0.809	0.935
TLC (VLAD) [433]	0.763	0.562	0.705	0.988	0.831	0.935	0.980	0.807	0.933

An important advantage of local descriptors is their flexibility in terms of the type of analysis that can be performed with. They can be used as local features for shape matching and recognition as shown in Chapters 10 and 11. They can also be aggregated over the entire shape to form global descriptors as shown in Table 12.4. Also, local descriptors characterize only the local geometry of the 3D shape, and thus they are usually robust to some nonrigid deformations, such as bending and stretching, and to geometric and topological noise.

12.4.2 Deep Learning-Based Methods

The main issue with the handcrafted (global or local) descriptors is that they do not scale well with the size of the datasets. Table 12.5 shows that the LFD and the SHD descriptors, which are the best performing global descriptors on the PSB, achieved only 33.3% and 40.9%, respectively, of mAP on the ModelNet dataset [73], which contains 127 915 models divided into 662 shape categories. Large 3D shape datasets are in fact very challenging. They often exhibit high inter-class similarities and also high intraclass variability. Deep learning techniques perform better in such situations since, instead of using handcrafted features, they learn, automatically from training data, the features which achieve the best performance.

Table 12.5 summarizes the performance, in terms of the classification accuracy and the mAP on the ModelNet benchmark [73], of multiview Convolutional Neural Network-based global descriptors [68]. It also compares them to the standard global descriptors such as SHD and LFD. In this table,

Table 12.5 Comparison between handcrafted descriptors and Multi-View CNN-based descriptors [68] on ModelNet [73].

| Method | Training configuration | | | Test config. | Classif. accuracy | Retrieval (mAP) |
	Pretrain	Fine-tune	#Views	#Views		
SHD [60, 76]	—	—	—	—	68.2	33.3
LFD [52]	—	—	—	—	75.5	40.9
CNN without f.t.	ImageNet1K	—	—	1	83.0	44.1
CNN, with f.t.	ImageNet1K	ModelNet40	12	1	85.1	61.7
MVCNN without f.t.	ImageNet1K	—	—	12	88.1	49.4
MVCNN, with f.t.	ImageNet1K	ModelNet40	12	12	89.9	70.1
MVCNN without f.t.	ImageNet1K	—	80	80	84.3	36.8
MVCNN with f.t.	ImageNet1K	ModelNet40	80	80	90.1	70.4

The pretraining is performed on ImageNet [70]. "f.t." refers to fine-tuning.
Source: The table is adapted from [68].

CNN refers to the descriptor obtained by training a CNN on a (large) image database, and using it to analyze views of 3D shapes. MVCNN refers to the multi-view CNN presented in Section 4.5.2. At runtime, one can render a 3D shape from any arbitrary viewpoint, pass it through the pretrained CNN, and take the neuron activations in the second-last layer as the descriptor. The descriptors are then compared using the Euclidean distance.

To further improve the accuracy, one can fine-tune these networks (CNN and MVCNN) on a training set of rendered shapes before testing. This is done by learning, in a supervised manner, the dissimilarity function which achieves the best performance. For instance, Su et al. [68] learned a Mahalanobis metric W which directly projects multiview CNN descriptors $h \in \mathbb{R}^d$ to $Wh \in \mathbb{R}^p, p < d$, such that the $\mathbb{L}^2$ distances in the projected space are small between shapes of the same category, and large between 3D shapes that belong to different categories. Su et al. [68] used the large-margin metric learning algorithm with $p < d$, ($p = 128$) to make the final descriptor compact.

With this fine-tuning, CNN descriptors, which use a single view (from an unknown direction) of the 3D shape, outperform the LFD. In fact, the mAP on the 40 classes of the ModelNet [73] improves from 40.9% (LFD) to 61.7%. This performance is further boosted to 70.4% when using the MultiView CNN, with 80 views and with fine-tuning the CNN training on ImageNet1K.

Table 12.6 summarizes the performance, in terms of the F-measure, mAP, and NDCG, of some other deep learning-based methods. Overall, the

Table 12.6 Performance comparison on ShapeNet Core55 normal dataset of some learning-based methods.

Method	Micro			Macro			Micro + Macro		
	F-	mAP	NDCG	F-	mAP	NDCG	F-	mAP	NDCG
Kanezaki [434]	0.798	0.772	0.865	0.590	0.583	0.650	0.694	0.677	0.757
Tatsuma_ReVGG [434]	0.772	0.749	0.828	0.519	0.496	0.559	0.645	0.622	0.693
Zhou [434]	0.767	0.722	0.827	0.581	0.575	0.657	0.674	0.648	0.742
MVCNN [68]	0.764	0.873	0.899	0.575	0.817	0.880	0.669	0.845	0.890
Furuya_DLAN [434]	0.712	0.663	0.762	0.505	0.477	0.563	0.608	0.570	0.662
Thermos_ MVFusionNet [434]	0.692	0.622	0.732	0.484	0.418	0.502	0.588	0.520	0.617
GIFT [82]	0.689	0.825	0.896	0.454	0.740	0.850	0.572	0.783	0.873
Li and coworkers [419]	0.582	0.829	0.904	0.201	0.711	0.846	0.392	0.770	0.875
Deng_CM- VGG5-6DB [434]	0.479	0.540	0.654	0.166	0.339	0.404	0.322	0.439	0.529
Tatsuma and Aono [435]	0.472	0.728	0.875	0.203	0.596	0.806	0.338	0.662	0.841
Wang and coworkers [419]	0.391	0.823	0.886	0.286	0.661	0.820	0.338	0.742	0.853
Mk_DeepVoxNet [434]	0.253	0.192	0.277	0.258	0.232	0.337	0.255	0.212	0.307

Here, F-refers to the F-measure; VLAD stands for vector of locally aggregated descriptors.

retrieval performance of these techniques is fairly high. In fact, most of the learning-based methods outperform traditional handcrafted features-based approaches.

12.5 Summary and Further Reading

In this chapter, we discussed how shape descriptors and similarity measures are used for 3D shape retrieval. We focused on the case where the user specifies a complete 3D model as a query. The retrieval system then finds, from a collection of 3D models, the models that are the most relevant to the query.

The field of 3D model retrieval has a long history, which we can summarize in three major eras. The methods in the first era used the handcrafted global descriptors described in Chapter 4. Their main focus was on designing descriptors that capture the overall properties of shapes while achieving invariance to rigid transformations (translation, scaling, and rotation).

Methods in the second era used local descriptors aggregated into global descriptors using aggregation techniques such as the BoF. Their main focus was on achieving (i) invariance to rigid transformations and isometric (bending) deformations, (ii) robustness to geometrical and topological noise, and incomplete data (as is the case with 3D scans), and (iii) better retrieval performance.

The new era of 3D shape retrieval tackled the scalability problem. Previous methods achieved relatively acceptable results only on small datasets. Their performance drops significantly when used on large datasets such as ShapeNet [75]. This new generation of methods is based on deep learning techniques that achieved remarkable performances on large-scale image retrieval. Their main limitation, however, is that they require large amounts of annotated training data and a substantial computational power. Although datasets such as the ShapeNet [75] and ModelNet [73] provide annotated data for training, their sizes are still too small compared to existing image collections (in the 100K while image collections are well above millions). Thus, we expect that the next generation of retrieval methods would focus on handling weakly annotated data. We also expect that very large datasets of 3D models will start to emerge in the near future.

13

Cross-domain Retrieval

13.1 Introduction

In Chapter 12, we emphasized on the need for efficient 3D search tools and presented the most important techniques for 3D shape retrieval using 3D models as queries. Ideally, however, a search engine should provide a mechanism that enables users to retrieve relevant information, whether it is an image, a 3D model, or a 2D sketch, using queries which can also be images, sketches, text, or 3D models. For instance, the user may draw a 2D sketch depicting an airplane and queries a collection of images looking for related photos. Other users may want to search for sketches that are similar to a given photo or 3D model. This is referred to as *cross-domain retrieval*. It aims at providing the users with an efficient way to search for shapes independently of their representations.

The main difficulty with cross-domain retrieval is that different modalities lie in different (disjoint) spaces or domains. The challenge is then how to narrow the modality gap between these different types of representations. Early works that attempted to solve this problem focused solely on two modalities, e.g. photos vs. 3D shapes, or hand-drawn sketches vs. 3D shapes. For instance, in photo-based 3D shape retrieval, the signature of the photo is compared to the signatures of the 2D projections of the 3D model, see view-based 3D shape descriptors described in Section 4.3. Similarly, sketch-based 3D shape retrieval methods extract 2D silhouettes from the 3D models by projecting them onto a binary image. The projected silhouettes are then compared to the 2D sketches using view-based descriptors, see for example the works of Kanai [436], Li et al. [437], and Aono and Iwabuchi [438]. Their main limitation is that, in general, hand-drawn sketches are significantly different from silhouettes extracted from 2D projections of 3D models. Some works, e.g. [438–440], have leveraged contour representations to build more realistic sketch-based 3D model retrieval systems [437, 441–444]. The solution, however, remains limited to only this type of modality.

Instead of trying to narrow the gap between different modalities, another class of methods formulate the cross-domain retrieval problem as a *joint*

3D Shape Analysis: Fundamentals, Theory, and Applications, First Edition.
Hamid Laga, Yulan Guo, Hedi Tabia, Robert B. Fisher, and Mohammed Bennamoun.
© 2019 John Wiley & Sons, Inc. Published 2019 by John Wiley & Sons, Inc.

embedding problem. They seek to embed all these modalities into a common domain space so that one can compare them directly using standard distance measures. These methods have been first investigated for joint image and text search where semantically related elements, whether they are text or images, are mapped onto points which are close to each other in the embedding space [445–447]. They have been later extended to photo-to-3D shape and sketch-to-3D shape comparison by using deep learning techniques, e.g. Siamese networks [448], which jointly learn shapes from hand-drawn sketches and rendered sketches from 3D shapes, and deep CNN [449], which jointly learn shape similarities from photo and 3D shapes.

In this chapter, we focus on three types of shape modalities, i.e. photos, hand-drawn sketches and 3D shapes (although the approach is general and can be used with other modalities). We start by outlining the challenges and the datasets. We then review the commonly proposed solutions that handle this new research problem. As shown in Table 13.1, these solutions, which mostly use deep learning techniques, aimed at finding a function that maps the original modalities into a shared space such that the Euclidean distance between elements in the shared space equals to the *semantic* distance in the original space. From Table 13.1, we can observe three main points:

- Deep learning techniques are extensively used. In fact, several recent works [405, 460, 461] have demonstrated the superiority of these algorithms, which are particularly able to learn complex visual information from raw

Table 13.1 An example of the state-of-the-art methods that are proposed for cross-domain comparison.

Works	Type of cross-domain	Techniques
Rasiwasia et al. [450]	Text-to-photo	Correlation model
Jia et al. [451]	Text-to-photo	Markov random field
Chen et al. [452]	Sketch-to-photo	Meta data
Eitz et al. [453]	Sketch-to-photo	BoF descriptors
Russell et al. [454]	Painting-to-photo	ICP-like registration
Liu et al. [455]	Sketch-to-photo	Deep Siamese CNN
Qi et al. [456]	Sketch-to-photo	Deep Siamese CNN
Song et al. [457]	Sketch-to-photo	Deep Siamese CNN
Li et al. [80]	Sketch-to-3D shape	Variety of tools
Wang et al. [448]	Sketch-to-3D shape	Deep Siamese CNN
Dai et al. [458]	Sketch-to-3D shape	Deep Siamese CNN
Li et al. [449]	Photo-to-3D shape	3D centric Deep CNN
Tabia and Laga [459]	Sketch-to-3D shape-to-photo	3D centric Deep CNN

data. The performance of deep learning-based methods, however, highly depends on the availability of large amounts of annotated training data, which are difficult to obtain. Fortunately, 3D shapes can be used to synthesize photos and sketches. Section 13.2.2 will discuss, in detail, how additional training data can be synthesized from 3D shapes.

- The use of the Siamese architecture, which jointly embeds two different modalities into the same space, has been proposed in [448, 458] (see Section 13.3).
- The use of a 3D shape centric method, which uses the 3D shapes as the center of interest and constructs an embedding space around them. Section 13.4 details this method.

Before discussing in details these approaches, we will first review in Section 13.2 the specific challenges and the main datasets which have been used in cross-domain 3D shape retrieval.

13.2 Challenges and Datasets

Cross-domain shape retrieval is a very challenging task, not only because of the different types of shape representations but also due to the level of details and complexity in each representation. Below, we summarize a few of these challenges.

- **Training data**. Probably, one of the biggest challenges currently faced by cross-domain retrieval is the lack of datasets and benchmarks that are specifically designed for this task. In particular, taking each modality alone, one can find many datasets and benchmarks for evaluating retrieval algorithms which operate on that modality. These datasets, however, are not interrelated across modalities. For instance, one cannot easily find a dataset that contains for each 3D model, its 2D images taken under various conditions, and its 2D sketches drawn in different ways and by multiple users.
- **Clutter and complex backgrounds in 2D images**. In general, images contain cluttered objects, with complex backgrounds. This makes their comparison with hand-drawn sketches or with 3D models very challenging.
- **Multiple objects in 2D images**. A photo may contain more than one object that also equivocates the comparison process.
- **Complexity of and ambiguity in hand-drawn sketches**. 3D objects and hand-drawn sketches can be complex. Hand-drawn sketches, for instance, are usually noisy as lines and strokes may be ambiguous and may not correspond to the real contour of the 3D shapes.
- **User perception**. Sketches depend highly on the user perception. They generally involve different levels of detail (e.g. contours [444] or skeleton sketches [463]). Figure 13.1 shows some examples of hand-drawn sketches

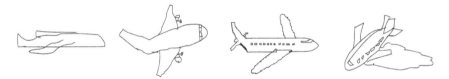

Figure 13.1 Sample of sketches taken from the SHREC 2013 dataset [462]. Observe that the same object can be sketched differently by different users.

taken from the SHREC 2013 dataset [462]. One can see from this example that the same object can be sketched differently by different users. Also, subject to the user imagination, sketches of the same object may have different levels of details.

13.2.1 Datasets

Since cross-domain shape retrieval is a relatively recent research area, there are no datasets that have been specifically designed for this task. Authors generally mix datasets from different domains to evaluate the performance of their proposed methods. The most commonly used datasets, from which shapes are mixed, are:

- **ShapeNet [401] and the Toyohashi Shape Benchmark (TSB) [418] datasets**. These are popular 3D shape datasets. They are initially released for 3D shape retrieval contests. A detailed description of these datasets can be found in Chapter 12.
- **ImageNet [70]**. This dataset contains more than 15 million labeled high-resolution images belonging to roughly 22 000 categories. The images were collected from the web and labeled by humans using Amazon's Mechanical Turk crowd-sourcing tool. Starting from 2010, as part of the Pascal Visual Object Challenge, an annual competition called the ImageNet Large-Scale Visual Recognition Challenge has been held. The challenge uses a subset of ImageNet with roughly 1000 image categories with 1000 images per category. Overall, there are roughly 1.2 million images for training, 50 000 images for validation, and 150 000 images for testing. ImageNet has been used by Li et al. [449] and Tabia and Laga [459] to learn joint embedding of photos and 3D shapes.
- **Sketch dataset [444]**. This dataset has been collected to explore how humans draw sketches. Eitz et al. [444] have collected 20 000 human drawn sketches, categorized into 250 classes, with 80 sketches per class. The dataset has been mixed with 3D models from the Princeton Shape Benchmark [53] to form the shape retrieval contest SHREC 2013 [462]. The contest benchmark contains 1258 shapes and 7200 sketches, which are grouped into 90 classes. The number of shapes in each class is about 14 on

average, while the number of sketches for each class is equal to 80 in total. For each class, there are 50 sketches for training and 30 for testing. The sketch dataset has also been used in the SHREC 2014 [80] contest, which is a benchmark for large-scale sketch-based 3D shape retrieval. It consists of shapes from various datasets, such as SHREC 2012 [437] and the Toyohashi Shape Benchmark (TSB) [418]. The dataset has about 13 680 sketches and 8987 3D models in total, grouped into 171 classes. The SHREC 2014 dataset is quite challenging due to its diversity of classes and the presence of large variations within the classes. The number of shapes in each class varies from less than 10 to more than 300, while the number of sketches for each class is equal to 80 (50 sketches for training and 30 for testing).

13.2.2 Training Data Synthesis

As deep learning-based methods require data augmentation, several authors [464, 465] have used 3D models to render large amounts of training data.

13.2.2.1 Photo Synthesis from 3D Models

Using 3D models to generate large amounts of training images has recently been used by several authors [465–467]. Most of them use the rendered images to train object detectors or viewpoint classifiers. A variety of images is generated from a single 3D shape by varying a set of parameters (including the position of the light, the camera, and the pose of the 3D shape). Su et al. [464] and Peng and Saenko [468] proposed to increase the number of synthesized images and their variations by acting on other parameters (including changing background patterns, illumination, and viewpoint). Su et al. [464] inject randomness in the three basic steps of their pipeline that includes rendering, background synthesis, and cropping. For image rendering, the lighting conditions and the camera configurations have been tuned. For the lighting conditions, the number of light sources, their positions and energies have also been varied and tuned. For the camera extrinsics, azimuth, elevation and in-plane rotation can be sampled from distributions estimated from a training set composed of real images. Figure 13.2 shows a few examples of images synthesized from a 3D model.

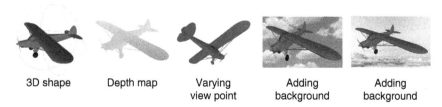

| 3D shape | Depth map | Varying view point | Adding background | Adding background |

Figure 13.2 Photo synthesis from a 3D model.

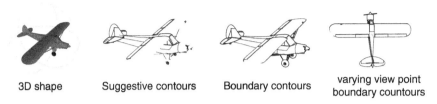

| 3D shape | Suggestive contours | Boundary contours | varying view point boundary countours |

Figure 13.3 Sketch synthesis from a 3D model.

13.2.2.2   2D Sketch Synthesis from 3D Models

Similarly to synthesized photos, a large amount of sketches can be generated from 3D models. Synthesizing sketches from a 3D shape is a very active area of research in the computer vision and graphic communities [439, 469–471]. A large set of parameters, including the viewpoint and the contour type, can be tuned to generate a diversity of sketches. Figure 13.3 shows four sketches generated from one 3D shape. The synthesized sketches correspond, respectively, from left to right to (i) suggestive contour [439], (ii) outer edges (boundary) sketch, and (iii) outer edges with different viewpoint.

13.3   Siamese Network for Cross-domain Retrieval

One of the common approaches used for cross-domain shape retrieval is based on Siamese network, a special type of neural network architectures. It has been proposed to differentiate between two inputs [472]. As illustrated in Figure 13.4, the Siamese network consists of two identical subnetworks, one per modality, joined at their outputs with a joining neuron. During the training, the two subnetworks extract features from the two different modalities, while the joining neuron measures the distance between the two feature vectors computed by the two subnetworks. The two subnetworks are identical and have the same weights.

Figure 13.4 shows an example of a Siamese network designed to learn the similarity between 3D models and 2D sketches. The first subnetwork (CNN1) takes as input a sketch synthesized from a 3D model (see Section 13.2.2) while the second one takes real sketches. During the training phase, the weights of the two networks are iteratively optimized.

Let us denote by S_1 a real (hand drawn) sketch and by S_2 a sketch rendered from a 3D shape. Let c be a binary label indicating whether S_1 and S_2 correspond to the same shape. Let $f_\theta(S_1)$, respectively $f_\theta(S_2)$, be the feature vector produced by the Siamese CNN for S_1, respectively S_2. The goal is to train the network so that it can learn the compatibility between the two sketches S_1 and

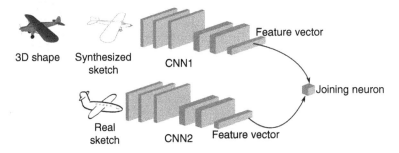

Figure 13.4 Joint embedding using Siamese network.

S_2. This compatibility can be defined as the $\mathbb{L}^2$ distance between the two feature vectors, i.e.:

$$E_\theta(S_1, S_2) = \|f_\theta(S_1) - f_\theta(S_2)\|_2. \tag{13.1}$$

The aim of the training phase is to compute the parameters θ of the Siamese network. Given a set of training pairs $((S_1, S_2), c)$, composed of positive $(c = 1)$ and negative $(c = -1)$ examples, the loss function of the network can be expressed as:

$$L(\theta) = \sum_{i=1}^{n} l(\theta, (S_1, S_2, c)_i), \tag{13.2}$$

where

$$l(\theta, (S_1, S_2, c)_i) = (1 - c)L^+(E_\theta(S_1, S_2)_i) + cL^-(E_\theta(S_1, S_2)_i). \tag{13.3}$$

Here, $(S_1, S_2)_i$ is the ith training sample, and n is the number of samples. L^+ is the partial loss function for a positive pair, and L^- is the partial loss function for an negative pair, see [473] (pages 142–143) for the definition of these partial lose functions.

The Siamese network has been used for the sketch-to-3D shape retrieval by Wang et al. [448]. The authors proposed to scale the input sketches to images of 100×100 for both sources. The two subnetworks of the Siamese network are two CNNs (see Chapter 4 for CNN networks). Each single CNN has

- Three convolutional layers, each with a max pooling.
- One fully connected layer to generate the features, and
- One output layer to compute the loss.

The first convolutional layer followed by a 4×4 pooling generates 32 response maps, each of size 22×22. The second layer and pooling outputs 64 maps of size 8×8 each. The third layer has 256 response maps, each pooled to a size

Table 13.2 An example of some Siamese networks based method and their performance on SHREC 2013 [462] and SHREC 2014 [80].

Dataset	Works	NN	FT	ST	E	DCG	mAP
SHREC 2013 [462]	Wang et al. [448]	0.405	0.403	0.548	0.287	0.607	0.469
	Dai et al. [458]	0.650	0.634	0.719	0.348	0.766	0.674
SHREC 2014 [80]	Wang et al. [448]	0.239	0.212	0.316	0.140	0.495	0.228
	Dai et al. [458]	0.272	0.275	0.345	0.171	0.498	0.286

of 3×3. The 2304 features, which are generated by the final pooling operation, are linearly transformed, in the last layer, to feature vectors of size 64×1. Rectified Linear Units are used in all layers. Dai et al. [458] extended the work of Wang et al. [448] by adding a marginal distance between a set of positive and negative pairs across different domains. This increases the discrimination of the learned features from the Siamese network.

Table 13.2 shows the performance, in terms of nearest neighbor (NN), first tier (FT), second tier (ST), discounted cumulative gain (DCG), and mean average precision (mAP), of some Siamese-based methods. We refer to reader to Chapter 12 for a detailed description of these evaluation metrics. The reported results have been achieved on the SHREC 2013 and the SHREC 2014 datasets. As we can see, the two methods achieved good results on the SHREC 2013 [462] datasets, with the approach of Dai et al. [458] performing significantly better than Wang et al. [448]. The two methods, however, perform equally low on the SHREC 2014 dataset [80].

The Siamese network efficiently learns the similarities between cross-domain modalities. However, the Siamese network, as its name implies, allows the joint learning of only two modalities (e.g. sketch-to-3D shape). Adding other modalities requires further considerations.

13.4   3D Shape-centric Deep CNN

The main idea behind 3D shape-centric approaches is to place the 3D representation of the shape in the center, while the other representations are set around, as shown in Figure 13.5. Placing 3D shapes in the center can have several advantages. **First**, the 3D representation provides the most complete visual information about the shape being analyzed. **Second**, as described in Section 13.2.2, 3D shapes can be used in a rendering process from which large amounts of 2D views can be generated. Learning algorithms can then benefit

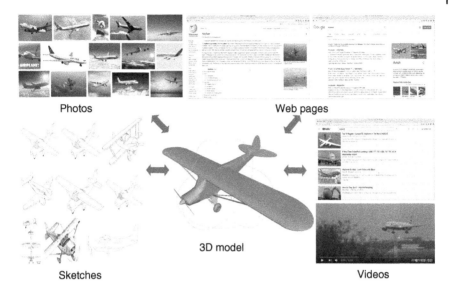

Photos

Web pages

3D model

Sketches

Videos

Figure 13.5 Example of five modalities from the same shape class (airplane in this case). The 3D shape is set in the center of interest. The comparison between the other modalities is feasible through the 3D shape.

from these data and bridge the gap between the different shape representations. By jointly learning the 3D and any other representations, the interaction between other modalities becomes possible by a simple transitivity.

The general pipeline of a 3D centric cross-domain shape retrieval method, which is illustrated in Figure 13.6, can be divided into three main steps; (i) embedding space construction from the collection of 3D shapes, (ii) photo and sketch synthesizing, and (iii) learning shape from the synthesized data. The following sections describe in detail each of these steps.

13.4.1 Embedding Space Construction

The embedding space can be defined as a topological shape space where each point corresponds to one shape represented by any modality. Since we consider 3D shapes as the central representation, the embedding space is first constructed using only the relations between 3D shapes. Let $\{M_i\}$ be a collection of 3D shapes and $\{\mathbf{x}_i\}$ their corresponding shape descriptors, which can be any of the descriptors presented in Chapters 4 and 5. Once the embedding space is constructed, all the other modalities can be mapped onto it, such that the different modalities which represent the same 3D shape will surround that shape.

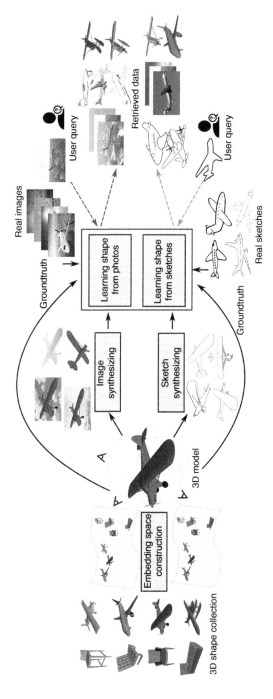

Figure 13.6 Cross-domain shape retrieval pipeline using 3D centric method.

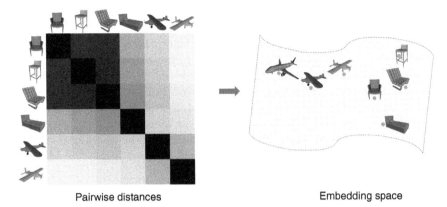

Pairwise distances   Embedding space

Figure 13.7 Embedding space construction. Given the pairwise similarities between 3D shapes in a collection, the aim is to find an (Euclidean) embedding space where the distances between shapes are preserved.

The embedding process starts by building a similarity graph such that each 3D shape is represented with a node in the graph. An edge in the graph connects two nodes (and thus shapes) that are similar to each other. The length of each edge is defined as the dissimilarity between the shape of the nodes it connects. In general, one can use any type of dissimilarity measure. However, for simplicity, one can assume that the shape descriptors are elements of a Euclidean space and use the $\mathbb{L}^2$ metric for their comparison. That is, $d(M_i, M_j) = \|\mathbf{x}_i - \mathbf{x}_j\|_2$ for any given pair of shapes M_i and M_j. The embedding process then aims to find an embedding space where the distances between shape are preserved. This is illustrated in Figure 13.7.

Under this setup, there are several ways of building the embedding space. The simplest way is to assume that the embedding space is the same as the space of the signatures of the 3D models. In general, however, these signatures are not compact and sometimes nonlinear. This may impact on the embedding space structure and scale. There are several efficient approaches for handling this issue. Below, we describe a few of them.

13.4.1.1 Principal Component Analysis

A simple way to build an embedding shape space is to apply principal component analysis (PCA) to the set of the 3D shape signatures. The PCA allows to normalize and to reduce the dimension of each signature. The reduced features can then be used as coordinates to represent 3D shapes in the embedding space. We refer the reader to Chapter 2 for more details on the application of PCA to a set of points in $\mathbb{R}^d$, where d is the dimension of the descriptors.

13.4.1.2 Multi-dimensional Scaling

The aim of the multi-dimensional scaling (MDS) algorithm [474] is to find a space that precisely represents the given relations between the data, while reducing its dimensionality (see Figure 13.7). The original MDS algorithm computes, from the matrix of dissimilarities between the 3D shapes, a coordinate matrix whose configuration minimizes a cost function. Let $\mathbf{D} = [d(\mathbf{x}_i, \mathbf{x}_j)]$ be a distance matrix computed for all pairs of descriptors $\mathbf{x}_i$ and $\mathbf{x}_j$. The goal is to find the new coordinates p_i, p_j of $\mathbf{x}_i, \mathbf{x}_j$ in the embedding space. If we consider a square error cost, the objective function to be minimized can be written as:

$$E_M = \sum_{i \neq j} [d(\mathbf{x}_i, \mathbf{x}_j) - d(p_i, p_j)]^2. \tag{13.4}$$

Let $\mathbf{P}$ be the matrix whose columns are the new points p_i. The goal is to find the coordinate matrix $\mathbf{P}$, which minimizes the cost function of Eq. (13.4). The MDS algorithm, which solves this optimization problem, can be summarized as follows:

- Setup the matrix of squared distances $\mathbf{D} = [d^2(\mathbf{x}_i, \mathbf{x}_j)]$.
- Apply the double centering: $\mathbf{B} = -\frac{1}{2}\mathbf{JDJ}$, where $\mathbf{J} = \mathbf{I} - n^{-1}\mathbf{1}\mathbf{1}^T$, n is the number of 3D shapes, $\mathbf{I}$ is the identity matrix of size $n \times n$, and $\mathbf{1}$ is a matrix of size $n \times n$ and who elements are all equal to 1.
- Compute the m largest positive eigenvalues $\lambda_1, \dots, \lambda_m$ of $\mathbf{B}$ and their corresponding eigenvectors $\Lambda_1, \dots, \Lambda_m$.
- An m-dimensional spatial configuration of the n 3D shapes is derived from the coordinate matrix $\mathbf{P} = \mathbf{E}_m \Lambda_m^{1/2}$, where $\mathbf{E}_m$ is the matrix of m eigenvectors and Λ_m is the diagonal matrix of m eigenvalues of $\mathbf{B}$.

In the literature, many variants of the original MDS with slightly different cost functions and optimization algorithms have been proposed. An interesting nonlinear variant, proposed by Sammon [475], uses the following cost function:

$$E_s = \frac{1}{\sum_{i \neq j} [d(p_i, p_j)]^2} \sum_{i \neq j} \frac{[d(\mathbf{x}_i, \mathbf{x}_j) - d(p_i, p_j)]^2}{d(\mathbf{x}_i, \mathbf{x}_j)}. \tag{13.5}$$

This is a weighted sum of differences between the original pairwise distances and the embedding pairwise distances.

13.4.1.3 Kernel-Based Analysis

Another approach is to compute a mapping function f to some specific space such that the dot product between the mapped elements $f(\mathbf{x}_i)^T f(\mathbf{x}_j)$ is as close as possible to a kernel $\mathbf{k}(\mathbf{x}_i, \mathbf{x}_j)$ associated with the $\mathbb{L}^2$ distance $d(\mathbf{x}_i, \mathbf{x}_j)$ [476].

Let us consider the expression of the triangular kernel, $k(\mathbf{x}_i, \mathbf{x}_j)$ associated to $d(\mathbf{x}_i, \mathbf{x}_j)$ as follows:

$$k(\mathbf{x}_i, \mathbf{x}_j) = 1 - \frac{d(\mathbf{x}_i, \mathbf{x}_j)^2}{2}. \tag{13.6}$$

With this expression for k, the projection onto a lower dimensional space, which preserves most of the metric proprieties, can easily be derived. Let us consider a training set $\mathcal{A} = \{\mathbf{x}_i\}$ used for the training of the projection $f(\cdot)$, which has to satisfy $\forall \mathbf{x}_i, \mathbf{x}_j \in \mathcal{A}, f(\mathbf{x}_i)^\top f(\mathbf{x}_j) = k(\mathbf{x}_i, \mathbf{x}_j)$. The Gram matrix of the kernel k on $\mathcal{A}$ is given by $K = [k(\mathbf{x}_i, \mathbf{x}_j)]_{\mathbf{x}_i, \mathbf{x}_j \in \mathcal{A}}$. K can then be approximated by a low rank version and the corresponding projection can be easily computed. This is known as the Nyström approximation for kernels, and has been used to speed up large scale kernel based classifiers in machine learning [477, 478]. Since k is a Mercer kernel, and thus positive semi-definite, its eigenvalue decomposition is given by: $K = V\Lambda V^\top$. The nonlinear projection $f(\cdot)$ is then given by:

$$f(X) = \Lambda^{-\frac{1}{2}} V^\top [k(\mathbf{x}_i, \mathbf{x})]_{\mathbf{x}_i \in \mathcal{A}}, \tag{13.7}$$

where $[k(\mathbf{x}_i, \mathbf{x})]_{\mathbf{x}_i \in \mathcal{A}}$ is the vector of entry-wise computation of the kernel between $\mathbf{x}$ and the elements of the training set. Let us consider the matrix Y of the projected elements of $\mathcal{A}$: $Y = [f(\mathbf{x})]_{\mathbf{x} \in \mathcal{A}}$. The Gram matrix in the embedding space using the linear kernel is thus:

$$Y^\top Y = (\Lambda^{-\frac{1}{2}} V^\top K)^\top \Lambda^{-\frac{1}{2}} V^\top K = KV\Lambda^{-1}V^\top K = KK^{-1}K = K. \tag{13.8}$$

The Euclidean distance between the elements projected onto this space is equal to the original distance on the training set, i.e. $\forall \mathbf{x}_i, \mathbf{x}_j \in \mathcal{A}$:

$$\begin{aligned}
\|f(\mathbf{x}_i) - f(\mathbf{x}_j)\|^2 &= f(\mathbf{x}_i)^\top f(\mathbf{x}_i) + f(\mathbf{x}_j)^\top f(\mathbf{x}_j) - 2f(\mathbf{x}_i)^\top f(\mathbf{x}_j) \\
&= k(\mathbf{x}_i, \mathbf{x}_i) + k(\mathbf{x}_j, \mathbf{x}_j) - 2k(\mathbf{x}_i, \mathbf{x}_j) \\
&= 1 - \frac{d(\mathbf{x}_i, \mathbf{x}_i)^2}{2} + 1 - \frac{d(\mathbf{x}_j, \mathbf{x}_j)^2}{2} - 2 + 2\frac{d(\mathbf{x}_i, \mathbf{x}_j)^2}{2} \\
&= d(\mathbf{x}_i, \mathbf{x}_j)^2. \tag{13.9}
\end{aligned}$$

The more $\mathcal{A}$ is a good sampling of the original space, the closer the linearization f along with the dot product is to the original kernel k, and thus the better the distance $d(\cdot, \cdot)$ in the embedding space $\mathbb{E}$ is approximated.

Finally, some of the eigenvalues of the Kernel matrix are often very small compared to the other ones and can be safely discarded to further gain in computational efficiency. Moreover, since the associated projectors often encode noise in the data, discarding them can act as a denoising process and may improve the discrimination power of the representation in $\mathbb{E}$.

13.4.2 Learning Shapes from Synthesized Data

In cross-domain shape retrieval, our aim is to place the different modalities depicting similar shapes closely in the embedding space. The intuitive solution is to train a model to project these modalities onto the embedding space. For instance, since the points in the embedding space that correspond to given 3D shapes are known, a model can be trained to map photos and sketches, which belong to the same category of a 3D shape into the coordinate induced by the 3D shape.

13.4.3 Image and Sketch Projection

Given the large amount of generated data from 3D shapes and their coordinates in the embedding space, learning algorithms (e.g. deep CNN) can be trained (in a regression fashion) to map the generated data onto the corresponding coordinates induced by the 3D shapes [449]. Two different CNNs (one for images and one for sketches) can be trained to learn the mapping between an image I (respectively a sketch S), to the point $f_X \in \mathbb{E}$ corresponding to the coordinates of a 3D shape X from which I and S have been generated. Figure 13.8 illustrates this process for the case of 2D images.

The mapping CNN can be expressed as a function f_I (respectively f_S for sketches) that receives as input an image I_i (respectively a sketch S_i) and is expected to output f_X. The actual output of f_I (respectively f_S) depends on the parameters of the network θ_I (respectively θ_S) that are tuned during training. Hence, these mapping can be written as $f_I(I_i; \theta_I) \mapsto f_X$ for images (respectively $f_S(S_i; \theta_S) \mapsto f_X$ for sketches). A mapping error can be defined with a Euclidean

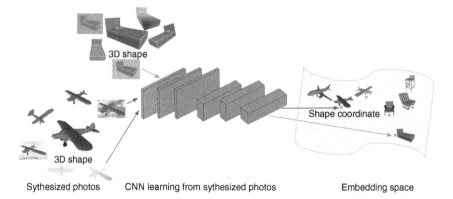

3D shape

3D shape

Shape coordinate

Sythesized photos   CNN learning from sythesized photos   Embedding space

Figure 13.8 Learning shapes from synthesized photos. The learning is based on a regression model. A CNN is given a large amount of training data and the coordinates of the points induced by each shape. By doing so, the CNN learns to map images onto the embedding space.

loss function for each CNN as follows:

$$L(\theta_I) = \sum_{X \in \mathcal{A}} \sum_i \|f_I(I_i, \theta_I) - f_X\|_2 \tag{13.10}$$

for image CNN, and

$$L(\theta_S) = \sum_{X \in \mathcal{A}} \sum_i \|f_S(S_i, \theta_S) - f_X\|_2 \tag{13.11}$$

for sketch mapping. Here, $\mathcal{A}$ denotes the shape collection. Li et al. [449] and Tabia and Laga [459] use the same architecture for joint shape and image embeddings. They use a CNN containing eight layers with weights initialized using ImageNet [460]:

- The first convolutional layer contains 96 kernels of size $11 \times 11 \times 3$ with a stride of 4 pixels.
- The second convolutional layer takes as input the (response-normalized and pooled) output of the first convolutional layer and filters it with 256 kernels of size $5 \times 5 \times 96$.
- The third, fourth, and fifth convolutional layers are connected together without any intervening pooling or normalization layers.
- The third convolutional layer has 384 kernels of size $3 \times 3 \times 256$ connected to the (normalized, pooled) outputs of the second convolutional layer.
- The fourth convolutional layer has 384 kernels of size $3 \times 3 \times 384$.
- The fifth convolutional layer has 256 kernels of size $3 \times 3 \times 384$.
- The fully-connected layers have 4096 neurons each.

The output of the last fully-connected layer is fed to a 128-dimensional vector, which corresponds to one point f_X in the embedding space $\mathbb{E}$.

Once learned, both image and sketch CNNs can project new images (respectively new sketches) into the embedding space $\mathbb{E}$. In order to do so, a new image I (respectively a new sketch S) passes through the image CNN (respectively the sketch, CNN) which produces an output vector corresponding to the image coordinates $f_i \in \mathbb{E}$ (respectively the sketch coordinates $f_S \in \mathbb{E}$). Figure 13.9 illustrates this process.

Figure 13.10 shows how these points are distributed in the embedding space. Note that Tabia and Laga [459] used a kernel embedding and for visualization purposes, they reduced the dimensionality of the embedding space to two. The red (dark gray) symbols in Figure 13.10 denote images downloaded from the Internet, and the green (light gray) ones denote hand-drawn sketches. The symbols indicate the class membership of the objects. As one can see, entities of the same class, independently whether they are 3D models, sketches or 2D images, form clusters. This suggests that distances in this space can be used for the classification and retrieval of 3D shapes by using queries that can be 3D shapes, images, or 2D sketches.

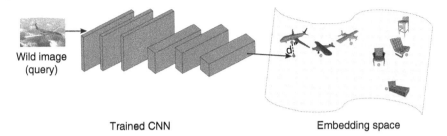

Trained CNN   Embedding space

Figure 13.9 The testing phase. The CNN maps input images onto the embedding space. The dissimilarity between different entities is computed as a distance in the embedding space.

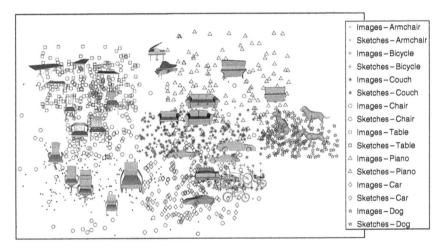

Figure 13.10 A visualization of some 3D shapes in the embedding space and a set of projected images and sketches [459]. The red (dark gray) symbols denote images from the wild, and the green (light gray) ones denote hand-drawn sketches. The symbols indicate the class membership of the objects.

13.5 Summary and Further Reading

We focused, in this chapter, on cross-domain shape retrieval where queries as well as queried objects can be images, sketches, or 3D models. This is an emerging field of research, which aims to bridge the gap between different modalities. This chapter only reviewed some of the most recent approaches that dealt with this problem. Other approaches include the work of Castrejon et al. [479], which proposed to learn an align cross modal scene representations from hundreds of natural scene types captured from five different modalities, excluding 3D shapes. One can expect in the near future an extension of these works to handle multiple other domains including text and video.

14

Conclusions and Perspectives

In this book, we have covered a broad range of 3D shape analysis techniques, their foundations, and their applications. Starting with the basics, we have seen how 3D objects are acquired and represented both mathematically and numerically. We have exposed a range of mathematical concepts that are essential for most of the 3D shape analysis tasks. This includes the mathematical representations of surfaces and the concepts of differential geometry.

Our journey in the 3D shape analysis world started with the preprocessing techniques, which take raw 3D data and transform them into a state that facilitates their analysis. This includes smoothing, parameterization, and normalization. Note that not all of these preprocessing steps are necessary for every 3D shape analysis task. For instance, some shape descriptors are invariant to some similarity-preserving transformations, such as translation, rotation, and scaling, and thus not all the normalization steps are required. Since many 3D shape analysis tasks are based on computing and comparing descriptors, we have looked at various global and local descriptors, which have been proposed in the literature, and studied their properties. Understanding the properties of each type of descriptors is important since different applications may have different requirements in terms of invariance to shape-preserving transformations; the type of shape features which need to be encoded in the descriptors; and the computational and storage constraints. We have covered hand-crafted descriptors as well as descriptors computed using Deep Learning techniques. As the availability of labeled 3D models on the Internet and in domain-specific repositories continues to increase, the latter approaches are gaining momentum and will continue to have a major impact on the 3D shape analysis field.

In this book, we have also looked at the correspondence and registration problem, which is an important aspect of 3D shape analysis. It is central to many applications including 3D reconstruction, deformation and interpolation, statistical analysis, and growth modeling and analysis. Despite the significant amount of research, this problem still remains unsolved especially when dealing with 3D models that deform elastically, and when putting in correspondence 3D models which are semantically similar but geometrically and topologically

3D Shape Analysis: Fundamentals, Theory, and Applications, First Edition.
Hamid Laga, Yulan Guo, Hedi Tabia, Robert B. Fisher, and Mohammed Bennamoun.
© 2019 John Wiley & Sons, Inc. Published 2019 by John Wiley & Sons, Inc.

very different. In fact, the gap between the human notion of similarity and the mathematical definition of similarity is still large.

The four applications that we covered, namely, 3D face recognition, object recognition in 3D scenes, 3D model retrieval, and cross-domain 3D retrieval, are just a few examples of what can be done with the techniques presented in the first and second parts of this book. Other applications include 3D growth and deformation pattern analysis (e.g. aging of 3D human faces and 3D human bodies, disease progression in 3D anatomical organs, etc.), 3D modeling, and 3D reconstruction. Given the decreasing price of 3D sensors, there will likely be a large growth in the variety of applications that use 3D data. This includes autonomous vehicles and road scanning, architectural modeling of current cities and buildings, the monitoring of humans and animals for their health and safety, factory automation and inspection, and human-robot interaction.

Given the breadth of material that we have covered in this book, what new developments are we likely to see in the future? One of the recent trends in image analysis is the use of the massive amount of images, which are currently available on the Internet for learning mathematical models of objects and scenes. This has been very successful, especially with the emergence of deep learning techniques. Although growing, the amount of 3D data currently available in online repositories created by the general public is still far less than their image counterpart. This is in part due to the fact that 3D acquisition systems are still not as simple and ubiquitous as 2D imaging systems. Nevertheless, we expect that in the near future, we will see massive 3D data repositories at a scale that is similar to what we are currently seeing with images and videos. We will also see the emergence of 3D video, or more precisely, the wide adoption of 3D video that will enable users and viewers to fully immerse inside 3D animated movies. Thus, 3D video analysis problems will start to attract more research.

3D models are just one type of digital representations of the world. People take images and videos, draw sketches, and describe the world with words and text. Traditionally, these modalities are treated independently, with most research focused on one modality (or just a few). One trend that we foresee in the future is how to jointly analyze and process multi-modal and heterogeneous visual and textual data. By doing so, one can search and retrieve relevant information independently whether it is a text document, a 3D model, a depth scan, a sketch, an image, or a video. This, however, would require embedding visual information in a common space independently of the type of their representation.

References

1 Botsch, M., Kobbelt, L., Pauly, M. et al. (2010). *Polygon Mesh Processing*. CRC Press.

2 Bernardini, F., Mittleman, J., Rushmeier, H. et al. (1999). The ball-pivoting algorithm for surface reconstruction. *IEEE Transactions on Visualization and Computer Graphics* 5 (4): 349–359.

3 Edelsbrunner, H. and Mücke, E.P. (1994). Three-dimensional alpha shapes. *ACM Transactions on Graphics (TOG)* 13 (1): 43–72.

4 Carr, J.C., Beatson, R.K., Cherrie, J.B. et al. (2001). Reconstruction and representation of 3D objects with radial basis functions. In: *Proceedings of the 28th Annual Conference on Computer Graphics and Interactive Techniques*, 67–76. ACM.

5 Kettner, L. (1999). Using generic programming for designing a data structure for polyhedral surfaces. *Computational Geometry* 13 (1): 65–90.

6 Campagna, S., Kobbelt, L., and Seidel, H.P. (1998). Directed edges-a scalable representation for triangle meshes. *Journal of Graphics Tools* 3 (4): 1–11.

7 Kendall, D.G. (1977). The diffusion of shape. *Advances in Applied Probability* 9 (3): 428–430.

8 Godil, A., Dutagaci, H., Bustos, B. et al. (2015). SHREC'15: range scans based 3D shape retrieval. *Proceedings of the Eurographics Workshop on 3D Object Retrieval, Zurich, Switzerland*, pp. 2–3.

9 Podolak, J., Shilane, P., Golovinskiy, A. et al. (2006). A planar-reflective symmetry transform for 3D shapes. *ACM Transactions on Graphics (TOG)* 25 (3): 549–559.

10 Mitra, N.J., Pauly, M., Wand, M., and Ceylan, D. (2013). Symmetry in 3D geometry: extraction and applications. *Computer Graphics Forum* 32 (6): 1–23.

11 Heeren, B., Rumpf, M., Schröder, P. et al. (2016). Splines in the space of shells. *Computer Graphics Forum* 35 (5): 111–120.

3D Shape Analysis: Fundamentals, Theory, and Applications, First Edition.
Hamid Laga, Yulan Guo, Hedi Tabia, Robert B. Fisher, and Mohammed Bennamoun.
© 2019 John Wiley & Sons, Inc. Published 2019 by John Wiley & Sons, Inc.

12 Windheuser, T., Schlickewei, U., Schmidt, F.R., and Cremers, D. (2011). Geometrically consistent elastic matching of 3D shapes: a linear programming solution. In: *2011 International Conference on Computer Vision*, 2134–2141. IEEE.

13 Heeren, B., Rumpf, M., Schröder, P. et al. (2014). Exploring the geometry of the space of shells. *Computer Graphics Forum* 33 (5): 247–256.

14 Zhang, C., Heeren, B., Rumpf, M., and Smith, W.A. (2015). Shell PCA: statistical shape modelling in shell space. *Proceedings of the IEEE International Conference on Computer Vision*, pp. 1671–1679.

15 Wirth, B., Bar, L., Rumpf, M., and Sapiro, G. (2011). A continuum mechanical approach to geodesics in shape space. *International Journal of Computer Vision* 93 (3): 293–318.

16 Heeren, B., Rumpf, M., Wardetzky, M., and Wirth, B. (2012). Time-discrete geodesics in the space of shells. *Computer Graphics Forum* 31 (5): 1755–1764.

17 Berkels, B., Fletcher, T., Heeren, B. et al. (2013). Discrete geodesic regression in shape space. *9th International Conference on Energy Minimization Methods in Computer Vision and Pattern Recognition*.

18 Jermyn, I.H., Kurtek, S., Klassen, E., and Srivastava, A. (2012). Elastic shape matching of parameterized surfaces using square root normal fields. In: *Computer Vision - ECCV 2012. ECCV 2012*, Lecture Notes in Computer Science, vol. 7576 (ed. A. Fitzgibbon, S. Lazebnik, P. Perona et al.), 804–817. Springer-Verlag, Berlin, Heidelberg.

19 Jermyn, I.H., Kurtek, S., Laga, H., and Srivastava, A. (2017). Elastic shape analysis of three-dimensional objects. *Synthesis Lectures on Computer Vision* 12 (1): 1–185.

20 Cohen-Or, D., Greif, C., Ju, T. et al. (2015). *A Sampler of Useful Computational Tools for Applied Geometry, Computer Graphics, and Image Processing: Foundations for Computer Graphics, Vision, and Image Processing*. CRC Press.

21 Hasler, N., Stoll, C., Sunkel, M. et al. (2009). A statistical model of human pose and body shape. *Computer Graphics Forum* 28 (2): 337–346.

22 Bronstein, A.M., Bronstein, M.M., and Kimmel, R. (2008). *Numerical Geometry of Non-Rigid Shapes*. Springer Science & Business Media.

23 Hartley, R. and Zisserman, A. (2003). *Multiple View Geometry in Computer Vision*. Cambridge University Press.

24 Guo, Y., Zhang, J., Lu, M. et al. (2014). Benchmark datasets for 3D computer vision. *The 9th IEEE Conference on Industrial Electronics and Applications*.

25 Gomes, L., Bellon, O.R.P., and Silva, L. (2014). 3D reconstruction methods for digital preservation of cultural heritage: a survey. *Pattern Recognition Letters* 50: 3–14.

26 Curless, B. (1999). From range scans to 3D models. *ACM SIGGRAPH Computer Graphics* 33 (4): 38–41.

27 Lanman, D. and Taubin, G. (2009). Build your own 3D scanner: 3D photography for beginners. In: *ACM SIGGRAPH 2009 Courses*, 8. ACM.

28 Scharstein, D. and Szeliski, R. (2002). A taxonomy and evaluation of dense two-frame stereo correspondence algorithms. *International Journal of Computer Vision* 47 (1–3): 7–42.

29 Furukawa, Y. and Hernández, C. (2015). *Multi-View Stereo: A Tutorial.* Citeseer.

30 Tombari, F., Mattoccia, S., Di Stefano, L., and Addimanda, E. (2008). Classification and evaluation of cost aggregation methods for stereo correspondence. In: *IEEE Conference on Computer Vision and Pattern Recognition*, 1–8. IEEE.

31 Geiger, A., Lenz, P., and Urtasun, R. (2012). Are we ready for autonomous driving? The KITTI vision benchmark suite. In: *2012 IEEE Conference on Computer Vision and Pattern Recognition (CVPR)*, 3354–3361. IEEE.

32 Menze, M. and Geiger, A. (2015). Object scene flow for autonomous vehicles. *Proceedings of the IEEE Conference on Computer Vision and Pattern Recognition*, pp. 3061–3070.

33 Liang, Z., Feng, Y., Guo, Y. et al. (2018). Learning for disparity estimation through feature constancy. *Proceedings of the IEEE Conference on Computer Vision and Pattern Recognition*, pp. 2811–2820.

34 Geng, J. (2011). Structured-light 3D surface imaging: a tutorial. *Advances in Optics and Photonics* 3 (2): 128–160.

35 Salvi, J., Pages, J., and Batlle, J. (2004). Pattern codification strategies in structured light systems. *Pattern Recognition* 37 (4): 827–849.

36 Brenner, C., Boehm, J., and Guehring, J. (1998). Photogrammetric calibration and accuracy evaluation of a cross-pattern stripe projector. In: *Videometrics VI*, vol. 3641, 164–173. International Society for Optics and Photonics.

37 MacWilliams, F.J. and Sloane, N.J. (1976). Pseudo-random sequences and arrays. *Proceedings of the IEEE* 64 (12): 1715–1729.

38 Zhang, L., Curless, B., and Seitz, S.M. (2002). Rapid shape acquisition using color structured light and multi-pass dynamic programming. In: *Proceedings, 1st International Symposium on 3D Data Processing Visualization and Transmission, 2002*, 24–36. IEEE.

39 Griffin, P.M., Narasimhan, L.S., and Yee, S.R. (1992). Generation of uniquely encoded light patterns for range data acquisition. *Pattern Recognition* 25 (6): 609–616.

40 Yous, S., Laga, H., Kidode, M., and Chihara, K. (2007). GPU-based shape from silhouettes. In: *Proceedings of the 5th International Conference on Computer Graphics and Interactive Techniques in Australia and Southeast Asia*, 71–77. ACM.

41 Taubin, G. (1995). Curve and surface smoothing without shrinkage. In: *Proceedings, 5th International Conference on Computer Vision, 1995*, 852–857. IEEE.

42 Desbrun, M., Meyer, M., Schröder, P., and Barr, A.H. (1999). Implicit fairing of irregular meshes using diffusion and curvature flow. In: *Proceedings of the 26th Annual Conference on Computer Graphics and Interactive Techniques*, 317–324. ACM Press/Addison-Wesley Publishing Co.

43 Floater, M.S. and Hormann, K. (2005). Surface parameterization: a tutorial and survey. In: *Advances in Multiresolution for Geometric Modelling*, 157–186. Springer-Verlag.

44 Gu, X., Wang, Y., Chan, T.F. et al. (2004). Genus zero surface conformal mapping and its application to brain surface mapping. *IEEE Transactions on Medical Imaging* 23 (8): 949–958.

45 Szeliski, R. (2010). *Computer Vision: Algorithms and Applications*. Springer Science & Business Media.

46 Praun, E. and Hoppe, H. (2003). Spherical parametrization and remeshing. *ACM Transactions on Graphics* 22 (3): 340–349.

47 Sheffer, A., Praun, E., and Rose, K. (2006). Mesh parameterization methods and their applications. *Foundations and Trends® in Computer Graphics and Vision* 2 (2): 105–171.

48 Osada, R., Funkhouser, T., Chazelle, B., and Dobkin, D. (2002). Shape distributions. *ACM Transactions on Graphics (TOG)* 21 (4): 807–832.

49 Paquet, E. and Rioux, M. (1997). Nefertiti: a query by content software for three-dimensional models databases management. In: *Proceedings, International Conference on Recent Advances in 3-D Digital Imaging and Modeling, 1997*, 345–352. IEEE.

50 Manjunath, B.S., Salembier, P., and Sikora, T. (2002). *Introduction to MPEG-7: Multimedia Content Description Interface*, vol. 1. Wiley.

51 Ankerst, M., Kastenmüller, G., Kriegel, H.P., and Seidl, T. (1999). Nearest neighbor classification in 3D protein databases. *ISMB-99 Proceedings*, pp. 34–43.

52 Chen, D.Y., Tian, X.P., Shen, Y.T., and Ouhyoung, M. (2003). On visual similarity based 3D model retrieval. *Computer Graphics Forum* 22 (3): 223–232.

53 Shilane, P., Min, P., Kazhdan, M., and Funkhouser, T. (2004). The Princeton Shape Benchmark. *Proceedings of Shape Modeling International*, pp. 167–178.

54 Laga, H., Takahashi, H., and Nakajima, M. (2004). Geometry image matching for similarity estimation of 3D shapes. *Proceedings of Computer Graphics International, 2004*, 490–496. IEEE.

55 Laga, H., Takahashi, H., and Nakajima, M. (2006). Spherical parameterization and geometry image-based 3D shape similarity estimation (CGS 2004 special issue). *The Visual Computer* 22 (5): 324–331.

56 Horn, B.K.P. (1984). Extended Gaussian images. *Proceedings of the IEEE* 72 (12): 1671–1686.

57 Kang, S.B. andIkeuchi, K. (1993). The complex EGI: a new representation for 3-D pose determination. *IEEE Transactions on Pattern Analysis and Machine Intelligence* 15 (7): 707–721.

58 Saupe, D. and Vranić, D.V. (2001). 3D model retrieval with spherical harmonics and moments. In: *Joint Pattern Recognition Symposium*, 392–397. Springer-Verlag.

59 Vranic, D.V. (2003). An improvement of rotation invariant 3D-shape based on functions on concentric spheres. In: *Proceedings. 2003 International Conference on Image Processing, 2003. ICIP 2003, vol. 3*, III–757. IEEE.

60 Kazhdan, M., Funkhouser, T., and Rusinkiewicz, S. (2003). Rotation invariant spherical harmonic representation of 3D shape descriptors. *Symposium on Geometry Processing, vol. 6*, pp. 156–164.

61 Laga, H., Takahashi, H., and Nakajima, M. (2006). Spherical wavelet descriptors for content-based 3D model retrieval. *IEEE International Conference on Shape Modeling and Applications 2006*, pp. 15–25.

62 Laga, H. and Nakajima, M. (2007). Statistical spherical wavelet moments for content-based 3D model retrieval. *Computer Graphics International*.

63 Laga, H., Nakajima, M., and Chihara, K. (2007). Discriminative spherical wavelet features for content-based 3D model retrieval. *International Journal of Shape Modeling* 13 (01): 51–72.

64 Schroder, P. and Sweldens, W. (1995). Spherical wavelets: efficiently representing functions on the sphere. In: *SIGGRAPH '95: Proceedings of the 22nd Annual Conference on Computer Graphics and Interactive Techniques*, 161–172. ACM Press.

65 Hoppe, H. and Praun, E. (2003). Shape compression using spherical geometry images. In: *MINGLE 2003 Workshop. Advances in Multiresolution for Geometric Modelling, vol. 2* (ed. N. Dodgson, M. Floater, and M. Sabin), 27–46. Springer-Verlag.

66 Laga, H. and Nakajima, M. (2007). A boosting approach to content-based 3D model retrieval. In: *Proceedings of the 5th International Conference on Computer Graphics and Interactive Techniques in Australia and Southeast Asia*, 227–234. ACM.

67 Laga, H. and Nakajima, M. (2008). Supervised learning of similarity measures for content-based 3D model retrieval. *Large-Scale Knowledge Resources*, pp. 210–225.

68 Su, H., Maji, S., Kalogerakis, E., and Learned-Miller, E. (2015). Multi-view convolutional neural networks for 3D shape recognition. *Proceedings of the IEEE International Conference on Computer Vision*, pp. 945–953.

69 Chatfield, K., Simonyan, K., Vedaldi, A., and Zisserman, A. (2014). Return of the devil in the details: delving deep into convolutional nets. arXiv preprint arXiv:1405.3531.

70 Deng, J., Dong, W., Socher, R. et al. (2009). ImageNet: a large-scale hierarchical image database. In: *CVPR 2009. IEEE Conference on Computer Vision and Pattern Recognition, 2009*, 248–255. IEEE.

71 Wu, Z., Song, S., Khosla, A. et al. (2015). 3D ShapeNets: a deep representation for volumetric shapes. *Proceedings of the IEEE Conference on Computer Vision and Pattern Recognition*, 1912–1920.

72 Maturana, D. and Scherer, S. (2015). VoxNet: a 3D convolutional neural network for real-time object recognition. In: *2015 IEEE/RSJ International Conference on Intelligent Robots and Systems (IROS)*, 922–928. IEEE.

73 The Princeton ModelNet. http://modelnet.cs.princeton.edu/ (accessed 29 August 2017).

74 Qi, C.R., Su, H., Nießner, M. et al. (2016). Volumetric and multi-view CNNs for object classification on 3D data. *Proceedings of the IEEE Conference on Computer Vision and Pattern Recognition*, 5648–5656.

75 ShapeNet. https://www.shapenet.org/ (accessed 29 August 2017).

76 Funkhouser, T., Min, P., Kazhdan, M. et al. (2003). A search engine for 3D models. *ACM Transactions on Graphics (TOG)* 22 (1): 83–105.

77 Kurtek, S., Shen, M., and Laga, H. (2014). Elastic reflection symmetry based shape descriptors. *2014 IEEE Winter Conference on Applications of Computer Vision (WACV)*, 293–300. IEEE.

78 Sundar, H., Silver, D., Gagvani, N., and Dickinson, S. (2003). Skeleton based shape matching and retrieval. In: *Shape Modeling International, 2003*, 130–139. IEEE Computer Society.

79 Tangelder, J.W. and Veltkamp, R.C. (2008). A survey of content based 3D shape retrieval methods. *Multimedia Tools and Applications* 39 (3): 441.

80 Li, B., Lu, Y., Li, C. et al. (2015). A comparison of 3D shape retrieval methods based on a large-scale benchmark supporting multimodal queries. *Computer Vision and Image Understanding* 131: 1–27.

81 Benhabiles, H. and Tabia, H. (2016). Convolutional neural network for pottery retrieval. *Journal of Electronic Imaging* 26 (1): 011005.

82 Bai, S., Bai, X., Zhou, Z. et al. (2017). Gift: towards scalable 3D shape retrieval. *IEEE Transactions on Multimedia* 19 (6): 1257–1271.

83 Ioannidou, A., Chatzilari, E., Nikolopoulos, S., and Kompatsiaris, I. (2017). Deep learning advances in computer vision with 3D data: a survey. *ACM Computing Surveys (CSUR)* 50 (2): 20.

84 Tombari, F., Salti, S., and Di Stefano, L. (2013). Performance evaluation of 3D keypoint detectors. *International Journal of Computer Vision* 102 (1): 198–220.

85 Guo, Y., Bennamoun, M., Sohel, F. et al. (2014). 3D object recognition in cluttered scenes with local surface features: a survey. *IEEE Transactions on Pattern Analysis and Machine Intelligence* 36 (11): 2270–2287.

86 Guo, Y., Bennamoun, M., Sohel, F. et al. (2016). A comprehensive performance evaluation of 3D local feature descriptors. *International Journal of Computer Vision* 116 (1): 66–89.

87 Guo, Y., Sohel, F., Bennamoun, M. et al. (2014). An accurate and robust range image registration algorithm for 3D object modeling. *IEEE Transactions on Multimedia* 16 (5): 1377–1390.

88 Lei, Y., Bennamoun, M., Hayat, M., and Guo, Y. (2014). An efficient 3D face recognition approach using local geometrical signatures. *Pattern Recognition* 47 (2): 509–524.

89 Furrer, F., Wermelinger, M., Yoshida, H. et al. (2017). Autonomous robotic stone stacking with online next best object target pose planning. In: *2017 IEEE International Conference on Robotics and Automation (ICRA)*, 2350–2356. IEEE.

90 Limberger, F.A., Wilson, R.C., Aono, M. et al. (2017). SHREC'17 track: point-cloud shape retrieval of non-rigid toys. *10th Eurographics Workshop on 3D Object Retrieval*, pp. 1–11.

91 Zaharescu, A., Boyer, E., and Horaud, R. (2012). Keypoints and local descriptors of scalar functions on 2D manifolds. *International Journal of Computer Vision* 100: 78–98.

92 Bronstein, A., Bronstein, M., and Ovsjanikov, M. (2010). 3D features, surface descriptors, and object descriptors. *3D Imaging, Analysis, and Applications*, pp. 1–27.

93 Mian, A., Bennamoun, M., and Owens, R. (2010). On the repeatability and quality of keypoints for local feature-based 3D object retrieval from cluttered scenes. *International Journal of Computer Vision* 89 (2): 348–361.

94 Ho, H. and Gibbins, D. (2008). Multi-scale feature extraction for 3D surface registration using local shape variation. *23rd International Conference on Image and Vision Computing, New Zealand*, pp. 1–6.

95 Johnson, A.E. and Hebert, M. (1999). Using spin images for efficient object recognition in cluttered 3D scenes. *IEEE Transactions on Pattern Analysis and Machine Intelligence* 21 (5): 433–449.

96 Frome, A., Huber, D., Kolluri, R. et al. (2004). Recognizing objects in range data using regional point descriptors. *8th European Conference on Computer Vision*, pp. 224–237.

97 Mian, A., Bennamoun, M., and Owens, R.A. (2006). A novel representation and feature matching algorithm for automatic pairwise registration of range images. *International Journal of Computer Vision* 66 (1): 19–40.

98 Bronstein, A., Bronstein, M., Bustos, B. et al. (2010). SHREC 2010: robust feature detection and description benchmark. *Eurographics Workshop on 3D Object Retrieval*.

99 Boyer, E., Bronstein, A., Bronstein, M. et al. (2011). SHREC 2011: robust feature detection and description benchmark. *Eurographics Workshop on Shape Retrieval*, pp. 79–86.

100 Mokhtarian, F., Khalili, N., and Yuen, P. (2001). Multi-scale free-form 3D object recognition using 3D models. *Image and Vision Computing* 19 (5): 271–281.

101 Chen, H. and Bhanu, B. (2007). 3D free-form object recognition in range images using local surface patches. *Pattern Recognition Letters* 28 (10): 1252–1262.

102 Matei, B., Shan, Y., Sawhney, H. et al. (2006). Rapid object indexing using locality sensitive hashing and joint 3D-signature space estimation. *IEEE Transactions on Pattern Analysis and Machine Intelligence* 28 (7): 1111–1126.

103 Zhong, Y. (2009). Intrinsic shape signatures: a shape descriptor for 3D object recognition. *IEEE International Conference on Computer Vision Workshops*, pp. 689–696.

104 Sipiran, I. and Bustos, B. (2011). Harris 3D: a robust extension of the harris operator for interest point detection on 3D meshes. *The Visual Computer* 27 (11): 963–976.

105 Bariya, P., Novatnack, J., Schwartz, G., and Nishino, K. (2012). 3D geometric scale variability in range images: features and descriptors. *International Journal of Computer Vision* 99 (2): 232–255.

106 Castellani, U., Cristani, M., Fantoni, S., and Murino, V. (2008). Sparse points matching by combining 3D mesh saliency with statistical descriptors. *Computer Graphics Forum* 27 (2): 643–652.

107 Lowe, D. (2004). Distinctive image features from scale-invariant keypoints. *International Journal of Computer Vision* 60 (2): 91–110.

108 Itti, L., Koch, C., and Niebur, E. (1998). A model of saliency-based visual attention for rapid scene analysis. *IEEE Transactions on Pattern Analysis and Machine Intelligence* 20 (11): 1254–1259.

109 Unnikrishnan, R. and Hebert, M. (2008). Multi-scale interest regions from unorganized point clouds. *IEEE Conference on Computer Vision and Pattern Recognition Workshops*, pp. 1–8.

110 Novatnack, J., Nishino, K., and Shokoufandeh, A. (2006). Extracting 3D shape features in discrete scale-space. *3rd International Symposium on 3D Data Processing, Visualization, and Transmission*, pp. 946–953.

111 Hua, J., Lai, Z., Dong, M. et al. (2008). Geodesic distance-weighted shape vector image diffusion. *IEEE Transactions on Visualization and Computer Graphics* 14 (6): 1643–1650.

112 Koenderink, J. (1984). The structure of images. *Biological Cybernetics* 50 (5): 363–370.

113 Pauly, M., Keiser, R., and Gross, M. (2003). Multi-scale feature extraction on point-sampled surfaces. *Computer Graphics Forum* 22 (3): 281–289.

114 Zhong, C., Sun, Z., and Tan, T. (2007). Robust 3D face recognition using learned visual codebook. In: *CVPR'07. IEEE Conference on Computer Vision and Pattern Recognition, 2007*, 1–6. IEEE.

115 Harris, C. and Stephens, M. (1988). A combined corner and edge detector. *Alvey Vision Conference, vol. 15, Manchester, UK*, p. 50.

116 Dutagaci, H., Cheung, C., and Godil, A. (2012). Evaluation of 3D interest point detection techniques via human-generated ground truth. *The Visual Computer* 28 (9): 901–917.

117 Ho, H. and Gibbins, D. (2009). Curvature-based approach for multi-scale feature extraction from 3D meshes and unstructured point clouds. *IET Computer Vision* 3 (4): 201–212.

118 Hou, T. and Qin, H. (2010). Efficient computation of scale-space features for deformable shape correspondences. *European Conference on Computer Vision*, pp. 384–397.

119 Zou, G., Hua, J., Dong, M., and Qin, H. (2008). Surface matching with salient keypoints in geodesic scale space. *Computer Animation and Virtual Worlds* 19 (3–4): 399–410.

120 Zou, G., Hua, J., Lai, Z. et al. (2009). Intrinsic geometric scale space by shape diffusion. *IEEE Transactions on Visualization and Computer Graphics* 15 (6): 1193–1200.

121 Sun, J., Ovsjanikov, M., and Guibas, L. (2009). A concise and provably informative multi-scale signature based on heat diffusion. *Computer Graphics Forum* 28 (5): 1383–1392.

122 Hu, J. and Hua, J. (2009). Salient spectral geometric features for shape matching and retrieval. *The Visual Computer* 25 (5): 667–675.

123 Bronstein, A.M., Bronstein, M.M., Guibas, L.J., and Ovsjanikov, M. (2011). Shape google: geometric words and expressions for invariant shape retrieval. *ACM Transactions on Graphics* 30 (1): 1:1–1:20.

124 Salti, S., Petrelli, A., Tombari, F., and Di Stefano, L. (2012). On the affinity between 3D detectors and descriptors. *2nd International Conference on 3D Imaging, Modeling, Processing, Visualization, and Transmission*, pp. 424–431.

125 Alexandre, L.A. (2012). 3D descriptors for object and category recognition: a comparative evaluation. *Workshop on Color-Depth Camera Fusion in Robotics at the IEEE/RSJ International Conference on Intelligent Robots and Systems (IROS)*.

126 Tombari, F., Salti, S., and Di Stefano, L. (2010). Unique signatures of histograms for local surface description. *European Conference on Computer Vision*, pp. 356–369.

127 Stein, F. and Medioni, G. (1992). Structural indexing: efficient 3D object recognition. *IEEE Transaction on Pattern Analysis and Machine Intelligence* 14 (2): 125–145.

128 Chua, C.S. and Jarvis, R. (1997). Point signatures: a new representation for 3D object recognition. *International Journal of Computer Vision* 25 (1): 63–85.

129 Guo, Y., Sohel, F., Bennamoun, M. et al. (2013). Rotational projection statistics for 3D local surface description and object recognition. *International Journal of Computer Vision* 105 (1): 63–86.

130 Carmichael, O., Huber, D., and Hebert, M. (1999). Large data sets and confusing scenes in 3-D surface matching and recognition. 2rd International Conference on 3-D Digital Imaging and Modeling, pp. 358–367.

131 Assfalg, J., Bertini, M., Bimbo, A., and Pala, P. (2007). Content-based retrieval of 3-D objects using spin image signatures. *IEEE Transactions on Multimedia* 9 (3): 589–599.

132 Dinh, H. and Kropac, S. (2006). Multi-resolution spin-images. In: *IEEE International Conference on Computer Vision and Pattern Recognition*, vol. 1, 863–870. IEEE.

133 Ruiz-Correa, S., Shapiro, L., and Melia, M. (2001). A new signature-based method for efficient 3-D object recognition. In: *IEEE Conference on Computer Vision and Pattern Recognition*, vol. 1, I–769. IEEE.

134 Darom, T. and Keller, Y. (2012). Scale invariant features for 3D mesh models. *IEEE Transactions on Image Processing* 21 (5): 2758–2769.

135 Belongie, S., Malik, J., and Puzicha, J. (2002). Shape matching and object recognition using shape contexts. *IEEE Transactions on Pattern Analysis and Machine Intelligence* 24 (4): 509–522.

136 Rusu, R.B., Blodow, N., Marton, Z.C., and Beetz, M. (2008). Aligning point cloud views using persistent feature histograms. *IEEE/RSJ International Conference on Intelligent Robots and Systems*, pp. 3384–3391.

137 Rusu, R.B., Blodow, N., and Beetz, M. (2009). Fast point feature histograms (FPFH) for 3D registration. *IEEE International Conference on Robotics and Automation*, pp. 3212–3217.

138 Tabia, H., Laga, H., Picard, D., and Gosselin, P.H. (2014). Covariance descriptors for 3D shape matching and retrieval. *Proceedings of the IEEE Conference on Computer Vision and Pattern Recognition*, pp. 4185–4192.

139 Tabia, H. and Laga, H. (2015). Covariance-based descriptors for efficient 3D shape matching, retrieval and classification. *IEEE Transactions on Multimedia* 17 (9): 1591–1603.

140 Cirujeda, P., Cid, Y.D., Mateo, X., and Binefa, X. (2015). A 3D scene registration method via covariance descriptors and an evolutionary stable strategy game theory solver. *International Journal of Computer Vision* 115 (3): 306–329.

141 Lian, Z., Godil, A., and Sun, X. (2010). Visual similarity based 3D shape retrieval using bag-of-features. In: *Shape Modeling International Conference (SMI), 2010*, 25–36. IEEE.

142 Tabia, H., Daoudi, M., Vandeborre, J.P., and Collot, O. (2010). Local visual patch for 3D shape retrieval. *ACM 3DOR'10*.

143 Ohbuchi, R., Osada, K., Furuya, T., and Banno, T. (2008). Salient local visual features for shape-based 3D model retrieval. In: *Shape Modeling International*, pp. 93–102.

144 Liu, Y., Zha, H., and Qin, H. (2006). Shape topics: a compact representation and new algorithms for 3D partial shape retrieval. *Proceedings of the*

2006 IEEE Computer Society Conference on Computer Vision and Pattern Recognition, CVPR '06, vol. 2, 2025–2032. IEEE Computer Society.

145 Li, X. and Godil, A. (2009). Exploring the bag-of-words method for 3D shape retrieval. In: *Proceedings of the 16th IEEE International Conference on Image Processing, ICIP'09*, 437–440. IEEE Press.

146 Tabia, H., Daoudi, M., Colot, O., and Vandeborre, J.P. (2012). Three-dimensional object retrieval based on vector quantization of invariant descriptors. *Journal of Electronic Imaging* 21 (2): 023011–1.

147 Toldo, R., Castelllani, U., and Fusiello, A. (2010). The bag of words approach for retrieval and categorization of 3D objects. *The Visual Computer* 26 (10): 1257–1268.

148 Picard, D. and Gosselin, P.H. (2011). Improving image similarity with vectors of locally aggregated tensors. In: *2011 18th IEEE International Conference on Image Processing (ICIP)*, 669–672. IEEE.

149 Tabia, H., Picard, D., Laga, H., and Gosselin, P.H. (2013). 3D shape similarity using vectors of locally aggregated tensors. In: *2013 20th IEEE International Conference on Image Processing (ICIP)*, 2694–2698. IEEE.

150 Taati, B. and Greenspan, M. (2011). Local shape descriptor selection for object recognition in range data. *Computer Vision and Image Understanding* 115 (5): 681–694.

151 Salti, S., Tombari, F., Spezialetti, R., and Di Stefano, L. (2015). Learning a descriptor-specific 3D keypoint detector. *Proceedings of the IEEE International Conference on Computer Vision*, pp. 2318–2326.

152 Zhi, S., Liu, Y., Li, X., and Guo, Y. (2017). LightNet: a lightweight 3D convolutional neural network for real-time 3D object recognition. *Eurographics Workshop on 3D Object Retrieval*.

153 Zhi, S., Liu, Y., Li, X., and Guo, Y. (2017). Toward real-time 3D object recognition: a lightweight volumetric cnn framework using multitask learning. *Computers & Graphics* 71: 199–207.

154 Lavoué, G. (2012). Combination of bag-of-words descriptors for robust partial shape retrieval. *The Visual Computer* 28 (9): 931–942.

155 Díez, Y., Roure, F., Lladó, X., and Salvi, J. (2015). A qualitative review on 3D coarse registration methods. *ACM Computing Surveys (CSUR)* 47 (3): 45.

156 Guo, Y., Bennamoun, M., Sohel, F. et al. (2015). An integrated framework for 3D modeling, object detection and pose estimation from point-clouds. *IEEE Transactions on Instrumentation and Measurement* 64 (3): 683–693.

157 Ma, Y., Guo, Y., Lei, Y. et al. (2017). Efficient rotation estimation for 3D registration and global localization in structured point clouds. *Image and Vision Computing* 67: 52–66.

158 Besl, P.J. and McKay, N.D. (1992). A method for registration of 3-D shapes. *IEEE Transactions on Pattern Analysis and Machine Intelligence* 14 (2): 239–256.

159 Myronenko, A. and Song, X. (2010). Point set registration: coherent point drift. *IEEE Transactions on Pattern Analysis and Machine Intelligence* 32 (12): 2262–2275.

160 Arun, K.S., Huang, T.S., and Blostein, S.D. (1987). Least-squares fitting of two 3-D point sets. *IEEE Transactions on Pattern Analysis and Machine Intelligence* 9 (5): 698–700.

161 Horn, B.K., Hilden, H.M., and Negahdaripour, S. (1988). Closed-form solution of absolute orientation using orthonormal matrices. *Journal of the Optical Society of America A* 5 (7): 1127–1135.

162 Yamany, S.M. and Farag, A.A. (2002). Surface signatures: an orientation independent free-form surface representation scheme for the purpose of objects registration and matching. *IEEE Transactions on Pattern Analysis and Machine Intelligence* 24 (8): 1105–1120.

163 Masuda, T. (2009). Log-polar height maps for multiple range image registration. *Computer Vision and Image Understanding* 113 (11): 1158–1169.

164 Fischler, M.A. and Bolles, R.C. (1981). Random sample consensus: a paradigm for model fitting with applications to image analysis and automated cartography. *Communications of the ACM* 24 (6): 381–395.

165 Malassiotis, S. and Strintzis, M. (2007). Snapshots: a novel local surface descriptor and matching algorithm for robust 3D surface alignment. *IEEE Transactions on Pattern Analysis and Machine Intelligence* 29 (7): 1285–1290.

166 Stamos, I. and Leordeanu, M. (2003). Automated feature-based range registration of urban scenes of large scale. In: *IEEE Conference on Computer Vision and Pattern Recognition*, vol. 2, II–555. IEEE.

167 Stamos, I. and Allen, P.K. (2002). Geometry and texture recovery of scenes of large scale. *Computer Vision and Image Understanding* 88 (2): 94–118.

168 Chao, C. and Stamos, I. (2005). Semi-automatic range to range registration: a feature-based method. *5th International Conference on 3D Digital Imaging and Modeling*, pp. 254–261.

169 Chung, D.H., Yun, I.D., and Lee, S.U. (1998). *Pattern Recognition* 31 (4): 457–464.

170 Salvi, J., Matabosch, C., Fofi, D., and Forest, J. (2007). A review of recent range image registration methods with accuracy evaluation. *Image and Vision Computing* 25 (5): 578–596.

171 Chen, C.S., Hung, Y.P., and Cheng, J.B. (1999). RANSAC-based DARCES: a new approach to fast automatic registration of partially overlapping range images. *IEEE Transactions on Pattern Analysis and Machine Intelligence* 21 (11): 1229–1234.

172 Aiger, D., Mitra, N.J., and Cohen-Or, D. (2008). 4-points congruent sets for robust pairwise surface registration. *ACM Transactions on Graphics (TOG)* 27 (3): 85.

173 Horn, B.K. (1987). Closed-form solution of absolute orientation using unit quaternions. *Journal of the Optical Society of America A* 4 (4): 629–642.

174 Chen, Y. and Medioni, G. (1992). Object modelling by registration of multiple range images. *Image and Vision Computing* 10 (3): 145–155.

175 Rusinkiewicz, S. and Levoy, M. (2001). Efficient variants of the ICP algorithm. In: *Proceedings, 3rd International Conference on 3-D Digital Imaging and Modeling, 2001*, 145–152. IEEE.

176 Weik, S. (1997). Registration of 3-D partial surface models using luminance and depth information. In: *Proceedings, International Conference on Recent Advances in 3-D Digital Imaging and Modeling, 1997*, 93–100. IEEE.

177 Godin, G., Rioux, M., and Baribeau, R. (1994). Three-dimensional registration using range and intensity information. In: *Proceedings of SPIE Videometric III, vol. 2350*, pp. 279–290.

178 Pulli, K. (1999). Multiview registration for large data sets. In: *2nd International Conference on 3-D Digital Imaging and Modeling*, 160–168. IEEE.

179 Dorai, C., Wang, G., Jain, A.K., and Mercer, C. (1998). Registration and integration of multiple object views for 3D model construction. *IEEE Transactions on Pattern Analysis and Machine Intelligence* 20 (1): 83–89.

180 Turk, G. and Levoy, M. (1994). Zippered polygon meshes from range images. *21st Annual Conference on Computer Graphics and Interactive Techniques*, pp. 311–318.

181 Chen, Y. and Medioni, G. (1991). Object modelling by registration of multiple range images. *IEEE International Conference on Robotics and Automation*, pp. 2724–2729.

182 Lu, M., Zhao, J., Guo, Y., and Ma, Y. (2016). Accelerated coherent point drift for automatic three-dimensional point cloud registration. *IEEE Geoscience and Remote Sensing Letters* 13 (2): 162–166.

183 van Kaick, O., Zhang, H., Hamarneh, G., and Cohen-Or, D. (2011). A survey on shape correspondence. *Comput. Graph Forum* 30 (6): 1681–1707.

184 Kurtek, S., Klassen, E., Ding, Z. et al. (2011). Parameterization-invariant shape comparisons of anatomical surfaces. *IEEE Transactions on Medical Imaging* 30 (3): 849–858.

185 Lipman, Y., Rustamov, R.M., and Funkhouser, T.A. (2010). Biharmonic distance. *ACM Transactions on Graphics* 29 (3): 27:1–27:11.

186 Kuhn, H.W. (1955). The hungarian method for the assignment problem. *Naval Research Logistics (NRL)* 2 (1–2): 83–97.

187 Lipman, Y. and Funkhouser, T. (2009). Mobius voting for surface correspondence. *ACM Transactions on Graphics* 28 (3): 72:1–72:12.

188 Kim, V.G., Lipman, Y., and Funkhouser, T. (2011). Blended intrinsic maps. *ACM Transactions on Graphics* 30 (4): 79:1–79:12.

189 Kurtek, S., Klassen, E., Ding, Z., and Srivastava, A. (2010). A novel Riemannain framework for shape analysis of 3D objects. *IEEE CVPR*, pp. 1625–1632.

190 Kurtek, S., Klassen, E., Ding, Z. et al. (2011). Parameterization-invariant shape statistics and probabilistic classification of anatomical surfaces. In: *Proceedings of IPMI*, vol. 6801, 147–158. Springer-Verlag.

191 Kurtek, S. and Srivastava, A. (2012). Elastic symmetry analysis of anatomical structures. *Mathematical Methods in Biomedical Image Analysis (MMBIA)*, pp. 33–38.

192 Kurtek, S., Klassen, E., Gore, J.C. et al. (2012). Elastic geodesic paths in shape space of parameterized surfaces. *IEEE Transactions on Pattern Analysis and Machine Intelligence* 34 (9): 1717–1730.

193 Laga, H., Xie, Q., Jermyn, I.H., and Srivastava, A. (2017). Numerical inversion of srnf maps for elastic shape analysis of genus-zero surfaces. *IEEE Transactions on Pattern Analysis and Machine Intelligence*. 39 (12): 2451–2464.

194 Xie, Q., Kurtek, S., Le, H., and Srivastava, A. (2013). Parallel transport of deformations in shape space of elastic surfaces. *IEEE International Conference on Computer Vision*.

195 Srivastava, A., Klassen, E., Joshi, S.H., and Jermyn, I.H. (2011). Shape analysis of elastic curves in Euclidean spaces. *IEEE Transactions on Pattern Analysis and Machine Intelligence* 33 (7): 1415–1428.

196 Laga, H., Kurtek, S., Srivastava, A. et al. (2012). A riemannian elastic metric for shape-based plant leaf classification. In: *2012 International Conference on Digital Image Computing Techniques and Applications (DICTA)*, 1–7. IEEE.

197 Laga, H., Kurtek, S., Srivastava, A., and Miklavcic, S.J. (2014). Landmark-free statistical analysis of the shape of plant leaves. *Journal of Theoretical Biology* 363: 41–52.

198 Kurtek, S., Srivastava, A., Klassen, E., and Laga, H. (2013). Landmark-guided elastic shape analysis of spherically-parameterized surfaces. *Computer Graphics Forum* 32 (2pt4): 429–438.

199 Kurtek, S., Laga, H., and Xie, Q. (2014). Elastic shape analysis of boundaries of planar objects with multiple components and arbitrary topologies. In: *Asian Conference on Computer Vision*, 424–439. Springer-Verlag.

200 Banerjee, I., Laga, H., Patané, G. et al. (2015). Generation of 3D canonical anatomical models: an experience on carpal bones. In: *International Conference on Image Analysis and Processing*, 167–174. Springer International Publishing.

201 Biasotti, S., Cerri, A., Bronstein, A., and Bronstein, M. (2014). Quantifying 3D shape similarity using maps: recent trends, applications and perspectives. In: *Eurographics 2014-State of the Art Reports*, 135–159. The Eurographics Association.

202 Laga, H. (2017). A survey on non-rigid 3D shape analysis. In: *Academic Press Library in Signal Processing: Image and Video Processing and Analysis and Computer Vision*, vol. 6, Chapter 7. Academic Press.

203 van Kaick, O., Tagliasacchi, A., Sidi, O. et al. (2011). Prior knowledge for part correspondence. *Computer Graphics Forum* 30 (2): 553–562.

204 Laga, H., Mortara, M., and Spagnuolo, M. (2013). Geometry and context for semantic correspondences and functionality recognition in man-made 3D shapes. *ACM Transactions on Graphics* 32 (5): 150:1–150:16.

205 Mitra, N.J., Wand, M., Zhang, H. et al. (2013). Structure-aware shape processing. *Eurographics State-of-the-Art Report*. The Eurographics Association.

206 Funkhouser, T., Kazhdan, M., Shilane, P. et al. (2004). Modeling by example. *ACM Transactions on Graphics* 23 (3): 652–663.

207 Zheng, Y., Cohen-Or, D., and Mitra, N.J. (2013). Smart variations: functional substructures for part compatibility. *Computer Graphics Forum* 32 (2pt2): 195–204.

208 Shamir, A. (2008). A survey on mesh segmentation techniques. *Computer Graphics Forum* 27 (6): 1539–1556.

209 Kalogerakis, E., Hertzmann, A., and Singh, K. (2010). Learning 3D mesh segmentation and labeling. *ACM Transactions on Graphics* 29 (4): 102.

210 Sidi, O., van Kaick, O., Kleiman, Y. et al. (2011). Unsupervised co-segmentation of a set of shapes via descriptor-space spectral clustering. *ACM Transactions on Graphics* 30 (6): 126.

211 Li, Z. and Chen, J. (2015). Superpixel segmentation using linear spectral clustering. *Proceedings of the IEEE Conference on Computer Vision and Pattern Recognition*, pp. 1356–1363.

212 Shapira, L., Shamir, A., and Cohen-Or, D. (2008). Consistent mesh partitioning and skeletonisation using the shape diameter function. *The Visual Computer* 24 (4): 249–259.

213 Hilaga, M., Shinagawa, Y., Kohmura, T., and Kunii, T.L. (2001). Topology matching for fully automatic similarity estimation of 3D shapes. In: *Proceedings of the 28th Annual Conference on Computer Graphics and Interactive Techniques*, 203–212. ACM.

214 Hearst, M.A., Dumais, S.T., Osuna, E. et al. (1998). Support vector machines. *IEEE Intelligent Systems and their Applications* 13 (4): 18–28.

215 Friedman, J., Hastie, T., Tibshirani, R. et al. (2000). Additive logistic regression: a statistical view of boosting (with discussion and a rejoinder by the authors). *The Annals of Statistics* 28 (2): 337–407.

216 Torralba, A., Murphy, K.P., and Freeman, W.T. (2007). Sharing visual features for multiclass and multiview object detection. *IEEE Transactions on Pattern Analysis and Machine Intelligence* 29 (5): 854–869.

217 Shapira, L., Shalom, S., Shamir, A. et al. (2010). Contextual part analogies in 3D objects. *Int. J. Comput. Vision* 89 (2–3): 309–326.

218 Golovinskiy, A. and Funkhouser, T. (2009). Consistent segmentation of 3D models. *Computers & Graphics* 33 (3): 262–269.

219 Comaniciu, D. and Meer, P. (2002). Mean shift: a robust approach toward feature space analysis. *IEEE Transactions on Pattern Analysis and Machine Intelligence* 24 (5): 603–619.

220 Shamir, A., Shapira, L., and Cohen-Or, D. (2006). Mesh analysis using geodesic mean-shift. *The Visual Computer* 22 (2): 99–108.

221 Boykov, Y., Veksler, O., and Zabih, R. (2001). Fast approximate energy minimization via graph cuts. *IEEE Transactions on Pattern Analysis and Machine Intelligence* 23 (11): 1222–1239.

222 Huang, Q., Koltun, V., and Guibas, L. (2011). Joint shape segmentation with linear programming. *ACM Transactions on Graphics* 30 (6): 125.

223 Laga, H. (2012). Graspable parts recognition in man-made 3D shapes. In: *Asian Conference on Computer Vision*, 552–564. Springer-Verlag.

224 Feragen, A., Lo, P., de Bruijne, M. et al. (2013). Toward a theory of statistical tree-shape analysis. *IEEE Transactions on Pattern Analysis and Machine Intelligence* 35 (8): 2008–2021.

225 Wang, G., Laga, H., Xie, N. et al. (2018). The shape space of 3D botanical tree models. *ACM Transactions on Graphics (TOG)* 37 (1): 7.

226 Liu, H., Vimont, U., Wand, M. et al. (2015). Replaceable substructures for efficient part-based modeling. *Computer Graphics Forum* 34 (2): 503–513.

227 Averkiou, M., Kim, V.G., Zheng, Y., and Mitra, N.J. (2014). ShapeSynth: parameterizing model collections for coupled shape exploration and synthesis. *Computer Graphics Forum* 33 (2): 125–134.

228 Zheng, Y., Cohen-Or, D., Averkiou, M., and Mitra, N.J. (2014). Recurring part arrangements in shape collections. *Computer Graphics Forum* 33 (2): 115–124.

229 Laga, H. and Tabia, H. (2017). Modeling and exploring co-variations in the geometry and configuration of man-made 3D shape families. *Computer Graphics Forum* 36 (5): 13–25.

230 Hansen, M., Smith, M., Smith, L. et al. (2015). Non-intrusive automated measurement of dairy cow body condition using 3D video. *Machine Vision of Animals and their Behaviour Workshop*.

231 Sternberg, S. (1986). Grayscale morphology. *Computer Vision, Graphics, and Image Processing* 35: 333–355.

232 Robinette, K.M., Daanen, H., and Paquet, E. (1999). The caesar project: a 3-D surface anthropometry survey. *2nd International Conference on 3-D Digital Imaging and Modeling*.

233 Domae, Y., Okuda, H., Taguchi, Y. et al. (2014). Fast graspability evaluation on single depth maps for bin picking with general grippers. *International Conference on Robotics and Automation*.

234 Hadsell, R., Sermanet, P., Scoffier, M. et al. (2009). Learning long-range vision for autonomous off-road driving. *Journal of Field Robotic* 26 (2): 120–144.

235 Zhang, J. and Fisher, R.B. (2018). Visual passphrase: behaviometrics via speech-related 3D facial dynamics. in review.

236 Sha, T., Song, M., Bu, J. et al. (2011). Feature level analysis for 3D facial expression recognition. *Neurocomputing* 74 (12-13): 2135–2141.

237 Fang, T., Zhao, X., Ocegueda, O. et al. (2012). 3D/4D facial expression analysis: an advanced annotated face model approach. *Image and Vision Computing* 30 (10): 738–749.

238 Aly, S., Trubanova, A., Abbott, L. et al. (2015). VT-KFER: a kinect-based RGBD+time dataset for spontaneous and non-spontaneous facial expression recognition. *International Conference on Biometrics*.

239 Barbosa, B.I., Cristani, M., Del Bue, A. et al. (2012). Re-identification with RGB-D sensors. *Proceedings of the 1st International Workshop on Re-Identification*.

240 Häne, C., Zach, C., Cohen, A., and Pollefeys, M. (2016). Dense semantic 3D reconstruction. *IEEE Transactions on Pattern Analysis and Machine Intelligence* 39 (9): 1730–1743.

241 Erdogmus, N. and Marcel, S. (2013). Spoofing in 2D face recognition with 3D masks and anti-spoofing with kinect. *6th International Conference on Biometrics: Theory, Applications, and Systems*.

242 Sung, J., Ponce, C., Selman, B., and Saxena, A. (2012). Unstructured human activity detection from RGBD images. *International Conference on Robotics and Automation*.

243 Liu, L. and Shao, L. (2013). Learning discriminative representations from RGB-D video data. *International Joint Conference on Artificial Intelligence*.

244 Xu, C. and Cheng, L. (2013). Efficient hand pose estimation from a single depth image. *International Conference on Computer Vision*.

245 Babaee, M. The TUM gait from audio, image and depth (GAID) database. https://www.mmk.ei.tum.de/verschiedenes/tum-gaid-database/ (accessed 11 September 2017).

246 Yun, K., Honorio, J., Chattopadhyay, D. et al. (2012). Two-person interaction detection using body-pose features and multiple instance learning. *International Workshop on Human Activity Understanding from 3D Data*.

247 Wikipedia. Simultaneous Localization and Mapping. https://en.wikipedia.org/wiki/Simultaneous_localization_and_mapping (accessed 10 September 2017).

248 Bloom, V., Argyriou, V., and Makris, D. (2016). Hierarchical transfer learning for online recognition of compound actions. *Computer Vision and Image Understanding* 144: 62–72.

249 Han J.J., Kurillo G., Abresch, R.T. et al. (2015). Upper extremity 3-D imensional reachable workspace analysis in dystrophinopathy using kinect: reachable workspace in DMD/BMD. *Muscle & Nerve* 52 (3): 344–55.

250 Verdone Sanchez, A. (2017). Tracking people and arms in robot surgery workcell using multiple RGB-D sensors. MSc Dissertation. School of Informatics, University of Edinburgh.

251 Al-Osaimi, F.R.M. and Bennammoun, M. (2013). 3D face surface analysis and recognition based on facial surface features. In: *3D Face Modeling, Analysis and Recognition*, 39–76. Wiley.

252 Phillips, P.J., Flynn, P.J., Scruggs, T. et al. (2005). Overview of the face recognition grand challenge. In: *CVPR 2005. IEEE Computer Society Conference on Computer Vision and Pattern Recognition, 2005*, vol. 1, 947–954. IEEE.

253 Moreno, A. and Sanchez, A. (2004). GavabDB: a 3D face database. *Proceedings of the 2nd COST275 Workshop on Biometrics on the Internet, Vigo (Spain)*, pp. 75–80.

254 Krishnan, P. and Naveen, S. (2015). RGB-D face recognition system verification using kinect and FRAV3D databases. *Procedia Computer Science* 46: 1653–1660.

255 Baocai, Y., Yanfeng, S., Chengzhang, W., and Yun, G. (2009). BJUT-3D large scale 3D face database and information processing. *Journal of Computer Research and Development* 6: 020.

256 Savran, A., Alyüz, N., Dibeklioğlu, H. et al. (2008). Bosphorus database for 3D face analysis. In: *European Workshop on Biometrics and Identity Management*, 47–56. Springer-Verlag.

257 Colombo, A., Cusano, C., and Schettini, R. (2011). UMB-DB: a database of partially occluded 3D faces. In: *2011 IEEE International Conference on Computer Vision Workshops (ICCV Workshops)*, 2113–2119. IEEE.

258 Bowyer, K.W., Chang, K., and Flynn, P. (2006). A survey of approaches and challenges in 3D and multi-modal 3D + 2D face recognition. *Computer Vision and Image Understanding* 101 (1): 1–15.

259 Abate, A.F., Nappi, M., Riccio, D., and Sabatino, G. (2007). 2D and 3D face recognition: a survey. *Pattern Recognition Letters* 28 (14): 1885–1906.

260 Bennamoun, M., Guo, Y., and Sohel, F. (2015). Feature selection for 2D and 3D face recognition. In: *Wiley Encyclopedia of Electrical and Electronics Engineering*. Wiley.

261 Kelkboom, E., Gökberk, B., Kevenaar, T. et al. (2007). 3D face: biometric template protection for 3D face recognition. In: *Advances in Biometrics. ICB 2007*, Lecture Notes in Computer Science, vol. 4642 (ed. S.W. Lee and S.Z. Li), 566–573. Berlin, Heidelberg: Springer-Verlag.

262 Wang, Y., Pan, G., and Wu, Z. (2007). 3D face recognition in the presence of expression: a guidance-based constraint deformation approach. In: *CVPR'07. IEEE Conference on Computer Vision and Pattern Recognition, 2007*, 1–7. IEEE.

263 Heseltine, T., Pears, N., and Austin, J. (2008). Three-dimensional face recognition using combinations of surface feature map subspace components. *Image and Vision Computing* 26 (3): 382–396.

264 Jahanbin, S., Choi, H., Liu, Y., and Bovik, A.C. (2008). Three dimensional face recognition using iso-geodesic and iso-depth curves. In: *BTAS 2008. 2nd IEEE International Conference on Biometrics: Theory, Applications, and Systems, 2008*, 1–6. IEEE.

265 Zhong, C., Sun, Z., and Tan, T. (2008). Learning efficient codes for 3D face recognition. In: *ICIP 2008. 15th IEEE International Conference on Image Processing, 2008*, 1928–1931. IEEE.

266 Kakadiaris, I.A., Passalis, G., Toderici, G. et al. (2006). 3D face recognition. In: *BMVC*, pp. 869–878.

267 Faltemier, T., Bowyer, K., and Flynn, P. (2006). 3D face recognition with region committee voting. In: *3rd International Symposium on 3D Data Processing, Visualization, and Transmission*, 318–325. IEEE.

268 Lin, W.Y., Wong, K.C., Boston, N., and Hu, Y.H. (2007). 3D face recognition under expression variations using similarity metrics fusion. In: *IEEE International Conference on Multimedia and Expo, 2007*, 727–730. IEEE.

269 Xu, C., Li, S., Tan, T., and Quan, L. (2009). Automatic 3D face recognition from depth and intensity gabor features. *Pattern Recognition* 42 (9): 1895–1905.

270 Jin, Y., Wang, Y., Ruan, Q., and Wang, X. (2011). A new scheme for 3D face recognition based on 2D gabor wavelet transform plus LBP. *2011 6th International Conference on Computer Science & Education (ICCSE)*, 860–865. IEEE.

271 Marras, I., Zafeiriou, S., and Tzimiropoulos, G. (2012). Robust learning from normals for 3D face recognition. In: *Computer Vision–ECCV 2012. Workshops and Demonstrations*, 230–239. Springer-Verlag.

272 Berretti, S., Del Bimbo, A., and Pala, P. (2010). 3D face recognition using isogeodesic stripes. *IEEE Transactions on Pattern Analysis and Machine Intelligence* 32 (12): 2162–2177.

273 Cook, J., Chandran, V., and Fookes, C. (2006). 3D face recognition using log-gabor templates. *Proceedings of the British Machine Conference*, pp. 769–778.

274 Cook, J., McCool, C., Chandran, V., and Sridharan, S. (2006). Combined 2D/3D face recognition using log-gabor templates. In: *AVSS'06. IEEE International Conference on Video and Signal Based Surveillance, 2006*, 83–83. IEEE.

275 Wang, Y., Pan, G., Wu, Z., and Wang, Y. (2006). Exploring facial expression effects in 3D face recognition using partial ICP. *Computer Vision–ACCV 2006*, pp. 581–590.

276 Mian, A., Bennamoun, M., and Owens, R. (2006). Automatic 3D face detection, normalization and recognition. In: *3rd International Symposium on 3D Data Processing, Visualization, and Transmission*, 735–742. IEEE.

277 Gokberk, B. and Akarun, L. (2006). Comparative analysis of decision-level fusion algorithms for 3D face recognition. In: *18th International Conference on Pattern Recognition, 2006. ICPR 2006*, vol. 3, 1018–1021. IEEE.

278 Zhang, L., Razdan, A., Farin, G. et al. (2006). 3D face authentication and recognition based on bilateral symmetry analysis. *The Visual Computer* 22 (1): 43–55.

279 Shin, H. and Sohn, K. (2006). 3D face recognition with geometrically localized surface shape indexes. In: *9th International Conference on Control, Automation, Robotics and Vision, 2006. ICARCV'06*, 1–6. IEEE.

280 Mian, A., Bennamoun, M., and Owens, R. (2007). An efficient multimodal 2D-3D hybrid approach to automatic face recognition. *IEEE Transactions on Pattern Analysis and Machine Intelligence* 29 (11): 1927–1943.

281 Kakadiaris, I.A., Passalis, G., Toderici, G. et al. (2007). Three-dimensional face recognition in the presence of facial expressions: an annotated deformable model approach. *IEEE Transactions on Pattern Analysis and Machine Intelligence* 29 (4): 640–649.

282 Feng, S., Krim, H., and Kogan, I. (2007). 3D face recognition using Euclidean integral invariants signature. In: *IEEE/SP 14th Workshop on Statistical Signal Processing, 2007. SSP'07*, 156–160. IEEE.

283 Li, X. and Zhang, H. (2007). Adapting geometric attributes for expression-invariant 3D face recognition. In: *IEEE International Conference on Shape Modeling and Applications, 2007. SMI'07*, 21–32. IEEE.

284 Mpiperis, I., Malassiotis, S. and Strintzis, M.G. (2007). 3-D face recognition with the geodesic polar representation. *IEEE Transactions on Information Forensics and Security* 2 (3): 537–547.

285 Mahoor, M.H. and Abdel-Mottaleb, M. (2007). 3D face recognition based on 3D ridge lines in range data. In: *IEEE International Conference on Image Processing, 2007. ICIP 2007*, vol. 1, I–137. IEEE.

286 Mpiperis, I., Malasiotis, S., and Strintzis, M.G. (2007). 3D face recognition by point signatures and iso-contours. In: *Proceedings of the 4th IASTED International Conference on Signal Processing, Pattern Recognition, and Applications*, 328–332. ACTA Press.

287 Mian, A.S., Bennamoun, M., and Owens, R. (2008). Keypoint detection and local feature matching for textured 3D face recognition. *International Journal of Computer Vision* 79 (1): 1–12.

288 Amberg, B., Knothe, R., and Vetter, T. (2008). Expression invariant 3D face recognition with a morphable model. In: *8th IEEE International Conference on Automatic Face & Gesture Recognition, 2008. FG'08*, 1–6. IEEE.

289 Alyuz, N., Gokberk, B., and Akarun, L. (2008). A 3D face recognition system for expression and occlusion invariance. In: *2nd IEEE International Conference on Biometrics: Theory, Applications and Systems, 2008. BTAS 2008*, 1–7. IEEE.

290 Faltemier, T.C., Bowyer, K.W., and Flynn, P.J. (2008). A region ensemble for 3-D face recognition. *IEEE Transactions on Information Forensics and Security* 3 (1): 62–73.

291 Al-Osaimi, F.R., Bennamoun, M., and Mian, A. (2008). Integration of local and global geometrical cues for 3D face recognition. *Pattern Recognition* 41 (3): 1030–1040.

292 Alyüz, N., Gökberk, B., Dibeklioğlu, H. et al. (2008). 3D face recognition benchmarks on the bosphorus database with focus on facial expressions. In: *European Workshop on Biometrics and Identity Management*, 57–66. Springer-Verlag.

293 Mpiperis, I., Malassiotis, S., and Strintzis, M.G. (2008). Bilinear models for 3-D face and facial expression recognition. *IEEE Transactions on Information Forensics and Security* 3 (3): 498–511.

294 Lu, X. and Jain, A. (2008). Deformation modeling for robust 3D face matching. *IEEE Transactions on Pattern Analysis and Machine Intelligence* 30 (8): 1346–1357.

295 Faltemier, T.C., Bowyer, K.W., and Flynn, P.J. (2008). Using multi-instance enrollment to improve performance of 3D face recognition. *Computer Vision and Image Understanding* 112 (2): 114–125.

296 Yang, W., Yi, D., Lei, Z. et al. (2008). 2D–3D face matching using CCA. *8th IEEE International Conference on Automatic Face & Gesture Recognition, 2008. FG'08*, 1–6. IEEE.

297 Theoharis, T., Passalis, G., Toderici, G., and Kakadiaris, I.A. (2008). Unified 3D face and ear recognition using wavelets on geometry images. *Pattern Recognition* 41 (3): 796–804.

298 Wang, Y., Tang, X., Liu, J. et al. (2008). 3D face recognition by local shape difference boosting. In: *Computer Vision–ECCV 2008*, 603–616.

299 Harguess, J., Gupta, S., and Aggarwal, J.K. (2008). 3D face recognition with the average-half-face. In: *19th International Conference on Pattern Recognition, 2008. ICPR 2008*, 1–4. IEEE.

300 Gokberk, B., Dutagaci, H., Ulas, A. et al. (2008). Representation plurality and fusion for 3-D face recognition. *IEEE Transactions on Systems, Man, and Cybernetics Part B: Cybernetics* 38 (1): 155–173.

301 Fabry, T., Vandermeulen, D., and Suetens, P. (2008). 3D face recognition using point cloud kernel correlation. In: *2nd IEEE International Conference on Biometrics: Theory, Applications and Systems, 2008. BTAS 2008*, 1–6. IEEE.

302 Zhong, C., Sun, Z., Tan, T., and He, Z. (2008). Robust 3D face recognition in uncontrolled environments. In: *IEEE Conference on Computer Vision and Pattern Recognition, 2008. CVPR 2008*, 1–8. IEEE.

303 Yan-Feng, S., Heng-Liang, T., and Bao-Cai, Y. (2008). The 3D face recognition algorithm fusing multi-geometry features. *Acta Automatica Sinica* 34 (12): 1483–1489.

304 Mousavi, M.H., Faez, K., and Asghari, A. (2008). Three dimensional face recognition using SVM classifier. In: *7th IEEE/ACIS International Conference on Computer and Information Science. 2008. ICIS 08*, 208–213. IEEE.

305 ter Haar, F. and Veltkamp, R. (2008). 3D face model fitting for recognition. In: *Computer Vision–ECCV 2008*, Lecture Notes in Computer Science, vol. 5305 (ed. D. Forsyth, P. Torr, and A. Zisserman), 652–664. Berlin, Heidelberg: Springer-Verlag.

306 Llonch, R.S., Kokiopoulou, E., Tosic, I., and Frossard, P. (2008). 3D face recognition using sparse spherical representations. In: *19th International Conference on Pattern Recognition, 2008. ICPR 2008*, 1–4. IEEE.

307 Al-Osaimi, F., Bennamoun, M., and Mian, A. (2008). Expression invariant non-rigid 3D face recognition: a robust approach to expression aware morphing. In: *International Symposium on 3D Data Processing Visualization and Transmission (3DPVT)*, pp. 19–26.

308 Paysan, P., Knothe, R., Amberg, B. et al. (2009). A 3D face model for pose and illumination invariant face recognition. In: *6th IEEE International Conference on Advanced Video and Signal Based Surveillance, 2009. AVSS'09*, 296–301. IEEE.

309 Al-Osaimi, F., Bennamoun, M., and Mian, A. (2009). An expression deformation approach to non-rigid 3D face recognition. *International Journal of Computer Vision* 81 (3): 302–316.

310 Li, X., Jia, T., and Zhang, H. (2009). Expression-insensitive 3D face recognition using sparse representation. In: *IEEE Conference on Computer Vision and Pattern Recognition, 2009. CVPR 2009*, 2575–2582. IEEE.

311 Boehnen, C., Peters, T., and Flynn, P. (2009). 3D signatures for fast 3D face recognition. In: *Advances in Biometrics. ICB 2009*, Lecture Notes in Computer Science, vol. 5558 (ed. M. Tistarelli and M.S. Nixon), 12–21. Berlin, Heidelberg: Springer-Verlag.

312 Huang, D., Ardabilian, M., Wang, Y., and Chen, L. (2009). Asymmetric 3D/2D face recognition based on LBP facial representation and canonical correlation analysis. In: *2009 16th IEEE International Conference on Image Processing (ICIP)*, 3325–3328. IEEE.

313 Dibeklioğlu, H., Gökberk, B., and Akarun, L. (2009). Nasal region-based 3D face recognition under pose and expression variations. In: *Advances in Biometrics. ICB 2009*, Lecture Notes in Computer Science, vol. 5558 (ed. M. Tistarelli and M.S. Nixon), 309–318. Berlin, Heidelberg: Springer-Verlag.

314 Cadoni, M., Bicego, M., and Grosso, E. (2009). 3D face recognition using joint differential invariants. In: *Advances in Biometrics. ICB 2009*, Lecture Notes in Computer Science, vol. 5558 (ed. M. Tistarelli and M.S. Nixon), 279–288. Berlin, Heidelberg: Springer-Verlag.

315 Daniyal, F., Nair, P., and Cavallaro, A. (2009). Compact signatures for 3D face recognition under varying expressions. In: *6th IEEE International Conference on Advanced Video and Signal Based Surveillance, 2009. AVSS'09*, 302–307. IEEE.

316 Yunqi, L., Dongjie, C., Meiling, Y. et al. (2009). 3D face recognition by surface classification image and PCA. In: *2nd International Conference on Machine Vision, 2009. ICMV'09*, 145–149. IEEE.

317 ter Haar, F.B. and Veltkamp, R.C. (2009). A 3D face matching framework for facial curves. *Graphical Models* 71 (2): 77–91.

318 Queirolo, C.C., Silva, L., Bellon, O.R., and Segundo, M.P. (2010). 3D face recognition using simulated annealing and the surface interpenetration measure. *IEEE Transactions on Pattern Analysis and Machine Intelligence* 32 (2): 206–219.

319 Wang, Y., Liu, J., and Tang, X. (2010). Robust 3D face recognition by local shape difference boosting. *IEEE Transactions on Pattern Analysis and Machine Intelligence* 32 (10): 1858–1870.

320 Maes, C., Fabry, T., Keustermans, J. et al. (2010). Feature detection on 3D face surfaces for pose normalisation and recognition. In: *2010 4th IEEE International Conference on Biometrics: Theory Applications and Systems (BTAS)*, 1–6. IEEE.

321 Drira, H., Amor, B.B., Daoudi, M., and Srivastava, A. (2010). Pose and expression-invariant 3D face recognition using elastic radial curves. *British Machine Vision Conference*, 1–11.

322 Llonch, R.S., Kokiopoulou, E., Tošić, I., and Frossard, P. (2010). 3D face recognition with sparse spherical representations. *Pattern Recognition* 43 (3): 824–834.

323 Huang, D., Zhang, G., Ardabilian, M. et al. (2010). 3D face recognition using distinctiveness enhanced facial representations and local feature hybrid matching. In: *2010 4th IEEE International Conference on Biometrics: Theory Applications and Systems (BTAS)*, 1–7. IEEE.

324 Huang, D., Ardabilian, M., Wang, Y., and Chen, L. (2010). Automatic asymmetric 3D-2D face recognition. In: *2010 20th International Conference on Pattern Recognition (ICPR)*, 1225–1228. IEEE.

325 Tang, H., Sun, Y., Yin, B., and Ge, Y. (2010). Expression-robust 3D face recognition using LBP representation. In: *2010 IEEE International Conference on Multimedia and Expo (ICME)*, 334–339. IEEE.

326 Smeets, D., Fabry, T., Hermans, J. et al. (2010). Fusion of an isometric deformation modeling approach using spectral decomposition and a region-based approach using ICP for expression-invariant 3D face recognition. In: *2010 20th International Conference on Pattern Recognition (ICPR)*, 1172–1175. IEEE.

327 Li, Y.A., Shen, Y.J., Zhang, G.D. et al. (2010). An efficient 3D face recognition method using geometric features. In: *2010 2nd International Workshop on Intelligent Systems and Applications (ISA)*, 1–4. IEEE.

328 Wang, X., Ruan, Q., and Ming, Y. (2010). 3D face recognition using corresponding point direction measure and depth local features. In: *2010 IEEE 10th International Conference on Signal Processing (ICSP)*, 86–89. IEEE.

329 Zhang, C., Uchimura, K., Zhang, C., and Koutaki, G. (2010). 3D face recognition using multi-level multi-feature fusion. In: *2010 Fourth Pacific-Rim Symposium on Image and Video Technology (PSIVT)*, 21–26. IEEE.

330 Miao, S. and Krim, H. (2010). 3D face recognition based on evolution of iso-geodesic distance curves. In: *2010 IEEE International Conference on Acoustics Speech and Signal Processing (ICASSP)*, 1134–1137. IEEE.

331 Ming, Y., Ruan, Q., and Ni, R. (2010). Learning effective features for 3D face recognition. In: *2010 17th IEEE International Conference on Image Processing (ICIP)*, 2421–2424. IEEE.

332 Passalis, G., Perakis, P., Theoharis, T., and Kakadiaris, I.A. (2011). Using facial symmetry to handle pose variations in real-world 3D face recognition. *IEEE Transactions on Pattern Analysis and Machine Intelligence* 33 (10): 1938–1951.

333 Spreeuwers, L. (2011). Fast and accurate 3D face recognition. *International Journal of Computer Vision* 93 (3): 389–414.

334 Zhang, G. and Wang, Y. (2011). Robust 3D face recognition based on resolution invariant features. *Pattern Recognition Letters* 32 (7): 1009–1019.

335 Li, H., Huang, D., Lemaire, P. et al. (2011). Expression robust 3D face recognition via mesh-based histograms of multiple order surface differential quantities. In: *2011 18th IEEE International Conference on Image Processing (ICIP)*, 3053–3056. IEEE.

336 Huang, D., Ouji, K., Ardabilian, M. et al. (2011). 3D face recognition based on local shape patterns and sparse representation classifier. In: *Advances in Multimedia Modeling. MMM 2011*, Lecture Notes in Computer Science, vol. 6523 (ed. K.T. Lee, W.H. Tsai, H.Y.M. Liao et al.), 206–216. Berlin, Heidelberg: Springer-Verlag.

337 Ocegueda, O., Passalis, G., Theoharis, T. et al. (2011). UR3D-C: linear dimensionality reduction for efficient 3D face recognition. In: *2011 International Joint Conference on Biometrics (IJCB)*, 1–6. IEEE.

338 Sharif, M., Mohsin, S., Hanan, R.A. et al. (2011). 3D face recognition using horizontal and vertical marked strips. *Sindh University Research Journal-SURJ (Science Series)* 43 (1-A): 63–68.

339 Smeets, D., Keustermans, J., Hermans, J. et al. (2011). Symmetric surface-feature based 3D face recognition for partial data. *2011 International Joint Conference on Biometrics (IJCB)*, 1–6. IEEE.

340 Tang, H., Sun, Y., Yin, B., and Ge, Y. (2011). 3D face recognition based on sparse representation. *The Journal of Supercomputing* 58 (1): 84–95.

341 Li, H., Huang, D., Morvan, J.M., and Chen, L. (2011). Learning weighted sparse representation of encoded facial normal information for expression-robust 3D face recognition. In: *2011 International Joint Conference on Biometrics (IJCB)*, 1–7. IEEE.

342 Ming, Y. and Ruan, Q. (2012). Robust sparse bounding sphere for 3D face recognition. *Image and Vision Computing* 30 (8): 524–534.

343 Huang, D., Ardabilian, M., Wang, Y., and Chen, L. (2012). 3-D face recognition using eLBP-based facial description and local feature hybrid matching. *IEEE Transactions on Information Forensics and Security* 7 (5): 1551–1565.

344 Li, X. and Da, F. (2012). Efficient 3D face recognition handling facial expression and hair occlusion. *Image and Vision Computing* 30 (9): 668–679.

345 Taghizadegan, Y., Ghassemian, H., and Naser-Moghaddasi, M. (2012). 3D face recognition method using 2DPCA -Euclidean distance classification. *ACEEE International Journal on Control System and Instrumentation* 3 (1): 5.

346 Ballihi, L., Amor, B.B., Daoudi, M. et al. (2012). Boosting 3-D-geometric features for efficient face recognition and gender classification. *IEEE Transactions on Information Forensics and Security* 7 (6): 1766–1779.

347 Zhang, Y.N., Guo, Z., Xia, Y. et al. (2012). 2D representation of facial surfaces for multi-pose 3D face recognition. *Pattern Recognition Letters* 33 (5): 530–536.

348 Inan, T. and Halici, U. (2012). 3-D face recognition with local shape descriptors. *IEEE transactions on Information Forensics and Security* 7 (2): 577–587.

349 Belghini, N., Zarghili, A., and Kharroubi, J. (2012). 3D face recognition using Gaussian-Hermite moments. *Special Issue of International Journal of Computer Application on Software Engineering, Databases and Expert Systems* 1: 1–4.

350 Drira, H., Amor, B.B., Srivastava, A. et al. (2013). 3D face recognition under expressions, occlusions, and pose variations. *IEEE Transactions on Pattern Analysis and Machine Intelligence* 35 (9): 2270–2283.

351 Lei, Y., Bennamoun, M., and El-Sallam, A.A. (2013). An efficient 3D face recognition approach based on the fusion of novel local low-level features. *Pattern Recognition* 46 (1): 24–37.

352 Smeets, D., Keustermans, J., Vandermeulen, D., and Suetens, P. (2013). meshSIFT: local surface features for 3D face recognition under expression variations and partial data. *Computer Vision and Image Understanding* 117 (2): 158–169.

353 Tang, H., Yin, B., Sun, Y., and Hu, Y. (2013). 3D face recognition using local binary patterns. *Signal Processing* 93 (8): 2190–2198.

354 Alyuz, N., Gokberk, B., and Akarun, L. (2013). 3-D face recognition under occlusion using masked projection. *IEEE Transactions on Information Forensics and Security* 8 (5): 789–802.

355 Zhang, L., Ding, Z., Li, H. et al. (2014). 3D face recognition based on multiple keypoint descriptors and sparse representation. *PLoS ONE* 9 (6): e100120.

356 Elaiwat, S., Bennamoun, M., Boussaïd, F., and El-Sallam, A. (2015). A curvelet-based approach for textured 3D face recognition. *Pattern Recognition* 48 (4): 1235–1246.

357 Emambakhsh, M. and Evans, A. (2017). Nasal patches and curves for expression-robust 3D face recognition. *IEEE Transactions on Pattern Analysis and Machine Intelligence* 39 (5): 995–1007.

358 Heseltine, T., Pears, N., and Austin, J. (2004). Three-dimensional face recognition: an eigensurface approach. In: *2004 International Conference on Image Processing, 2004. ICIP'04*, vol. 2, 1421–1424. IEEE.

359 Russ, T.D., Koch, M.W., and Little, C.Q. (2005). A 2D range hausdorff approach for 3D face recognition. In: *IEEE Computer Society Conference on Computer Vision and Pattern Recognition-Workshops, 2005. CVPR Workshops*, 169–169. IEEE.

360 Achermann, B. and Bunke, H. (2000). Classifying range images of human faces with hausdorff distance. In: *Proceedings, 15th International Conference on Pattern Recognition, 2000*, vol. 2, 809–813. IEEE.

361 Lee, Y.H. and Shim, J.C. (2004). Curvature based human face recognition using depth weighted hausdorff distance. In: *2004 International Conference on Image Processing, 2004. ICIP'04*, vol. 3, 1429–1432. IEEE.

362 Bronstein, A.M., Bronstein, M.M., and Kimmel, R. (2005). Three-dimensional face recognition. *International Journal of Computer Vision* 64 (1): 5–30.

363 Bronstein, A.M., Bronstein, M.M., and Kimmel, R. (2007). Expression-invariant representations of faces. *IEEE Transactions on Image Processing* 16 (1): 188–197.

364 Besl, P.J. and Jain, R.C. (1988). Segmentation through variable-order surface fitting. *IEEE Transactions on Pattern Analysis and Machine Intelligence* 10 (2): 167–192.

365 Lu, X., Colbry, D., and Jain, A. (2004). Matching 2.5D scans for face recognition. *Biometric Authentication*, pp. 9–33.

366 Perakis, P., Passalis, G., Theoharis, T., and Kakadiaris, I.A. (2013). 3D facial landmark detection under large yaw and expression variations. *IEEE Transactions on Pattern Analysis and Machine Intelligence* 35 (7): 1552–1564.

367 Koudelka, M.L., Koch, M.W., and Russ, T.D. (2005). A prescreener for 3D face recognition using radial symmetry and the hausdorff fraction. In: *IEEE Computer Society Conference on Computer Vision and Pattern Recognition-Workshops, 2005. CVPR Workshops*, 168–168. IEEE.

368 Creusot, C., Pears, N., and Austin, J. (2013). A machine-learning approach to keypoint detection and landmarking on 3D meshes. *International Journal of Computer Vision* 102 (1-3): 146–179.

369 Gupta, S., Markey, M.K., and Bovik, A.C. (2010). Anthropometric 3D face recognition. *International Journal of Computer Vision* 90 (3): 331–349.

370 Gordon, G.G. (1992). Face recognition based on depth and curvature features. In: *Proceedings Computer Vision and Pattern Recognition*. IEEE.

371 Moreno, A.B., Sánchez, A., Vélez, J.F., and Díaz, F.J. (2003). Face recognition using 3D surface-extracted descriptors. In: *Irish Machine Vision and Image Processing Conference*, vol. 2. Citeseer.

372 Xu, C., Wang, Y., Tan, T., and Quan, L. (2004). Automatic 3D face recognition combining global geometric features with local shape variation information. In: *Proceedings, 6th IEEE international conference on Automatic Face and Gesture Recognition, 2004*, 308–313. IEEE.

373 Samir, C., Srivastava, A., and Daoudi, M. (2006). Three-dimensional face recognition using shapes of facial curves. *IEEE Transactions on Pattern Analysis and Machine Intelligence* 28 (11): 1858–1863.

374 Samir, C., Srivastava, A., Daoudi, M., and Klassen, E. (2009). An intrinsic framework for analysis of facial surfaces. *International Journal of Computer Vision* 82 (1): 80–95.

375 Klassen, E., Srivastava, A., Mio, M., and Joshi, S.H. (2004). Analysis of planar shapes using geodesic paths on shape spaces. *IEEE Transactions on Pattern Analysis and Machine Intelligence* 26 (3): 372–383.

376 Soltanpour, S., Boufama, B., and Wu, Q.J. (2017). A survey of local feature methods for 3D face recognition. *Pattern Recognition* 72: 391–406.

377 Ojala, T., Pietikainen, M., and Maenpaa, T. (2002). Multiresolution gray-scale and rotation invariant texture classification with local binary patterns. *IEEE Transactions on Pattern Analysis and Machine Intelligence* 24 (7): 971–987.

378 Ahonen, T., Hadid, A., and Pietikainen, M. (2006). Face description with local binary patterns: application to face recognition. *IEEE Transactions on Pattern Analysis and Machine Intelligence* 28 (12): 2037–2041.

379 Li, S.Z., Zhao, C., Ao, M., and Lei, Z. (2005). Learning to fuse 3D + 2D based face recognition at both feature and decision levels. In: *International Workshop on Analysis and Modeling of Faces and Gestures*, 44–54. Springer-Verlag.

380 Li, H., Huang, D., Morvan, J.M. et al. (2014). Expression-robust 3D face recognition via weighted sparse representation of multi-scale and multi-component local normal patterns. *Neurocomputing* 133: 179–193.

381 Werghi, N., Berretti, S., and Del Bimbo, A. (2015). The mesh-LBP: a framework for extracting local binary patterns from discrete manifolds. *IEEE Transactions on Image Processing* 24 (1): 220–235.

382 Gupta, S., Markey, M.K., and Bovik, A.C. (2007). Advances and challenges in 3D and 2D + 3D human face recognition. In: *Pattern Recognition in Biology* (ed. M.S. Corrigan), 63–103. Nova Science Punlisher Inc.

383 Wang, Y., Chua, C.S., and Ho, Y.K. (2002). Facial feature detection and face recognition from 2D and 3D images. *Pattern Recognition Letters* 23 (10): 1191–1202.

384 Chua, C.S., Han, F., and Ho, Y.K. (2000). 3D human face recognition using point signature. In: *Proceedings, 4th IEEE International Conference on Automatic Face and Gesture Recognition, 2000*, 233–238. IEEE.

385 Guo, Y., Sohel, F.A., Bennamoun, M. et al. (2013). TriSI: a distinctive local surface descriptor for 3D modeling and object recognition. *GRAPP/IVAPP*, pp. 86–93.

386 Guo, Y., Sohel, F., Bennamoun, M. et al. (2015). A novel local surface feature for 3D object recognition under clutter and occlusion. *Information Sciences* 293 (2): 196–213.

387 Kakadiaris, I.A., Toderici, G., Evangelopoulos, G. et al. (2017). 3D-2D face recognition with pose and illumination normalization. *Computer Vision and Image Understanding* 154: 137–151.

388 Gilani, S.Z., Mian, A., Shafait, F., and Reid, I. (2018). Dense 3D face correspondence. *IEEE Transactions on Pattern Analysis and Machine Intelligence* 40 (7): 1584–1598.

389 Zulqarnain Gilani, S. and Mian, A. (2018). Learning from millions of 3D scans for large-scale 3D face recognition. *Proceedings of the IEEE Conference on Computer Vision and Pattern Recognition*, pp. 1896–1905.

390 Mikolajczyk, K. and Schmid, C. (2005). A performance evaluation of local descriptors. *IEEE Transactions on Pattern Analysis and Machine Intelligence* 27 (10): 1615–1630.

391 Mian, A., Bennamoun, M., and Owens, R. (2006). Three-dimensional model-based object recognition and segmentation in cluttered scenes. *IEEE Transactions on Pattern Analysis and Machine Intelligence* 28 (10): 1584–1601.

392 Johnson, A.E. and Hebert, M. (1998). Surface matching for object recognition in complex three-dimensional scenes. *Image and Vision Computing* 16 (9–10): 635–651.

393 Papazov, C., Haddadin, S., Parusel, S. et al. (2012). Rigid 3D geometry matching for grasping of known objects in cluttered scenes. *The International Journal of Robotics Research* 31 (4): 538–553.

394 Rodolà, E., Albarelli, A., Bergamasco, F., and Torsello, A. (2013). A scale independent selection process for 3D object recognition in cluttered scenes. *International Journal of Computer Vision* 102 (1–3): 129–145.

395 Zai, D., Li, J., Guo, Y. et al. (2017). Pairwise registration of TLS point clouds using covariance descriptors and a non-cooperative game. *ISPRS Journal of Photogrammetry and Remote Sensing* 134: 15–29.

396 Tombari, F. and Stefano, L.D. (2012). Hough voting for 3D object recognition under occlusion and clutter. *IPSJ Transactions on Computer Vision and Applications* 4: 20–29.

397 Huber, D. and Hebert, M. (2003). Fully automatic registration of multiple 3D data sets. *Image and Vision Computing* 21 (7): 637–650.

398 Aldoma, A., Tombari, F., Di Stefano, L., and Vincze, M. (2016). A global hypothesis verification framework for 3D object recognition in clutter. *IEEE Transactions on Pattern Analysis and Machine Intelligence* 38 (7): 1383–1396.

399 Wang, H., Wang, C., Luo, H. et al. (2015). 3-D point cloud object detection based on supervoxel neighborhood with hough forest framework. *IEEE Journal of Selected Topics in Applied Earth Observations and Remote Sensing* 8 (4): 1570–1581.

400 Papon, J., Abramov, A., Schoeler, M., and Worgotter, F. (2013). Voxel cloud connectivity segmentation-supervoxels for point clouds. In: *2013 IEEE Conference on Computer Vision and Pattern Recognition (CVPR)*, 2027–2034. IEEE.

401 Chang, A.X., Funkhouser, T., Guibas, L. et al. (2015). ShapeNet: an information-rich 3D model repository. arXiv preprint arXiv:1512.03012.

402 Hackel, T., Savinov, N., Ladicky, L. et al. (2017). Semantic3D. net: a new large-scale point cloud classification benchmark. arXiv preprint arXiv:1704.03847.

403 De Deuge, M., Quadros, A., Hung, C., and Douillard, B. (2013). Unsupervised feature learning for classification of outdoor 3D scans. *Australasian Conference on Robitics and Automation, volume 2*.

404 Bu, S., Liu, Z., Han, J. et al. (2014). Learning high-level feature by deep belief networks for 3-D model retrieval and recognition. *IEEE Transactions on Multimedia* 16 (8): 2154–2167.

405 Xie, J., Dai, G., Zhu, F. et al. (2017). DeepShape: deep-learned shape descriptor for 3D shape retrieval. *IEEE Transactions on Pattern Analysis and Machine Intelligence* 39 (7): 1335–1345.

406 Ma, Y., Zheng, B., Guo, Y. et al. (2017). Boosting multi-view convolutional neural networks for 3D object recognition via view saliency. *Chinese Conference on Image and Graphics Technologies*, pp. 199–209.

407 Shi, B., Bai, S., Zhou, Z., and Bai, X. (2015). DeepPano: deep panoramic representation for 3-D shape recognition. *IEEE Signal Processing Letters* 22 (12): 2339–2343.

408 Chen, X., Ma, H., Wan, J. et al. (2016). Multi-view 3D object detection network for autonomous driving. arXiv preprint arXiv:1611.07759.

409 Song, S. and Xiao, J. (2015). Deep sliding shapes for amodal 3D object detection in RGB-D images. arXiv preprint arXiv:1511.02300.

410 Riegler, G., Ulusoys, A.O., and Geiger, A. (2016). OctNet: learning deep 3D representations at high resolutions. arXiv preprint arXiv:1611.05009.

411 Ma, C., An, W., Lei, Y., and Guo, Y. (2017). BV-CNNs: binary volumetric convolutional networks for 3D object recognition. *British Machine Vision Conference*.

412 Qi, C.R., Su, H., Mo, K., and Guibas, L.J. (2016). PointNet: deep learning on point sets for 3D classification and segmentation. arXiv preprint arXiv:1612.00593.

413 Li, Y., Bu, R., Sun, M., and Chen, B. (2018). PointCNN. arXiv preprint arXiv:1801.07791.

414 Lähner, Z., Rodolà, E., Bronstein, M.M. et al. (2016). SHREC'16: Matching of deformable shapes with topological noise. *Proceedings of Eurographics Workshop on 3D Object Retrieval (3DOR)*.

415 Pauly, M., Mitra, N.J., Giesen, J. et al. (2005). Example-based 3D scan completion. *Symposium on Geometry Processing. EPFL-CONF-149337*, pp. 23–32.

416 Bronstein, A.M., Bronstein, M.M., and Kimmel, R. (2006). Efficient computation of isometry-invariant distances between surfaces. *SIAM Journal on Scientific Computing* 28 (5): 1812–1836.

417 Siddiqi, K., Zhang, J., Macrini, D. et al. (2008). Retrieving articulated 3-D models using medial surfaces. *Machine Vision and Applications* 19 (4): 261–275.

418 Tatsuma, A., Koyanagi, H., and Aono, M. (2012). A large-scale shape benchmark for 3D object retrieval: Toyohashi shape benchmark. In: *2012 Asia-Pacific Signal & Information Processing Association Annual Summit and Conference (APSIPA ASC)*, 1–10. IEEE.

419 Savva, M., Yu, F., Su, H. et al. (2016). SHREC'16 track: large-scale 3D shape retrieval from shapenet core55. *Proceedings of the Eurographics Workshop on 3D) Object Retrieval*.

420 Gong, Y., Lazebnik, S., Gordo, A., and Perronnin, F. (2013). Iterative quantization: a procrustean approach to learning binary codes for large-scale image retrieval. *IEEE Transactions on Pattern Analysis and Machine Intelligence* 35 (12): 2916–2929.

421 Mu, Y., Shen, J., and Yan, S. (2010). Weakly-supervised hashing in kernel space. In: *2010 IEEE Conference on Computer Vision and Pattern Recognition (CVPR)*, 3344–3351. IEEE.

422 Wang, J., Kumar, S., and Chang, S.F. (2012). Semi-supervised hashing for large-scale search. *IEEE Transactions on Pattern Analysis and Machine Intelligence* 34 (12): 2393–2406.

423 Liu, W., Wang, J., Ji, R. et al. (2012). Supervised hashing with kernels. In: *2012 IEEE Conference on Computer Vision and Pattern Recognition (CVPR)*, 2074–2081. IEEE.

424 Liu, X., He, J., Liu, D., and Lang, B. (2012). Compact kernel hashing with multiple features. In: *Proceedings of the 20th ACM International Conference on Multimedia*, 881–884. ACM.

425 Furuya, T. and Ohbuchi, R. (2014). Hashing cross-modal manifold for scalable sketch-based 3D model retrieval. In: *2014 2nd International Conference on 3D Vision (3DV)*, vol. 1, 543–550. IEEE.

426 Furuya, T. and Ohbuchi, R. (2017). Deep semantic hashing of 3D geometric features for efficient 3D model retrieval. In: *Proceedings of the Computer Graphics International Conference*, 8. ACM.

427 Tabia, H. and Laga, H. (2017). Multiple vocabulary coding for 3D shape retrieval using bag of covariances. *Pattern Recognition Letters*. 95: 78–84.

428 Zhou, D., Bousquet, O., Lal, T.N. et al. (2004). Learning with local and global consistency. *Advances in Neural Information Processing Systems*, 321–328.

429 Osada, R., Funkhouser, T., Chazelle, B., and Dobkin, D. (2001). Matching 3D models with shape distributions. In: *SMI 2001 International Conference On Shape Modeling and Applications*, 154–166. IEEE.

430 Giorgi, D., Biasotti, S., and Paraboschi, L. (2007). Shape retrieval contest 2007: Watertight models track. *SHREC Competition* 8 (7)

431 Tabia, H., Picard, D., Laga, H., and Gosselin, P.H. (2013). Compact vectors of locally aggregated tensors for 3D shape retrieval. *Eurographics workshop on 3D object retrieval.*

432 Papadakis, P., Pratikakis, I., Theoharis, T. et al. (2008). 3d object retrieval using an efficient and compact hybrid shape descriptor. *3DOR.*

433 Bai, X., Bai, S., Zhu, Z., and Latecki, L.J. (2015). 3D shape matching via two layer coding. *IEEE Transactions on Pattern Analysis and Machine Intelligence* 37 (12): 2361–2373.

434 Savva, M., Yu, F., Su, H. et al. (2017). Large-scale 3D shape retrieval from shapeNet core55. In: *Eurographics Workshop on 3D Object Retrieval* (ed. I. Pratikakis, F. Dupont, and M. Ovsjanikov). The Eurographics Association.

435 Tatsuma, A. and Aono, M. (2016). Food image recognition using covariance of convolutional layer feature maps. *IEICE Transactions on Information and Systems* 99 (6): 1711–1715.

436 Kanai, S. (2008). Content-based 3D mesh model retrieval from hand-written sketch. *International Journal on Interactive Design and Manufacturing (IJIDeM)* 2 (2): 87–98.

437 Li, B., Schreck, T., Godil, A. et al. (2012). SHREC'12 track: sketch-based 3D shape retrieval. *3DOR*, 109–118.

438 Aono, M. and Iwabuchi, H. (2012). 3D shape retrieval from a 2D image as query. In: *2012 Asia-Pacific Signal & Information Processing Association Annual Summit and Conference (APSIPA ASC)*, 1–10. IEEE.

439 DeCarlo, D., Finkelstein, A., Rusinkiewicz, S., and Santella, A. (2003). Suggestive contours for conveying shape. *ACM Transactions on Graphics (TOG)* 22 (3): 848–855.

440 Li, B. and Johan, H. (2013). Sketch-based 3D model retrieval by incorporating 2D-3D alignment. *Multimedia Tools and Applications* 65 (3): 363–385.

441 Yoon, S.M., Scherer, M., Schreck, T., and Kuijper, A. (2010). Sketch-based 3D model retrieval using diffusion tensor fields of suggestive contours. In: *Proceedings of the 18th ACM International Conference on Multimedia*, 193–200. ACM.

442 Yoon, S.M. and Kuijper, A. (2011). View-based 3D model retrieval using compressive sensing based classification. In: *2011 7th International Symposium on Image and Signal Processing and Analysis (ISPA)*, 437–442. IEEE.

443 Saavedra, J.M., Bustos, B., Schreck, T. et al. (2012). Sketch-based 3D model retrieval using keyshapes for global and local representation. In: *3DOR*, 47–50. Citeseer.

444 Eitz, M., Richter, R., Boubekeur, T. et al. (2012). Sketch-based shape retrieval. *ACM Transactions on Graphics* 31 (4): 1–10.

445 Norouzi, M., Mikolov, T., Bengio, S. et al. (2013). Zero-shot learning by convex combination of semantic embeddings. arXiv preprint arXiv:1312.5650.

446 Frome, A., Corrado, G.S., Shlens, J. et al. (2013). DeViSE: a deep visual-semantic embedding model. In: *Advances in Neural Information Processing Systems*, 2121–2129. NIPS.

447 Socher, R., Ganjoo, M., Manning, C.D., and Ng, A. (2013). Zero-shot learning through cross-modal transfer. In: *Advances in Neural Information Processing Systems*, 935–943. NIPS.

448 Wang, F., Kang, L., and Li, Y. (2015). Sketch-based 3D shape retrieval using convolutional neural networks. *Proceedings of the IEEE Conference on Computer Vision and Pattern Recognition*, pp. 1875–1883.

449 Li, Y., Su, H., Qi, C.R. et al. (2015). Joint embeddings of shapes and images via CNN image purification. *ACM Transactions on Graphics (TOG)* 34 (6): 234.

450 Rasiwasia, N., Costa Pereira, J., Coviello, E. et al. (2010). A new approach to cross-modal multimedia retrieval. In: *Proceedings of the 18th ACM International Conference on Multimedia*, 251–260. ACM.

451 Jia, Y., Salzmann, M., and Darrell, T. (2011). Learning cross-modality similarity for multinomial data. In: *2011 IEEE International Conference on Computer Vision (ICCV)*, 2407–2414. IEEE.

452 Chen, T., Cheng, M.M., Tan, P. et al. (2009). Sketch2Photo: internet image montage. *ACM Transactions on Graphics* 28 (5): 124.

453 Eitz, M., Hildebrand, K., Boubekeur, T., and Alexa, M. (2010). An evaluation of descriptors for large-scale image retrieval from sketched feature lines. *Computers & Graphics* 34 (5): 482–498.

454 Russell, B.C., Sivic, J., Ponce, J., and Dessales, H. (2011). Automatic alignment of paintings and photographs depicting a 3D scene. In: *2011 IEEE International Conference on Computer Vision Workshops (ICCV Workshops)*, 545–552. IEEE.

455 Liu, L., Shen, F., Shen, Y. et al. (2017). Deep sketch hashing: fast free-hand sketch-based image retrieval. *Proceedings of CVPR*, pp. 2862–2871.

456 Qi, Y., Song, Y.Z., Zhang, H., and Liu, J. (2016). Sketch-based image retrieval via siamese convolutional neural network. In: *2016 IEEE International Conference on Image Processing (ICIP)*, 2460–2464. IEEE.

457 Song, J., Qian, Y., Song, Y.Z. et al. (2017). Deep spatial-semantic attention for fine-grained sketch-based image retrieval. *ICCV*.

458 Dai, G., Xie, J., Zhu, F., and Fang, Y. (2017). Deep correlated metric learning for sketch-based 3D shape retrieval. *AAAI*, pp. 4002–4008.

459 Tabia, H. and Laga, H. (2017). Learning shape retrieval from different modalities. *Neurocomputing* 253: 24–33.

460 Krizhevsky, A., Sutskever, I., and Hinton, G.E. (2012). ImageNet classification with deep convolutional neural networks. In: *Advances in Neural Information Processing Systems*, 1097–1105. ACM.

461 Luvizon, D.C., Tabia, H., and Picard, D. (2017). Human pose regression by combining indirect part detection and contextual information. arXiv preprint arXiv:1710.02322.

462 Li, B., Lu, Y., Godil, A. et al. (2013). SHREC'13 track: large scale sketch-based 3D shape retrieval. In: *Eurographics Workshop on 3D Object Retrieval*. The Eurographics Association.

463 Zhao, X., Gregor, R., Mavridis, P., and Schreck, T. (2017). Sketch-based 3D object retrieval with skeleton line views - initial results and research problems. In: *Eurographics Workshop on 3D Object Retrieval* (ed. I. Pratikakis, F. Dupont, and M. Ovsjanikov). The Eurographics Association.

464 Su, H., Qi, C.R., Li, Y., and Guibas, L.J. (2015). Render for CNN: viewpoint estimation in images using cnns trained with rendered 3D model views. *Proceedings of the IEEE International Conference on Computer Vision*, 2686–2694.

465 Peng, X., Sun, B., Ali, K., and Saenko, K. (2014). Exploring invariances in deep convolutional neural networks using synthetic images. 2 (4), CoRR, abs/1412.7122.

466 Stark, M., Goesele, M., and Schiele, B. (2010). Back to the future: learning shape models from 3D CAD data. In: *Proceedings of the British Machine Vision Conference*. Citeseer.

467 Liebelt, J. and Schmid, C. (2010). Multi-view object class detection with a 3D geometric model. In: *2010 IEEE Conference on Computer Vision and Pattern Recognition (CVPR)*, 1688–1695. IEEE.

468 Peng, X. and Saenko, K. (2017). Synthetic to real adaptation with deep generative correlation alignment networks. arXiv preprint arXiv:1701.05524.

469 DeCarlo, D., Finkelstein, A., and Rusinkiewicz, S. (2004). Interactive rendering of suggestive contours with temporal coherence. In: *Proceedings* of the 3rd International Symposium on Non-Photorealistic Animation and Rendering, 15–145. ACM.

470 DeCarlo, D. and Rusinkiewicz, S. (2007). Highlight lines for conveying shape. In: *Proceedings of the 5th International Symposium on Non-Photorealistic Animation and Rendering*, pp. 63–70. ACM.

471 Lee, Y., Markosian, L., Lee, S., and Hughes, J.F. (2007). Line drawings via abstracted shading. *ACM Transactions on Graphics (TOG)* 26 (3): 18.

472 Bromley, J., Guyon, I., LeCun, Y. et al. (1994). Signature verification using a "Siamese" time delay neural network. In: *Advances in Neural Information Processing Systems*, 737–744. World Scientific.

473 Khan, S., Rahmani, H., Shah, S.A.A., and Bennamoun, M. (2018). A guide to convolutional neural networks for computer vision. *Synthesis Lectures on Computer Vision* 8 (1): 1–207.

474 Kruskal, J.B. and Wish, M. (1978). *Multidimensional Scaling*, vol. 11. Sage.

475 Sammon, J.W. (1969). A nonlinear mapping for data structure analysis. *IEEE Transactions on computers* 100 (5): 401–409.

476 Schölkopf, B. (2000). The kernel trick for distances, TR MSR 2000-51. *Advances in Neural Information Processing Systems, 2001*. Redmond, WA: Microsoft Research.

477 Williams, C. and Seeger, M. (2001). Using the Nystroem method to speed up kernel machines. In: *Advances in Neural Information Processing Systems 13*, 682–688. MIT Press.

478 Drineas, P. and Mahoney, M.W. (2005). On the Nyström method for approximating a Gram matrix for improved kernel-based learning. *The Journal of Machine Learning Research* 6: 2153–2175.

479 Castrejon, L., Aytar, Y., Vondrick, C. et al. (2016). Learning aligned cross-modal representations from weakly aligned data. arXiv preprint arXiv:1607.07295.

Index

3D Shape Analysis: Fundamentals, Theory, and Applications, First Edition.
Hamid Laga, Yulan Guo, Hedi Tabia, Robert B. Fisher, and Mohammed Bennamoun.